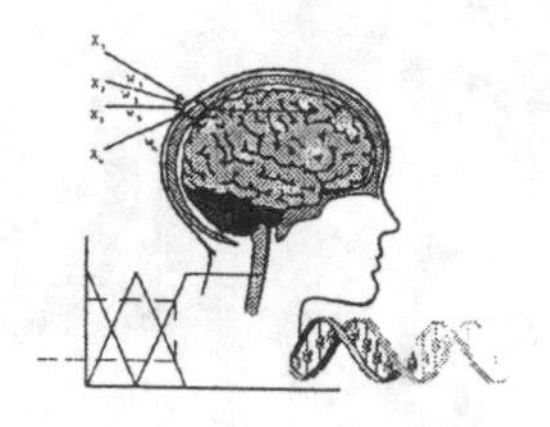

The CRC Press

International Series on Computational Intelligence

Series Editor
L.C. Jain, Ph.D.

L.C. Jain, R.P. Johnson, Y. Takefuji, and L.A. Zadeh
Knowledge-Based Intelligent Techniques in Industry

L.C. Jain and C.W. de Silva
Intelligent Adaptive Control: Industrial Applications

L.C. Jain and N.M. Martin
Fusion of Neural Networks, Fuzzy Systems, and Genetic Algorithms: Industrial Applications

H.N. Teodorescu, A. Kandel, and L.C. Jain
Fuzzy and Neuro-Fuzzy Systems in Medicine

C.L. Karr and L.M. Freeman
Industrial Applications of Genetic Algorithms

L.C. Jain and Beatrice Lazzerini
Knowledge-Based Intelligent Techniques in Character Recognition

L.C. Jain and V. Vemuri
Industrial Applications of Neural Networks

FUSION of NEURAL NETWORKS, FUZZY SETS, and GENETIC ALGORITHMS

Industrial Applications

Edited by
Lakhmi C. Jain
N. M. Martin

CRC

CRC Press
Boca Raton London New York Washington, D.C.

Library of Congress Cataloging-in-Publication Data

Catalog record is available from the Library of Congress.

This book contains information obtained from authentic and highly regarded sources. Reprinted material is quoted with permission, and sources are indicated. A wide variety of references are listed. Reasonable efforts have been made to publish reliable data and information, but the author and the publisher cannot assume responsibility for the validity of all materials or for the consequences of their use.

Direct all inquiries to CRC Press LLC, 2000 Corporate Blvd., N.W., Boca Raton, Florida 33431.

No claim to original U.S. Government works
International Standard Book Number 0-8493-9804-5
Printed in the United States of America 1 2 3 4 5 6 7 8 9 0
Printed on acid-free paper

PREFACE

The past two decades have seen an explosion of renewed interest in the areas of Artificial Intelligence and Information Processing. Much of this interest has come about with the successful demonstration of real-world applications of Artificial Neural Networks (ANNs) and their ability to learn. Initially proposed during the 1950s, the technology suffered a roller coaster development accompanied by exaggerated claims of their virtues, excessive competition between rival research groups, and the perils of boom and bust research funding. ANNs have only recently found a reasonable degree of respectability as a tool suitable for achieving a nonlinear mapping between an input and output space. ANNs have proved particularly valuable for applications where the input data set is of poor quality and not well characterized. At this stage, pattern recognition and control systems have emerged as the most successful ANN applications.

In more recent times, ANNs have been joined by other information processing techniques that come from a similar conceptual origin, with Genetic Algorithms, Fuzzy Logic, Chaos, and Evolutionary Computing the most significant examples. Together these techniques form what we refer to as the field of Knowledge-Based Engineering (KBE). For the most part, KBE techniques are those information and data processing techniques that were developed based on our understanding of the biological nervous system. In most cases the techniques used attempt, in some way, to mimic the manner in which a biological system might perform the task under consideration.

There has been intense interest in the development of Knowledge-Based Engineering as a research subject. Undergraduate course components in KBE were first conducted at the University of South Australia in 1992. Popularity of many aspects of Information Technology has been a world-wide phenomenon and, KBE as part of information technology, has followed accordingly. With a background of high demand from undergraduate and postgraduate students, the University of South Australia established a Research Centre in Knowledge-Based Engineering Systems in 1995. Since then the Centre has developed rapidly. Working in this rapidly evolving area of research has demanded a high degree of collaboration with researchers around the globe. The Centre has many international visitors each year and runs an annual international conference on KBE techniques. The Centre has also established industrial partners with some of the development projects. This book, therefore, is a natural progression in the Centre's activities. It represents a timely compilation of contributions from world-renowned practicing research engineers and scientists, describing the practical application of knowledge-based techniques to real-world problems.

Artificial neural networks can mimic the biological information processing mechanism in a very limited sense. The fuzzy logic provides a basis for representing uncertain and imprecise knowledge and forms a basis for human reasoning. The neural networks have shown real promise in solving problems, but there is not yet a definitive theoretical basis for their design. We see a need for integrating neural net, fuzzy system, and evolutionary computing in system design that can help us handle complexity. Evolutionary computation techniques possibly offer a method for doing that and, at the least, we would hope to gain some insight into alternative approaches to neural network design. The trend is to fuse these novel paradigms for offsetting the demerits of one paradigm by the merits of another. This book presents specific projects where fusion techniques have been applied. Overall, it covers a broad selection of applications that will serve to demonstrate the advantages of fusion techniques in industrial applications.

We see this book being of great value to the researcher and practicing engineer alike. The student of KBE will receive an in-depth tutorial on the KBE topics covered. The seasoned researcher will appreciate the practical applications and the gold mine of other possibilities for novel research topics. Most of all, however, this book aims to provide the practicing engineer and scientist with case studies of the application of a combination of KBE techniques to real-world problems.

We are grateful to the authors for preparing such interesting and diverse chapters. We would like to express our sincere thanks to Berend Jan van Zwaag, Ashlesha Jain, Ajita Jain and Sandhya Jain for their excellent help in the preparation of the manuscript. Thanks are due to Gerald T. Papke, Josephine Gilmore, Jane Stark, Dawn Mesa, Mimi Williams, Lourdes Franco, Tom O'Neill and Suzanne Lassandro for their editorial assistance

L.C. Jain
N.M. Martin
Adelaide,
AUSTRALIA

Contents

Chapter 1:

Introduction to Neural Networks, Fuzzy Systems, Genetic Algorithms, and their Fusion

INTRODUCTION TO NEURAL NETWORKS, FUZZY SYSTEMS, GENETIC ALGORITHMS, AND THEIR FUSION

N.M. Martin
Defence Science and Technology Organisation
P.O. Box 1500
Salisbury, Adelaide, S.A. 5108
Australia

L.C. Jain
Knowledge-Based Intelligent Engineering Systems Centre
University of South Australia
Adelaide, Mawson Lakes, S.A. 5095
Australia

This chapter presents an introduction to knowledge-based information systems which include artificial neural networks, evolutionary computing, fuzzy logic and their fusion. Knowledge-based systems are designed to mimic the performance of biological systems. Artificial neural networks can mimic the biological information processing mechanism in a very limited sense. Evolutionary computing algorithms are used for optimization applications, and fuzzy logic provides a basis for representing uncertain and imprecise knowledge. The trend is to fuse these novel paradigms in order that the demerits of one paradigm may be offset by the merits of another. These fundamental paradigms form the basis of the novel design and application related projects presented in the following chapters.

1. Knowledge-Based Information Systems

As is typical with a new field of scientific research, there is no precise definition for knowledge-based information systems. Generally speaking, however, so-called knowledge-based data and information processing techniques are those that are inspired by an understanding of information processing in biological systems. In some

0-8493-9804-5/99/$0.00+$.50

cases an attempt is made to mimic some aspects of biological systems. When this is the case, the process will include an element of adaptive or evolutionary behavior similar to biological systems and, like the biological model, there will be a very high level of connection between distributed processing elements.

Knowledge-based information (KBI) systems are being applied in many of the traditional rule-based Artificial Intelligence (AI) areas. Intelligence is also not easy to define, however, we can say that a system is intelligent if it is able to improve its performance or maintain an acceptable level of performance in the presence of uncertainty. The main attributes of intelligence are learning, adaptation, fault tolerance and self-organization. Data and information processing paradigms that exhibit these attributes can be referred to as members of the family of techniques that make up the knowledge-based engineering area. Researchers are trying to develop AI systems that are capable of performing, in a limited sense, "like a human being."

The popular knowledge-based paradigms are: artificial neural networks, evolutionary computing, of which genetic algorithms are the most popular example, chaos, and the application of data and information fusion using fuzzy rules. The chapters that follow in this book have concentrated on the application of artificial neural networks, genetic algorithms, and evolutionary computing. Overall, the family of knowledge-based information processing paradigms have recently generated tremendous interest among researchers. To date the tendency has been to concentrate on the fundamental development and application of a single paradigm. The thrust of the topics in this book is the application of the various paradigms to appropriate parts of real-world engineering problems. Emphasis is placed on examining the attributes of particular paradigms to particular problems, and combining them with the aim of achieving a systems solution to the engineering requirement. The process of coordinating the most appropriate paradigm for the task will be referred to as an hybrid approach to knowledge-based information systems. The greatest gains in the application of KBI systems will come from exploring the synergies that often exist when paradigms are used together.

The one KBI paradigm not reported in this book is chaos theory. From the point of view of engineering applications chaos stands as the most novel of several novel paradigms. In recent years chaos engineering has generated tremendous interest among application engineers. The word chaos refers to the complicated and noise-like phenomena originated from nonlinearities involved in deterministic dynamic systems. There is a growing interest to discover the law of nature hidden in these complicated phenomena and the attempt to use it to solve engineering problems is gaining momentum. A number of successful engineering applications of chaos engineering are reported in the literature [1]. These include suppression of vibrations and oscillations in mechanical and electrical systems, industrial plant control, adaptive equalization, data compression, dish washer control, washing machine control and heater control.

In the following paragraphs the main KBI paradigms used throughout the book are reviewed; these are artificial neural networks, evolutionary computing and fuzzy logic.

The review will serve to give the reader some insight into the fundamentals of the paradigms and their typical applications. The reader is referred to the reference list for further detailed reading.

2. Artificial Neural Networks

Artificial Neural Networks (ANNs) mimic biological information processing mechanisms. They are typically designed to perform a nonlinear mapping from a set of inputs to a set of outputs. ANNs are developed to try to achieve biological system type performance using a dense interconnection of simple processing elements analogous to biological neurons. ANNs are information driven rather than data driven. They are non-programmed adaptive information processing systems that can autonomously develop operational capabilities in response to an information environment. ANNs learn from experience and generalize from previous examples. They modify their behavior in response to the environment, and are ideal in cases where the required mapping algorithm is not known and tolerance to faulty input information is required.

ANNs contain electronic processing elements (PEs) connected in a particular fashion. The behavior of the trained ANN depends on the weights, which are also referred to as strengths of the connections between the PEs. ANNs offer certain advantages over conventional electronic processing techniques. These advantages are the generalization capability, parallelism, distributed memory, redundancy, and learning.

Artificial neural networks are being applied to a wide variety of automation problems including adaptive control, optimization, medical diagnosis, decision making, as well as information and signal processing, including speech processing. ANNs have proven to be very suitable for processing sensor data, in particular, feature extraction and automated recognition of signals and multi-dimensional objects. Pattern recognition has, however, emerged as a major application because the network structure is suited to tasks that biological systems perform well, and pattern recognition is a good example where biological systems out-perform traditional rule-based artificial intelligence approaches.

The name artificial neural network given to the study of these mathematical processes is, in a sense, unfortunate in that it creates a false impression which leads to the formation of unwarranted expectations. Despite some efforts to change to a less spectacular name such as connectionist systems, it seems that the title Artificial Neural Networks is destined to remain. At this time the performance of the best ANN is trivial when compared with even the simplest biological system.

The first significant paper on artificial neural networks is generally considered to be that of McCullock and Pitts [2] in 1943. This paper outlined some concepts concerning how biological neurons could be expected to operate. The neuron models proposed were modeled by simple arrangements of hardware that attempted to mimic the performance of the single neural cell. In 1949 Hebb [3] formed the basis of 'Hebbian learning' which is now regarded as an important part of ANN theory. The basic concept underlying 'Hebbian learning' is the principle that every time a neural connection is used, the pathway is strengthened. About this time of neural network development, the digital computer became more widely available and its availability proved to be of great practical value in the further investigation of ANN performance. In 1958 Neumann proposed modeling the brain performance using items of computer hardware available at that time. Rosenblatt [4] constructed neuron models in hardware during 1957. These models ultimately resulted in the concept of the Perceptron. This was an important development and the underlying concept is still in wide use today. Widrow and Hoff [5] were responsible for simplified artificial neuron development. First the ADALINE and then the MADALINE networks. The name 'ADALINE' comes from ADAptive LInear NEuron, and the name 'MADALINE' comes from Multiple ADALINE.

In 1969 Minsky and Pappert published [6] an influential book "Perceptrons" which showed that the Perceptron developed by Rosenblatt had serious limitations. He further contended that the Perceptron, at the time, suffered from severe limitations. The essence of the book "Perceptrons" was the assumption that the inability of the perception to be able to handle the 'exclusive or' function was a common feature shared by all neural networks. As a result of this assumption, interest in neural networks greatly reduced. The overall effect of the book was to reduce the amount of research work on neural networks for the next 10 years. The book served to dampen the unrealistically high expectations previously held for ANNs. Despite the reduction in ANN research funding, a number of people still persisted in ANN research work.

John Hopfield [7] produced a paper in 1982 that showed that the ANN had potential for successful operation, and proposed how it could be developed. This paper was timely as it marked a second beginning for the ANN. While Hopfield is the name frequently associated with the resurgence of interest in ANN it probably represented the culmination of the work of many people in the field. From this time onward the field of neural computing began to expand and now there is world-wide enthusiasm as well as a growing number of important practical applications.

Today there are two classes of ANN paradigm, supervised and unsupervised. The multilayer back-propagation network (MLBPN) is the most popular example of a supervised network. It results from work carried out in the mid-eighties largely by David Rumelhart [8] and David Parker [9]. It is a very powerful technique for constructing nonlinear transfer functions between several continuous valued inputs and one or more continuously valued outputs. The network basically uses a multilayer perceptron architecture and gets its name from the manner in which it processes errors during training.

Adaptive Resonance Theory (ART) is an example of an unsupervised or self-organizing network and was proposed by Carpenter and Grossberg [10]. Its architecture is highly adaptive and evolved from the simpler adaptive pattern recognition networks known as the competitive learning models. Kohonen's Learning vector quantiser [11] is another popular unsupervised neural network that learns to form activity bubbles through the actions of competition and cooperation when the feature vectors are presented to the network. A feature of biological neurons, such as those in the central nervous system, is their rich interconnections and abundance of recurrent signal paths. The collective behavior of such networks is highly dependent upon the activity of each individual component. This is in contrast to feed forward networks where each neuron essentially operates independent of other neurons in the network.

The underlying reason for using an artificial neural network in preference to other likely methods of solution is that there is an expectation that it will be able to provide a rapid solution to a non-trivial problem. Depending on the type of problem being considered, there are often satisfactory alternative proven methods capable of providing a fast assessment of the situation.

Artificial Neural Networks are not universal panaceas to all problems. They are really just an alternative mathematical device for rapidly processing information and data. It can be argued that animal and human intelligence is only a huge extension of this process. Biological systems learn and then interpolate and extrapolate using slowly propagated (100 m/s) information when compared to the propagation speed ($3\ 10^8$ m/s) of a signal in an electronic system. Despite this low signal propagation speed the brain is able to perform splendid feats of computation in everyday tasks. The reason for this enigmatic feat is the degree of parallelism that exists within the biological brain.

3. Evolutionary Computing

Evolutionary computation is the name given to a collection of algorithms based on the evolution of a population toward a solution of a certain problem. These algorithms can be used successfully in many applications requiring the optimization of a certain multi-dimensional function. The population of possible solutions evolves from one generation to the next, ultimately arriving at a satisfactory solution to the problem. These algorithms differ in the way a new population is generated from the present one, and in the way the members are represented within the algorithm. Three types of evolutionary computing techniques have been widely reported recently. These are Genetic Algorithms (GAs), Genetic Programming (GP) and Evolutionary Algorithms (EAs). The EAs can be divided into Evolutionary Strategies (ES) and Evolutionary Programming (EP). All three of these algorithms are modeled in some way after the evolutionary processes occurring in nature.

Genetic Algorithms were envisaged by Holland [12] in the 1970s as an algorithmic concept based on a Darwinian-type survival-of-the-fittest strategy with sexual reproduction, where stronger individuals in the population have a higher chance of creating an offspring. A genetic algorithm is implemented as a computerized search and optimization procedure that uses principles of natural genetics and natural selection. The basic approach is to model the possible solutions to the search problem as strings of ones and zeros. Various portions of these bit-strings represent parameters in the search problem. If a problem-solving mechanism can be represented in a reasonably compact form, then GA techniques can be applied using procedures to maintain a population of knowledge structure that represent candidate solutions, and then let that population evolve over time through competition (survival of the fittest and controlled variation). The GA will generally include the three fundamental genetic operations of selection, crossover and mutation. These operations are used to modify the chosen solutions and select the most appropriate offspring to pass on to succeeding generations. GAs consider many points in the search space simultaneously and have been found to provide a rapid convergence to a near optimum solution in many types of problems; in other words, they usually exhibit a reduced chance of converging to local minima. GAs show promise but suffer from the problem of excessive complexity if used on problems that are too large.

Generic algorithms are an iterative procedure that consists of a constant-sized population of individuals, each one represented by a finite linear string of symbols, known as the genome, encoding a possible solution in a given problem space. This space, referred to as the search space, comprises all possible solutions to the optimization problem at hand. Standard genetic algorithms are implemented where the initial population of individuals is generated at random. At every evolutionary step, also known as generation, the individuals in the current population are decoded and evaluated according to a fitness function set for a given problem. The expected number of times an individual is chosen is approximately proportional to its relative performance in the population. Crossover is performed between two selected individuals by exchanging part of their genomes to form new individuals. The mutation operator is introduced to prevent premature convergence.

Every member of a population has a certain fitness value associated with it, which represents the degree of correctness of that particular solution or the quality of solution it represents. The initial population of strings is randomly chosen. The strings are manipulated by the GA using genetic operators, to finally arrive at a quality solution to the given problem. GAs converge rapidly to quality solutions. Although they do not guarantee convergence to the single best solution to the problem, the processing leverage associated with GAs make them efficient search techniques. The main advantage of a GA is that it is able to manipulate numerous strings simultaneously, where each string represents a different solution to a given problem. Thus, the possibility of the GA getting stuck in local minima is greatly reduced because the whole space of possible solutions can be simultaneously searched. A basic genetic algorithm comprises three genetic operators.

- selection
- crossover, and
- mutation.

Starting from an initial population of strings (representing possible solutions), the GA uses these operators to calculate successive generations. First, pairs of individuals of the current population are selected to mate with each other to form the offspring, which then form the next generation. Selection is based on the survival-of-the-fittest strategy, but the key idea is to select the better individuals of the population, as in tournament selection, where the participants compete with each other to remain in the population. The most commonly used strategy to select pairs of individuals is the method of roulette-wheel selection, in which every string is assigned a slot in a simulated wheel sized in proportion to the string's relative fitness. This ensures that highly fit strings have a greater probability to be selected to form the next generation through crossover and mutation. After selection of the pairs of parent strings, the crossover operator is applied to each of these pairs.

The crossover operator involves the swapping of genetic material (bit-values) between the two parent strings. In single point crossover, a bit position along the two strings is selected at random and the two parent strings exchange their genetic material as illustrated below.

Parent A = $a_1\ a_2\ a_3\ a_4 \mid a_5\ a_6$
Parent B = $b_1\ b_2\ b_3\ b_4 \mid b_5\ b_6$

The swapping of genetic material between the two parents on either side of the selected crossover point, represented by "|", produces the following offspring:

Offspring A' $= a_1\ a_2\ a_3\ a_4 \mid b_5\ b_6$
Offspring B' $= b_1\ b_2\ b_3\ b_4 \mid a_5\ a_6$

The two individuals (children) resulting from each crossover operation will now be subjected to the mutation operator in the final step to forming the new generation.

The mutation operator alters one or more bit values at randomly selected locations in randomly selected strings. Mutation takes place with a certain probability, which, in accordance with its biological equivalent, typically occurs with a very low probability. The mutation operator enhances the ability of the GA to find a near optimal solution to a given problem by maintaining a sufficient level of genetic variety in the population, which is needed to make sure that the entire solution space is used in the search for the best solution. In a sense, it serves as an insurance policy; it helps prevent the loss of genetic material.

Genetic algorithms are most appropriate for optimization type problems, and have been applied successfully in a number of automation applications including job shop scheduling, proportional integral derivative (PID) control loops, and the automated design of fuzzy logic controllers and ANNs.

John Koza of Stanford University developed genetic programming (GP) techniques in the 1990s [13]. Generic programming is a special implementation of GAs. It uses hierarchical genetic material that is not limited in size. The members of a population or chromosomes are tree structured programs and the genetic operators work on the branches of these trees. The structures generally represent computer programs written in LISP.

Evolutionary algorithms do not require separation between a recombination and an evaluation space. The genetic operators work directly on the actual structure. The structures used in EAs are representations that are problem dependent and more natural for the task than the general representations used in GAs.

Evolutionary programming is currently experiencing a dramatic increase in popularity. Several examples have been successfully completed that indicate EP is full of potential. Koza and his students have used EP to solve problems in various domains including process control, data analysis, and computer modeling. Although at the present time the complexity of the problems being solved with EP lags behind the complexity of applications of various other evolutionary computing algorithms, the technique is promising. Because EP actually manipulates entire computer programs, the technique can potentially produce effective solutions to very large-scale problems. To reach its full potential, EP will likely require dramatic improvements in computer hardware.

4. Fuzzy Logic

Fuzzy logic was first developed by Zadeh [14] in the mid-1960s for representing uncertain and imprecise knowledge. It provides an approximate but effective means of describing the behavior of systems that are too complex, ill-defined, or not easily analyzed mathematically. Fuzzy variables are processed using a system called a fuzzy logic controller. It involves fuzzification, fuzzy inference, and defuzzification. The fuzzification process converts a crisp input value to a fuzzy value. The fuzzy inference is responsible for drawing conclusions from the knowledge base. The defuzzification process converts the fuzzy control actions into a crisp control action.

Zadeh argues that the attempts to automate various types of activities from assembling hardware to medical diagnosis have been impeded by the gap between the way human beings reason and the way computers are programmed. Fuzzy logic uses graded statements rather than ones that are strictly true or false. It attempts to incorporate the "rule of thumb" approach generally used by human beings for decision making. Thus, fuzzy logic provides an approximate but effective way of describing the behavior of systems that are not easy to describe precisely. Fuzzy logic controllers, for example, are extensions of the common expert systems that use production rules like "if-then." With fuzzy controllers, however, linguistic variables like "tall" and "very tall" might be incorporated in a traditional expert system. The result is that fuzzy logic can be used in controllers that are capable of making intelligent control decisions in sometimes volatile and rapidly changing problem environments.

Fuzzy logic techniques have been successfully applied in a number of applications: computer vision, decision making, and system design including ANN training. The most extensive use of fuzzy logic is in the area of control, where examples include controllers for cement kilns, braking systems, elevators, washing machines, hot water heaters, air-conditioners, video cameras, rice cookers, and photocopiers.

5. Fusion

Neural networks, fuzzy logic and evolutionary computing have shown capability on many problems, but have not yet been able to solve the really complex problems that their biological counterparts can (e.g., vision). It is useful to fuse neural networks, fuzzy systems and evolutionary computing techniques for offsetting the demerits of one technique by the merits of another techniques. Some of these techniques are fused as:

Neural networks for designing fuzzy systems
Fuzzy systems for designing neural networks
Evolutionary computing for the design of fuzzy systems
Evolutionary computing in automatically training and generating neural network architectures

6. Summary

The following chapters discuss specific projects where knowledge-based techniques have been applied. The chapters start with the design of a new fuzzy-neural controller. The remaining chapters show the application of expert systems, neural networks, fuzzy control and evolutionary computing techniques in modern engineering systems. These specific applications include direct frequency converters, electro-hydraulic systems, motor control, toaster control, speech recognition, vehicle routing, fault diagnosis, asynchronous transfer mode (ATM) communications networks, telephones for hard-of-hearing people, control of gas turbine aero-engines and telecommunications systems design. Overall, these chapters cover a broad selection of applications that will serve to demonstrate the advantages and disadvantages of the application of KBI paradigms. KBI paradigms are demonstrated to be very powerful tools when applied in an appropriate manner.

References

[1] Katayama, R., Kuwata, K. and Jain, L.C. (1996), Fusion Technology of Neuro, Fuzzy, Genetic and Chaos Theory and its Applications, in Hybrid Intelligent Engineering Systems, World Scientific Publishing Company, Singapore, pp. 167-186.

[2] McCullock, W.W, Pitts, W. (1943), A Logical Calculus of the Ideas Imminent in Nervous Activity, Bulletin of Mathematical Biophysics, Vol. 5, pp. 115-133.

[3] Hebb, D.O. (1949), The Organisation of Behaviours, John Wiley & Sons, New York.

[4] Rosenblatt, F. (1959), Principles of Neurocomputing, Addison-Wesley Publishing Co.

[5] Widrow, B., Hoff. M.E. (1960), Adaptive Switching Circuits, IRE WESTCON Convention Record, Part 4, pp. 96-104.

[6] Minsky, M.L., Papert, S. (1969), Perceptrons, MIT Press, Cambridge MA.

[7] Hopfield, J.J. (1982), Neural Networks and Physical Systems with Emergent Collective Computational Abilities, Proc. Nat. Acad. Sci, USA, Vol. 79, pp. 2554-2558.

[8] Rumelhart, D.E., McClelland J.L. (1986), Parallel Distributed Processing: Explorations in the Microstructure of Cognition, Vol. 1 Foundations. MIT Press, Cambridge MA.

[9] Parker, D.B. (1985), Learning-logic, Report TR-47, Massachusetts Institute of Technology, Centre for Computational Research in Economics and Management Science, Cambridge, MA.

[10] Carpenter, G.A., Grossberg, S. (1987), ART2 Self-Organisation of Stable Category Recognition Codes for Analog Input Patterns, Applied Optics, 26, pp. 4919-4930.

[11] Kohonen, T. (1989), Self-Organisation and Associative Memory, Third Edition, Springer Verlag, Heidelberg.

[12] Holland, J.H. (1975), Adaptation in Natural and Artificial Systems, MIT Press, Cambridge MA.

[13] Jain, L.C. (Editor) (1997), Soft Computing Techniques in Knowledge-Based Intelligent Engineering Systems, Springer-Verlag, Heidelberg.

[14] Zadeh, L.A. (1988), Fuzzy Logic, IEEE Computer, 1988, pp. 83-89.

Chapter 2:

A New Fuzzy-Neural Controller

A NEW FUZZY-NEURAL CONTROLLER

Koji Shimojima
Material Processing Department
National Research Institute of Nagoya
Japan

Toshio Fukuda
Department of Micro System Engineering
Nagoya University
Japan

In this chapter, we introduce a hierarchical control system for an unsupervised Radial Basis Function (RBF) fuzzy system. This hierarchical control system has the skill database, which controls fuzzy controllers acquired through the unsupervised learning process based on Genetic Algorithms. Thus, the control system can use the acquired fuzzy controller effectively and it leads to reducing the iteration time for a new target. The effectiveness of the proposed method is shown using the simulations of the cart-pole problem.

1. Introduction

In recent years, intelligent techniques such as fuzzy logic, fuzzy reasoning, and fuzzy modeling are used in many fields including engineering, medical, and social sciences. Some fuzzy control systems can be seen in home appliances, transportation systems, and manufacturing systems. We also successfully applied the fuzzy inference for the sensor integration system [1-3].

The fuzzy system has a characteristic to represent human knowledge or experiences using fuzzy rules; however, the fuzzy systems have some problems. In most fuzzy systems, the shape of membership functions of the antecedent and the consequent and fuzzy rules are determined using trial and error by operators. It is time consuming. The problem is more serious when the fuzzy logic is applied to a complex system.

In order to solve this problem, some self-tuning methods have been proposed such as Fuzzy Neural Network [4,5] using back propagation algorithm [6], fuzzy learning using the Radial Basis Function (RBF) [7,8], Genetic Algorithm (GA) for deciding the

0-8493-9804-5/99/$0.00+$.50

shapes of membership functions and fuzzy rules [9,10], and the gradient descent method [11].

These methods can learn faster than neural networks; however, the operator must determine the number and shapes of membership functions before learning. The learning ability and accuracy of approximation are related to the number or shape of membership functions. Fuzzy inference with more membership functions and fuzzy rules have higher learning ability, however, these may include some redundant or unlearned rules. The number of rules is the product of the number of membership functions for each input, and these increase with increase of the input dimension; therefore, operators need to pay attention when deciding the structure of the fuzzy systems.

The fuzzy inference based on the RBF that adds a new rule for the maximal error point through the learning process has been proposed. In this method, fuzzy rules depend on the learning data set. If the learning data is biased, there are some unlearned areas or redundant fuzzy rules. Furthermore, this method does not delete a fuzzy rule, instead it adds new fuzzy rules; therefore it poses a problem because addition of fuzzy rules causes problems in the calculation time and memories.

To solve these problems, we proposed a new type of self-tuning fuzzy inference [12]. The membership function of the antecedent is expressed by the RBF. The supervised/unsupervised learning algorithms are based on the genetic algorithm, and the supervised learning also utilizes the gradient descent method to tune the shape and position of membership functions and the consequent values. However, these systems do not use previous learning results effectively. Therefore, if the systems handle a new task, the systems need additional learning for the new task. The GA based learning takes a long time to learning.

In this chapter, we propose the hierarchical control system with unsupervised learning based on skill knowledge database. In this system, the skill knowledge database manages fuzzy controllers acquired through previous learning process as skills. Therefore the system can use previous learning results for control/learning a new task. The effectiveness of this system is shown through some simulation results.

2. RBF Based Fuzzy System with Unsupervised Learning

2.1 Fuzzy System Based on RBF

Several researchers have proposed automatic design (self-tuning) methods. Most of them focused on tuning membership functions. For example, neural networks are used as membership value's generator, fuzzy systems are treated as networks, and backpropagation techniques are used to adjust the shapes of membership functions. However, these tuning methods are weak, because the convergence of tuning depends on the initial conditions such as the number and shapes of membership function, and sometimes it converges to a local minimum.

We have proposed a new method for auto-tuning and optimization of the structure of the fuzzy model. The GA is one of the optimization methods using a stochastic search algorithm based on the biological evolution process. However, the GA is a coarse searching and not the best method to find the optimal value.

We have proposed a fuzzy system based on RBF and its tuning method based on the GA [12]. The tuning algorithm not only tunes the shapes of membership functions and the consequent value, but also optimizes the number of membership functions and the number of rules.

First, we present the equations of the fuzzy system between input and output variables. The fitness value μ_j of the rules and the output value Yp are expressed by Equations (1) and (2),

$$\mu_j = \prod_{i \in J} \mu_{ij} \,, \tag{1}$$

$$Y_p = \frac{\sum_{j=1}^{J} \mu_j \cdot w_{pj}}{\sum_{j=1}^{J} \mu_j} \,, \tag{2}$$

where i is the input variable number, j is the fuzzy rule's number, and p is the data set's number.

The shapes of the membership functions are expressed by RBF with a dead zone c that is useful for reducing the membership functions and fuzzy rules. The membership function in the i-th input value and the j-th fuzzy rule is expressed by:

$$f(I_i) = \begin{cases} -b_{ij}(|I_i - a_{ij}| - c_{ij})^2 & if \quad |I_i - a_{ij}| \geq c_{ij} \\ 0 & if \quad |I_i - a_{ij}| \leq c_{ij} \end{cases}, \tag{3}$$

$$\mu_{ij} = \exp\{f(I_i)\}, \tag{4}$$

where a, b, c are the coefficients that decide the shape of membership functions shown in Figure 1.

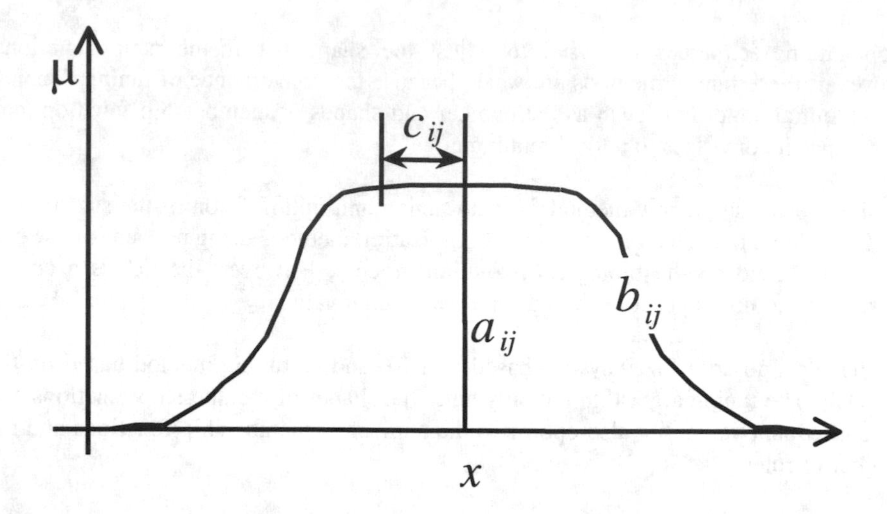

Figure 1 Membership function based on RBF.

2.2 Coding

To encode the information of membership functions, we use 31 bits memory for every membership function: each coefficient *a, b, c* used 10 bits; 1 bit is used as a flag of the membership function's validity. The consequent part is encoded into 16 bits memory in the case of unsupervised learning. Equations (5), (6), and (7) are used to decode the chromosome into the parameters of membership functions (see Figs. 1 and 2) in both unsupervised and supervised learning methods. Equation (8) is used to decode the value for the consequent parts in case of the unsupervised learning.

$$a_{ij} = \frac{2Sa_{ij}}{1023} - 0.5, \tag{5}$$

$$b_{ij} = 100\left(\frac{Sb_{ij}}{1023}\right)^3, \tag{6}$$

$$c_{ij} = \frac{Sc_{ij}}{4092}, \tag{7}$$

$$W_i = \frac{2w_i}{65535} - 0.5 \tag{8}$$

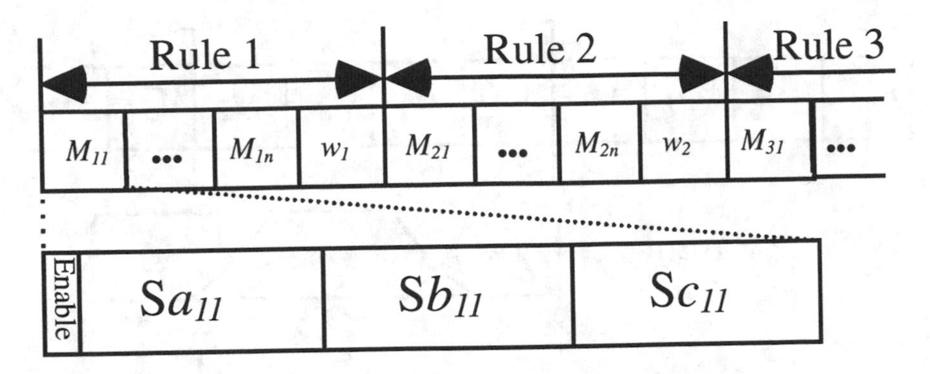

Figure 2 Coding scheme.

2.3 Selection

We rank each string of a society based on the fitness value F expressed in Equation (9), and the smallest value is the best string. We used the elite preserving strategy and the roulette wheel selection strategy to keep higher fitness chromosomes.

$$F(s_n) = \alpha P + \beta R_n + \gamma M_n, \tag{9}$$

where P, R_n, M_n, and α, β, γ means the performance index of the system, the number of the rules, the number of the membership functions, and the coefficients, respectively. In this equation, coefficients are classified into two types, one is the performance (α), the other is the size of fuzzy system (β and γ). The operator can acquire the preferable fuzzy system such as small fuzzy system (β and γ are larger than α), or highly accurate fuzzy system (α is larger than β and γ), by setting these coefficients.

2.4 Crossover Operator

In order to generate a new group of membership functions and rules, we apply the crossover operator. Crossover operator randomly selects the target chromosome. We adopt two points crossover operator as shown in Figure 3.

2.5 Mutation Operator

In this chapter, two types of mutation operators are utilized: (A) uniform distribution random set based mutation operator and (B) normal distribution random number based mutation operator. In both types, the target strings and mutation sites are randomly selected.

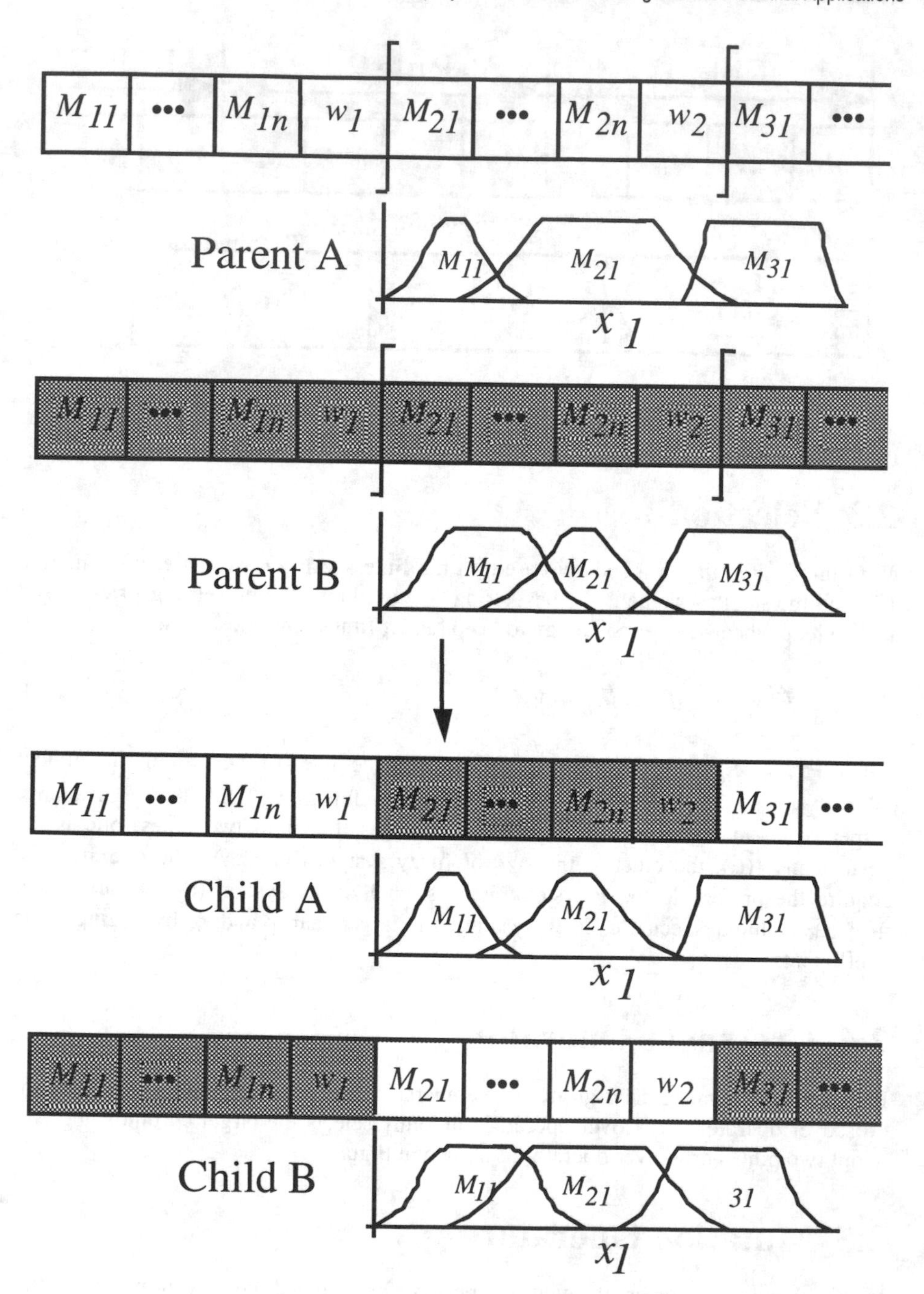

Figure 3 Crossover operator.

In the case of Mutation operator A shown in Figure 4(A), some bits of the strings are changed for global and rough search. This operator can change the enable/disable of the membership function.

The mutation operator with normal distribution in Figure 4(B) does not change the bits of the chromosomes directly, except the validity of the membership function, but adds (or subtracts) the random values to (from) the parameters of the membership functions, S_a, S_b, *and* S_c, and the consequent values *w*.

The random values are generated based on the age of the string. When the highest fitness value is improved, then the age is reset to zero; otherwise, the age is incremented. If the age is smaller, the random values are generated into a small region. On the contrary, if the age is large, the random values are generated in a large region. To change the region from small to large, the search space is changed from small to large. This mutation operator A also changes the validity of each membership function.

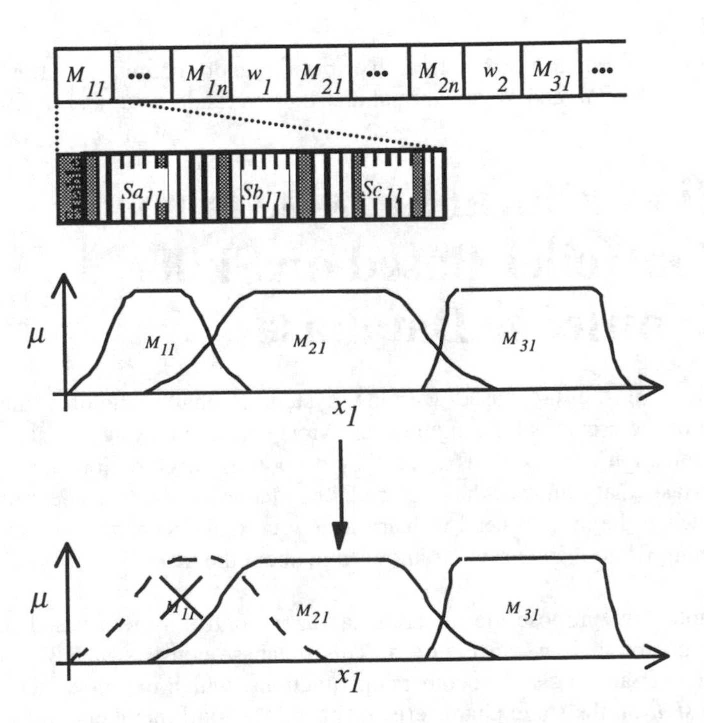

Figure 4(A) Mutation with uniform distribution random set for global search.

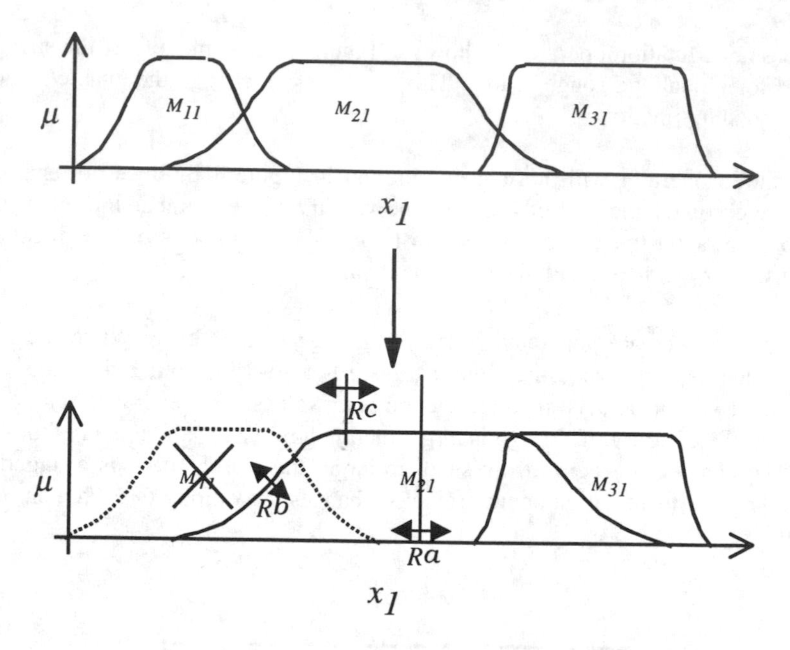

Figure 4(B) Mutation with normal distribution random number for fine search (*Ra*, *Rb*, and *Rc* are normal distribution random numbers).

3. Hierarchical Fuzzy-Neuro Controller Based on Skill Knowledge Database

One problem of the GA-based learning system is that it ignores the acquired knowledge of the previous learning process. Most studies utilizing GA are carried out as the optimization of a fixed task, and they do not use any previous learning results for a new task that can use the acquired knowledge of previous learning results. Therefore when the system need to learn a new task, the system must start on GA-based learning without any previous knowledge about the tasks.

In this chapter, we propose the hierarchical fuzzy control system based on the skill knowledge database shown in Figure 5. This database consists of RBF fuzzy-neuro controller (*skill*) and it's skill membership functions, which expresses the applicable area of the *skill* on the static characteristic space. The skill membership functions are expressed as shown in Figure1. This membership function is generated when the system learns a new task/target, and its location is decided by the static property of the task/target.

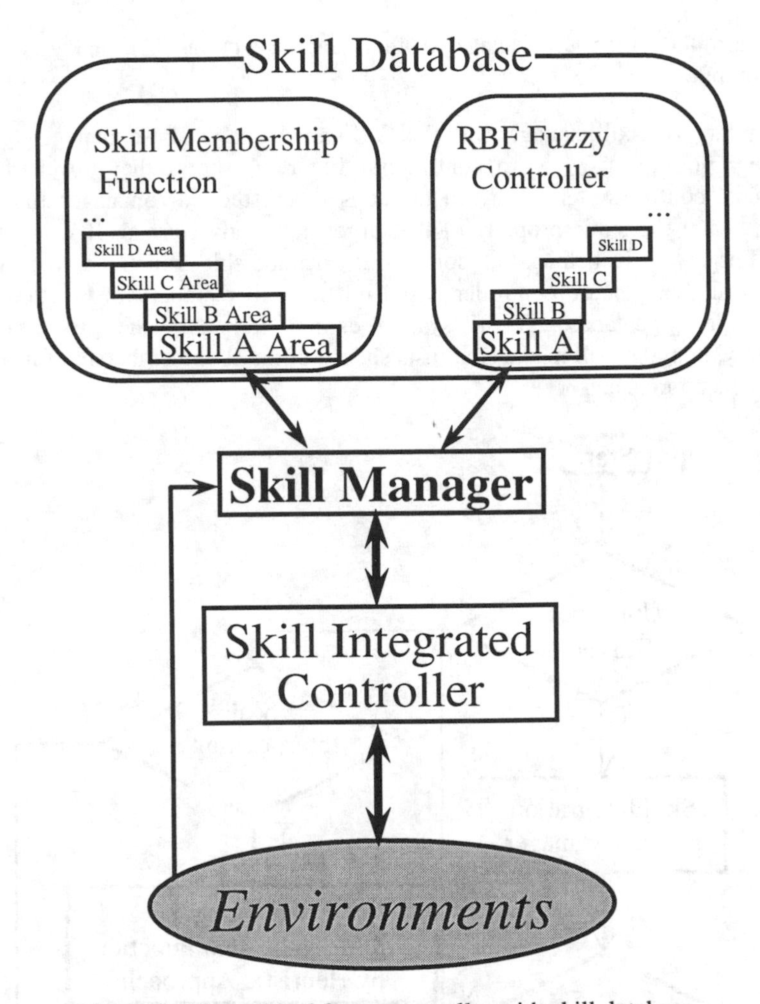

Figure 5 Hierarchical fuzzy controller with skill database.

Skill-Membership Functions express the applicable region of the acquired fuzzy-neuro controller. They are used for integration of controllers. Integration of controllers is done by the following equations:

$$y = \frac{\sum_{j=1}^{J} \mu_j \cdot Skill_j}{\sum_{j=1}^{J} \mu_j} \tag{10}$$

where μ_j is the applicable ability of the j-th fuzzy controller (j-th skill) and calculated from its skill membership function, the $Skill_j$ means the output of the j-th fuzzy controller which is calculated from Equations (1) and (2), and y is the total control output.

In this system, the skill manager with the skill knowledge database manage generation and integration of fuzzy-neuro controllers. Figure 6 shows the flowchart of the hierarchical control system. When a target is given, the skill manager first checks whether or not the static property of the target is already learned. If it is already a learned target or it belongs to some skill membership functions, the manager integrates all fuzzy-neuro controller based on the skill membership functions of the skill knowledge database. If the system cannot carry out the given new task sufficiently, then the skill manager adjusts shapes of the skill membership functions by the heuristic approach.

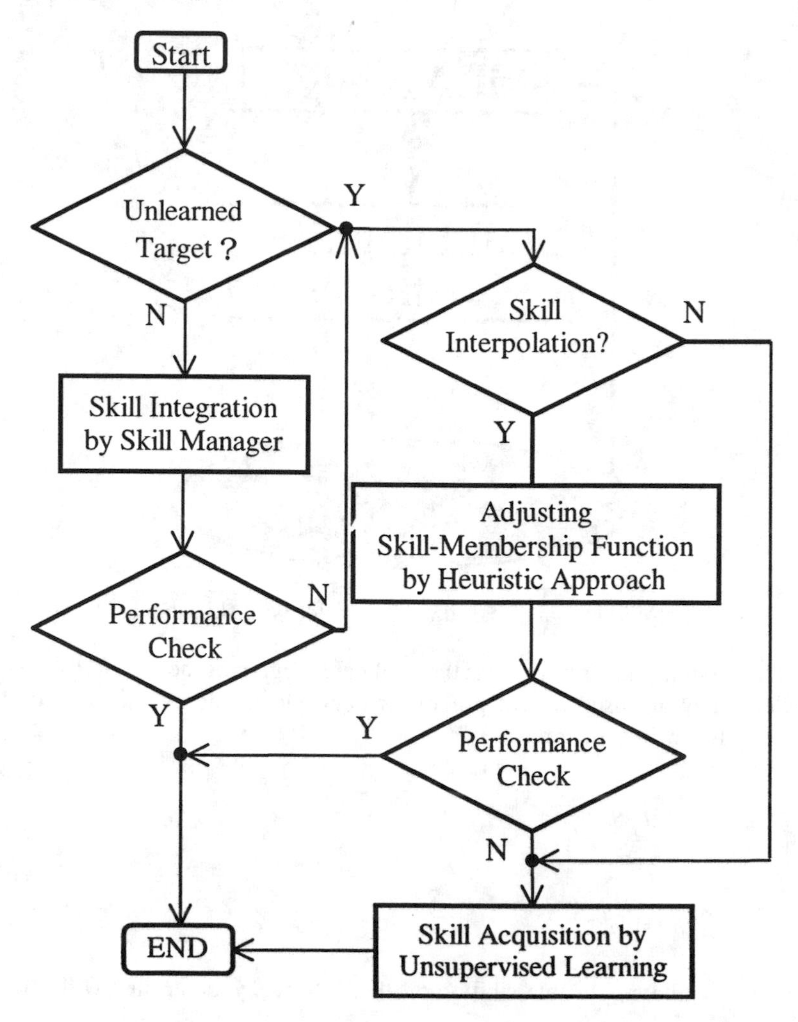

Figure 6 Learning flowchart.

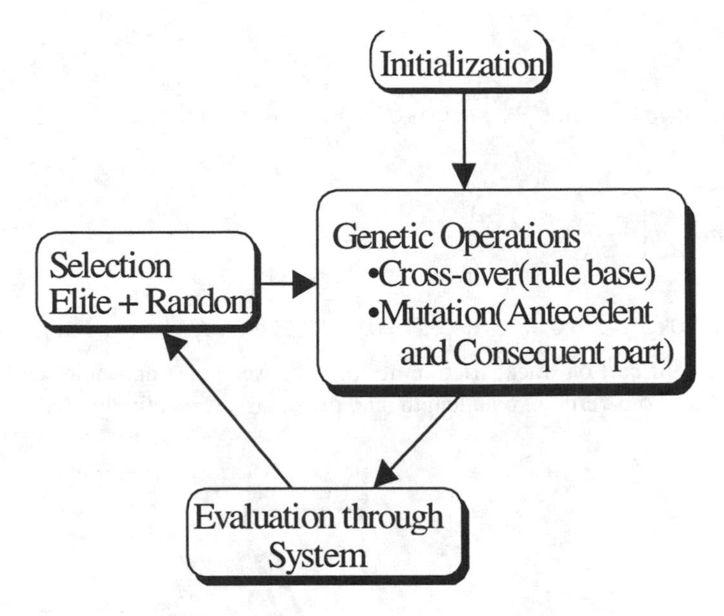

Figure 7 Unsupervised learning process based on Genetic Algorithm.

In the case of skill learning, the fuzzy-neuro controller is acquired through the unsupervised learning process as shown in Figure 7 and previously learned fuzzy-neuro controllers are encoded and set in the strings of the first generation.

4. Fuzzy-Neuro Controller for Cart-Pole System

Let us apply the proposed hierarchical fuzzy-neuro control system with the unsupervised learning method for the cart-pole system shown in Figure 8. The pole is controlled from a pendant position to an upright position and then kept it up.

The cart-pole system is described by the following equations:

$$\ddot{r} = \frac{F - \mu_c sign(\dot{r}) + \tilde{F}}{M + \tilde{m}}, \tag{11}$$

$$\ddot{\theta} = -\frac{3}{4l}\left(\ddot{r}\cos\theta + g\sin\theta + \frac{\mu_p\dot{\theta}}{ml}\right) \tag{12}$$

where

$$\tilde{F} = ml\dot{\theta}^2 \sin\theta + \frac{3}{4} m \cos\theta \left(\frac{\mu_p \dot{\theta}^2}{ml} + g \sin\theta \right) \tag{13}$$

$$\tilde{m} = m\left(1 - \frac{3}{4}\cos^2\theta\right) \tag{14}$$

where $M = 1.0$kg $\mu_c = 0.0005$N, $\mu_p = 0.000002$kg•m, r, θ, l, and m mean the cart mass, friction of cart on track, friction at hinge between cart and pole, cart position, pole deviation from vertical, pole length, and pole mass, respectively.

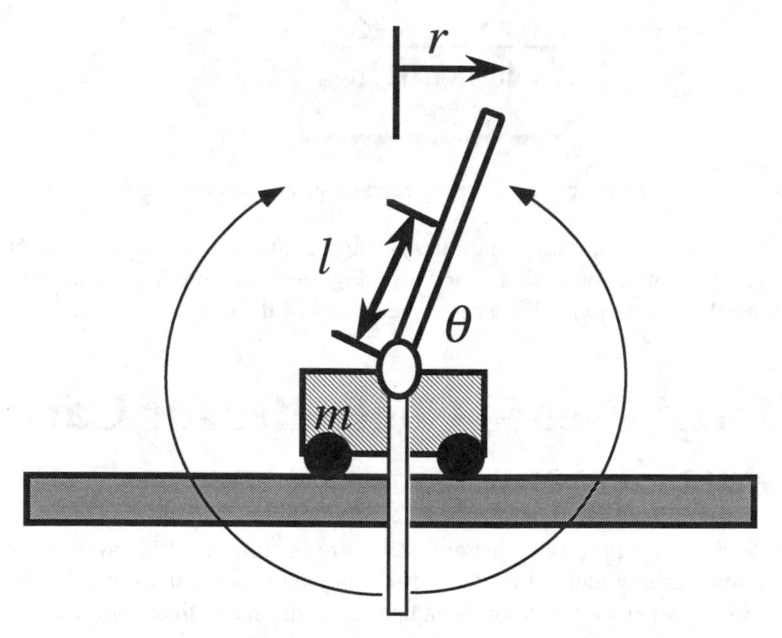

Figure 8 Cart-Pole system.

Inputs to the skill knowledge database are l and m, and inputs of the fuzzy-neuro controller shown in Figure 9 are r, $\dot{r}$, θ, and $\dot{\theta}$. The number of individuals is 50. Mutation rate is 0.5%. We use Equation (15) as the fitness function that is a modified Equation (9) for this simulation.

$$F(S_n) = \alpha \sum_{t=0}^{T} (\theta_d - \theta(t))^2 + \beta \sum_{t=0}^{T} (r_d - r(t))^2 + \gamma R_n + \delta M_n \tag{15}$$

Here, α, β, γ, and δ are equal to 0.0005, 1.0, 0.001, and 0.001, respectively.

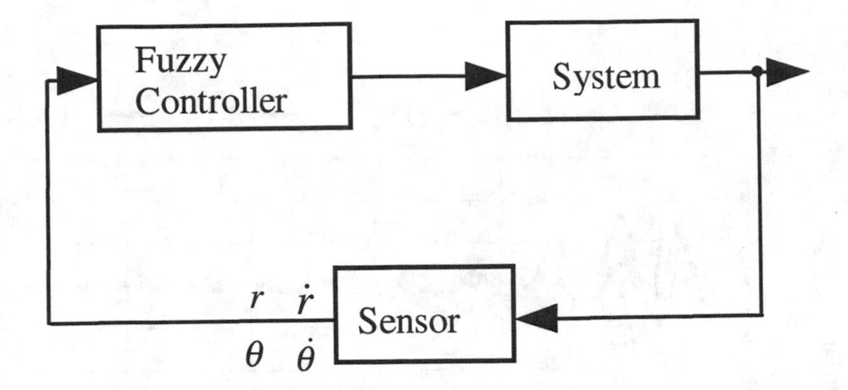

Figure 9 RBF Fuzzy-Neuro controller.

We apply the controller to four different poles for the learning process. Mass/length of four poles is 0.1kg/0.5m, 0.1kg/1.5m, 0.5kg/0.5m, and 0.5kg/1.5m. Sampling time is 20 ms and control time is 30 sec.. Iteration times of each pole are 300 times. Evaluation is carried out using the pole mass of 0.3kg and length of 1m.

Figures 10 and 11 show the simulation results. Here FS-n means the n-th fuzzy-neuro controller and FS-1-4 means the integration results of FS-1 to FS-4 based on the skill knowledge database adjusted by the heuristic approach. The integrated fuzzy-neuro controller has the best performance of all controllers. Figure 12 shows that this proposed system can learn a new fuzzy-neuro controller for unknown target faster by using the skill knowledge database; here the pole with 0.1kg/0.5m is first target and 0.1kg/1.5m is the next target.

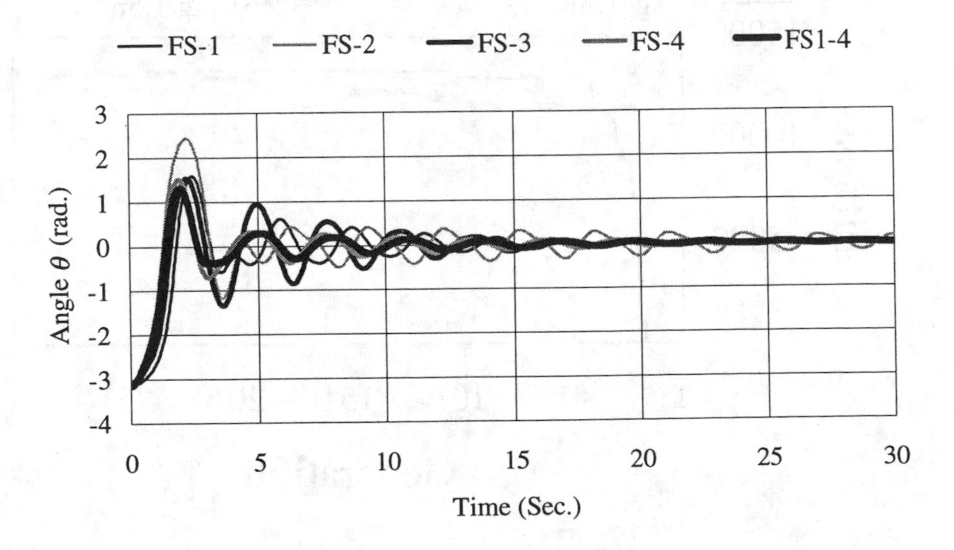

Figure 10 Simulation results of pole angle.

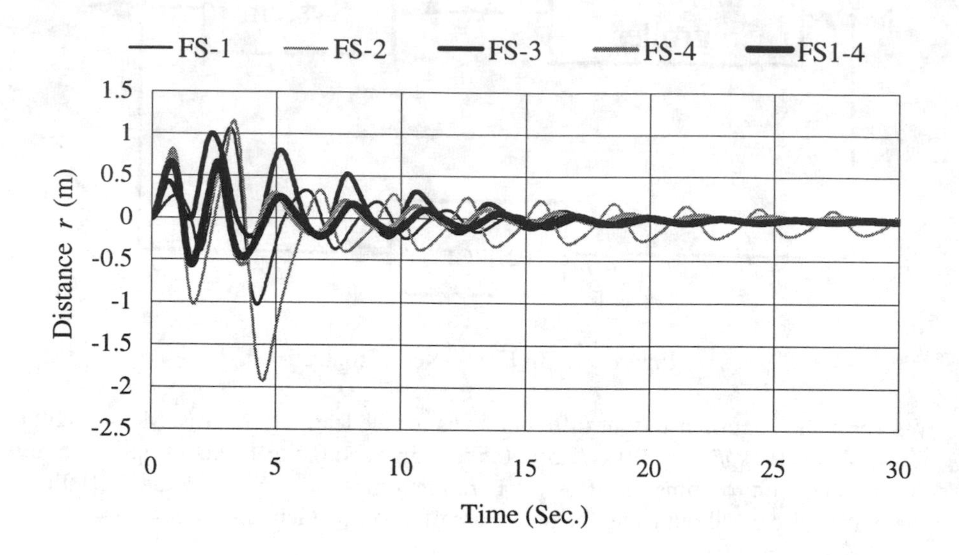

Figure 11 Simulation results of cart position.

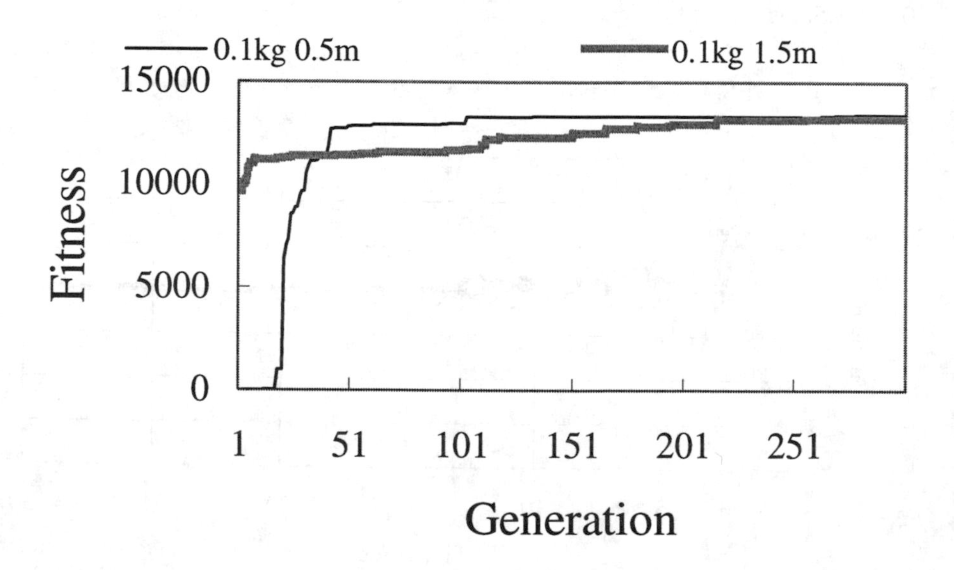

Figure 12 Learning results.

5. Conclusions

In this chapter, we proposed a new hierarchical fuzzy-neural control system based on the skill knowledge database. The skill knowledge database consists of the skills which are the fuzzy-neuro controller acquired through the GA based unsupervised learning and their membership functions. Membership functions of the skill database are used for integration of the skills. In this system, the skill database manages the skills in order to accomplish the given task.

We also show the effectiveness of the proposed system through simulations. These results show that the skill knowledge database can manage the skills to accomplish the given task with high performance.

References

[1] Fukuda, T., Shimojima, K., Arai, F., Matsuura, H. (1993) Multi-Sensor Integration System based on Fuzzy Inference and Neural Network, Journal of Information Sciences, Vol. 71, No. 1 and 2, pp. 27-41.

[2] Shimojima, K., Fukuda, T., Arai, F., Matsuura, H. (1992) Multi-Sensor Integration System utilizing Fuzzy Inference and Neural Network, Journal of Robotics and Mechatronics, Vol. 4, No. 5, pp. 416-421.

[3] Shimojima, K., Fukuda, T., Arai, F., Matsuura, H. (1993) Fuzzy Inference Integrated 3-D Measuring System with LED Displacement Sensor and Vision System, Journal of Intelligent and Fuzzy Systems, Vol. 1, No. 1, pp. 63-72.

[4] Higgins, C.M., Goodman, R.M. (1992) Learning Fuzzy Rule-Based Neural Networks for Function Approximation, Proc. of IJCNN, Vol.1, pp. 251-256.

[5] Horikawa, S., Furuhashi, T., Uchikawa, Y. (1992) On fuzzy modeling using fuzzy neural networks with the back-propagation algorithm, IEEE Trans. Neural Networks, Vol. 3, No. 5, pp. 801-806.

[6] Rumelhart, D.E., McClelland, J.L. and The PDP Research Group (1986) Parallel Distributed Processing, Vol. 1, 547; Vol. 2, 611, MIT Press.

[7] Katayama, R., Kajitani, Y., Kuwata, K., Nishida, Y. (1993) Self generating radial basis function as neuro-fuzzy model and its application to nonlinear prediction of chaotic time series, IEEE Intl. Conf. on Fuzzy Systems 1993, pp. 407- 414.

[8] Linkens, D.A., Nie, J. (1993) Fuzzified RBF Network-based learning control: structure and self-construction, IEEE Intl. Conf. on Neural Networks 1993, pp. 1016-1021.

[9] Lee, M.A, Takagi, H. (1993) Integrating Design Stages of fuzzy systems using genetic algorithms, Proc. of Second IEEE International Conference on Fuzzy System, pp. 612-617.

[10] Whitley, D., Strakweather, T., Bogart, C. (1990) Genetic Algorithms and Neural Networks: Optimizing Connection and Connectivity, Parallel Computing 14, pp. 347-361.

[11] Nomura, H., Hayashi, I., Wakami, N. (1991) A self-tuning method of fuzzy control by descent method, Proc. of 4th IFSA Congress, Engineering, pp. 155-158.

[12] Shimojima K., Fukuda T., Hasegawa Y. (1995) RBF-Fuzzy System with GA Based Unsupervised/Supervised Learning Method, Proc. of Intl. Joint Conf. of 4th Fuzz-IEEE/2nd IFES, pp. 253-258.

Chapter 3:

Expert Knowledge-Based Direct Frequency Converter Using Fuzzy Logic Control

EXPERT KNOWLEDGE-BASED DIRECT FREQUENCY CONVERTER USING FUZZY LOGIC CONTROL

E. Wiechmann and R. Burgos
Department of Electrical Engineering
University of Concepcion
Chile

This chapter presents the analysis and design of an eXpert knowledge based Direct Frequency Converter controlled by fuzzy logic (XDFC). Space vectors, knowledge based control, and fuzzy logic are combined to control the proposed converter. The XDFC main feature is the capability of achieving a unity ac-ac voltage gain, thus eliminating the need for coupling transformers, thereby enabling the straight use of XDFC driven standard voltage motors from a system of the same standard nominal voltage. The control system of the converter simultaneously controls the output and input currents. A set of rules is used by the expert knowledge based space vector modulation technique to significantly reduce processing time. While performing like a predictive current control-loop, this method is independent of the load's parameters. Finally, the converter operates with a maximum commutation frequency of 850 Hz throughout a wide output frequency range, thus reducing the converter's power losses, switches' stress, and increasing the operational power rating it can handle.

1. Introduction

The Direct Frequency Converter (DFC) is believed to be the natural evolution of the conventional ac drive, comprised of a diode bridge (ac-dc stage), a dc link filter, and a Voltage Source Inverter (VSI) (dc-ac stage). This evolution will depend on the scientific research semiconductors undergo during the years to come, as these devices seem to be the only restraint that has kept DFCs out of industrial production.

Direct frequency converters were envisioned by Gyugyi and Pelly in 1976 [1]. The authors conceived the idea of a static power converter capable of directly converting ac power. Later in 1979, Wood introduced a whole new concept and theory for designing and analyzing switching power converters [2][3]. Given a generalized converter structure (switches' array or matrix) with *n* input phases and *m* output

0-8493-9804-5/99/$0.00+$.50

phases, he stated that the converter could perform any type of power conversion, i.e., ac-dc, dc-dc, dc-ac, and ac-ac, if the proper switches and control technique were employed.

The matrix converter structure introduced by Wood in [2] triggered interest in this new promising ac-ac power conversion scheme. This was the case for Alesina and Venturini, as they presented the first real implementation of an ac-ac DFC [4]. The results they achieved were so encouraging that they named the converter the Generalized Transformer. This particular converter was capable of handling bidirectional power flow, it controlled the output frequency and voltage, and could even control the input power factor, producing sinusoidal input and output waveforms throughout its operational range. The only drawback was that it offered a maximum ac-ac voltage gain of 0.5.

The advantages offered by the Generalized Transformer remain the main characteristics of the DFC. However, this converter has other intrinsic advantages when compared to conventional ac drives; namely, a reduced size and weight, as it doesn't require a dc link filter. This lack of energy storage elements allows a better dynamic response of the converter. Working under machine regeneration is completely natural for the DFC, due to its bidirectionnal switches. This is not the case with conventional drives that can't work on regeneration mode unless they employ a dc chopper to burn the extra power returned by the machine. Also, the DFC doesn't use snubber circuits when using Staggered Commutation. This switching technique was introduced by Alesina and Venturini in [5], and basically emulates the commutation of VSI converters, thus producing a soft commutation between lines.

The allure of the DFC as the next generation converter has received the attention of a number of authors throughout the years. Although the converter's structure remains with the same switches' array proposed by Wood, the control techniques employed have evolved to offer an improved converter performance. Reference [5] introduced a closed loop control for the converter, achieving an ac-ac voltage gain of 0.867. This gain was proved to be the theoretical maximum when operating the converter with high frequency modulation techniques. In reference [6], a new modeling was introduced for the DFC, named Fictitious Link. With this approach the converter operation is split into a fictitious rectifying stage, that generates the fictitious dc link voltage, and a fictitious inverter stage, that inverts the fictitious dc link voltage at a desired voltage amplitude and frequency. This approach offered a higher voltage level that reached 0.95, with current and voltage waveforms similar to conventional ac drives.

The introduction of the Space Vector Modulation (SVM) for static power converters [7], together with the development of high speed processors suitable for on-line converter control (Digital Signal Processor, DSP) motivated the development of control techniques with enhanced characteristics for the DFC. These contributions can be clearly grouped under two different trends. The first group favours sinusoidal input currents with unity power factor, at the expense of a high commutation frequency (20 kHz) which restrains the converter's ac-ac voltage gain to 0.867 [8-11]. The second group uses lower commutation frequencies (less than 1 kHz), with a unity ac-ac voltage gain, suitable for higher power applications [12-14]. This at the expense of non-sinusoidal input currents similar to conventional ac drives.

The DFC presented in this chapter clearly belongs to the second trend described above. Fuzzy logic control [15][16], successfully used for various power electronics applications [17-19], is used to control this converter. Also, a new SVM technique is presented, named eXpert knowledge-based SVM (XSVM), whose combined action with fuzzy control originates the eXpert Direct Frequency Converter (XDFC). The XDFC is capable of working with a unity ac-ac voltage gain and a maximum commutation frequency of 850 Hz. The unity voltage gain allows XDFC driven motors rated at the system's nominal voltage to work at this voltage level without requiring transformers to counteract the voltage loss from other types of DFCs or conventional converters. The XDFC controls the load line currents, keeping their distortion below 6%, and reduces the harmonic distortion of the input line currents by simultaneously controlling them. To achieve these control goals the converter is required to measure the output line currents and input phase voltages. These are then used to realize a software waveform reconstruction of the output line voltages and input lines currents, thus reducing the measuring devices and circuitry required for controlling the XDFC.

2. XDFC Topology and Operation

The XDFC structure used is the one presented by Wood in [2], i.e., the nine bidirectional switches array or matrix. The converter is shown in Figure 1. Each input line *r*, *s*, *t* is connected through a switch to each output line *a*, *b*, and *c*. The input port has a capacitive input filter to provide a voltage source at this port, while the output port has an inductive filter which produces a current source. This filter is not required when the XDFC feeds a motor load, as the machine itself has a current source characteristic.

In order to produce a safe commutation of the converter switches, two conditions must be avoided. The first one is not to short the input voltage source, and the second one, not to open the output current source. These restraints can be stated mathematically as follows:

$$\sum_{j=1,4,7} S_j = 1 \wedge \sum_{j=2,5,8} S_j = 1 \wedge \sum_{j=3,6,9} S_j = 1 \tag{1}$$

$$S_j = \begin{cases} 1 \text{ if ON} \\ 0 \text{ if OFF} \end{cases}$$

Where S_j denotes switch j in Figure 1.

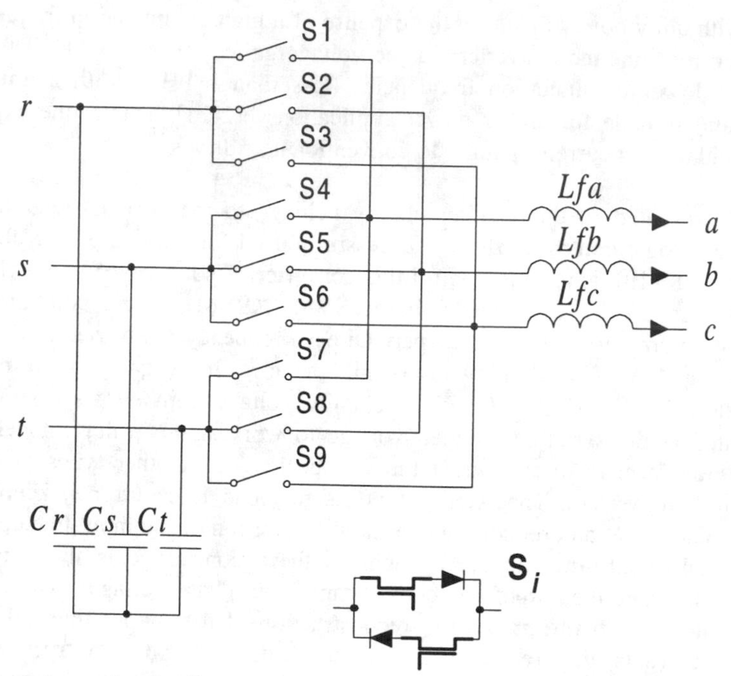

Figure 1: Schematic of a forced-commutated DFC using nine bidirectional switches.

These restraints imply that from the 512 (2^9) possible switching combinations of the converter structure, only 27 are valid and thus comply with Equation (1). These valid switching combinations are named the converter's Electric States, and are shown in Table I. This Table shows the converter switches which are on and off for every state, and the respective output voltages and input currents as a function of the input voltages and output currents, respectively.

In order to simplify the analysis and modeling of the DFC, the Generalized Transfer Function presented in [20] is used. Basically, the transfer function relates the converter's input and output electric variables, without actually dealing with the converter's structure, switches, drivers, etc. It is just a simple means of modeling highly complex converters. Table II shows various converters and their input and output dependent and independent variables, where the input port is connected to the power source, and the output port to the load. A converter transfer function can be employed to determine the dependent variable as a function of the independent variable. Thus, using the definition of dependent variables, the transfer function of a three-phase static power converter can be defined as

$$\text{conv. dep. electrical variable} = T.F. \times \text{conv. indep. electrical variable} \tag{2}$$

TABLE I
DFC ELECTRIC STATES

	Switch state (1=on, 0=off)									Connection			Voltage			Current		
Nº	S1	S2	S3	S4	S5	S6	S7	S8	S9	a	b	c	ab	bc	ca	r	s	t
1	1	0	0	0	1	1	0	0	0	r	s	s	rs	0	-rs	a	-a	0
2	0	0	0	1	0	0	0	1	1	s	t	t	st	0	-st	0	a	-a
3	0	1	1	0	0	0	1	0	0	t	r	r	tr	0	-tr	-a	0	a
4	1	1	0	0	0	1	0	0	0	r	r	s	0	rs	-rs	-c	c	0
5	0	0	0	1	1	0	0	0	1	s	s	t	0	st	-st	0	-c	c
6	0	0	1	0	0	0	1	1	0	t	t	r	0	tr	-tr	c	0	-c
7	0	1	0	1	0	1	0	0	0	s	r	s	-rs	rs	0	b	-b	0
8	0	0	0	0	1	0	1	0	1	t	s	t	-st	st	0	0	b	-b
9	1	0	1	0	0	0	0	1	0	r	t	r	-tr	tr	0	-b	b	0
10	0	1	1	1	0	0	0	0	0	s	r	r	-rs	0	rs	-a	a	0
11	0	0	0	0	1	1	1	0	0	t	s	s	-st	0	st	0	-a	a
12	1	0	0	0	0	0	0	1	1	r	t	t	-tr	0	tr	a	0	-a
13	0	0	1	1	1	0	0	0	0	s	s	r	0	-rs	rs	c	-c	0
14	0	0	0	0	0	1	1	1	0	t	t	s	0	-st	st	0	c	-c
15	1	1	0	0	0	0	0	0	1	r	r	t	0	-tr	tr	-c	0	c
16	1	0	1	0	1	0	0	0	0	r	s	r	rs	-rs	0	-b	b	0
17	0	0	0	1	0	1	0	1	0	s	t	s	st	-st	0	0	-b	b
18	0	1	0	0	0	0	1	0	1	t	r	t	tr	-tr	0	b	0	-b
19	1	0	0	0	1	0	0	0	1	r	s	t	rs	st	tr	a	b	c
20	0	0	1	1	0	0	0	1	0	s	t	r	st	tr	rs	c	a	b
21	0	1	0	0	0	1	1	0	0	t	r	s	tr	rs	st	b	c	a
22	0	1	0	1	0	0	0	0	1	s	r	t	-rs	-tr	-st	b	a	c
23	0	0	1	0	1	0	1	0	0	t	s	r	-st	-rs	-tr	c	b	a
24	1	0	0	0	0	1	0	1	0	r	t	s	-tr	-st	-rs	a	c	b
25	1	1	1	0	0	0	0	0	0	r	r	r	0	0	0	0	0	0
26	0	0	0	1	1	1	0	0	0	s	s	s	0	0	0	0	0	0
27	0	0	0	0	0	0	1	1	1	t	t	t	0	0	0	0	0	0

TABLE II
ELECTRICAL VARIABLE CLASSIFICATION FOR THREE-PHASE STATIC POWER CONVERTERS

	Input Voltage	Output Voltage	Input Current	Output Current
controlled rectifier	*independent*	*dependent*	*dependent*	*independent*
boost rectifier	*dependent*	*independent*	*independent*	*dependent*
current source inverter	*dependent*	*independent*	*independent*	*dependent*
voltage source inverter	*independent*	*dependent*	*dependent*	*independent*
DFC (matrix) converter	*independent*	*dependent*	*dependent*	*independent*

For example, a controlled rectifier's voltage independent variable is the input terminal's ac mains, and its dependent variable the output terminal's dc voltage. On the contrary, the independent current variable is the output terminal's dc current, and the dependent variable the input terminal's ac line current. Clearly, two transfer functions are defined. One relates input and output voltages, and the other one input and output currents. However, these voltage and current transfer functions are a single converter transfer function. This can be proved for every converter shown in Table II.

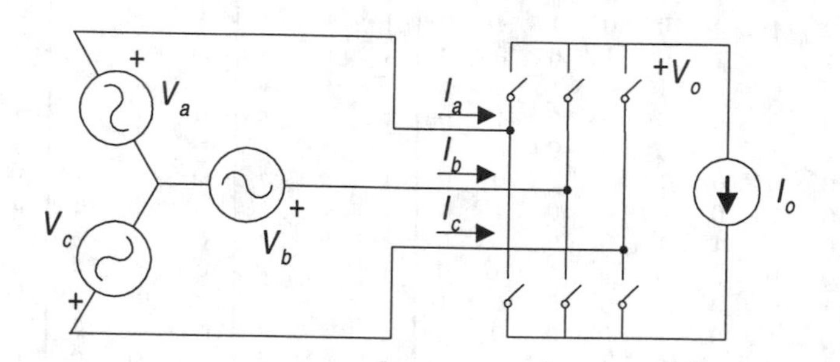

Figure 2. Schematic of a three-phase rectifier, including ac mains (V_a, V_b, V_c) and a current-source load (I_o). Input line currents (I_a, I_b, I_c) and output dc voltage (V_o) are also shown.

Let us consider the system shown in Figure 2, comprising the ac mains, a three-phase rectifier, and a current-source load. If an ideal converter is considered, i.e., with zero losses and no energy storage elements, then the following relation can be established considering the input and output instantaneous power.

$$P_{in} = P_{out} \tag{3}$$

Expanding (3) yields the next expression,

$$v_a \cdot i_a + v_b \cdot i_b + v_c \cdot i_c = v_o \cdot i_o. \tag{4}$$

Using matrix notation, (4) can be rearranged as

$$v_o = \begin{bmatrix} v_a & v_b & v_c \end{bmatrix} \cdot \frac{\begin{bmatrix} i_a & i_b & i_c \end{bmatrix}^T}{i_o} \tag{5}$$

By inspection of (5), the input *independent* voltages are multiplied by a matrix to obtain the output *dependent* voltage. Hence, (5) can be rewritten using definition (2) and Table II in the following way:

$$v_o = \begin{bmatrix} v_a & v_b & v_c \end{bmatrix} \cdot H \tag{6}$$

Where matrix H is defined by Equation (7a and b), and matches the voltage transfer function characteristics.

$$H = \frac{\left[i_a \ i_b \ i_c\right]^T}{i_o} \tag{7a}$$

$$H = \left[H_a \ H_b \ H_c\right]^T \tag{7b}$$

It should be noticed that matrix H is defined even if the load current i_o is zero (7a). Therefore, as the line currents i_a, i_b, and i_c approach zero, the load current i_o approaches zero, thus defining the limit shown in (8).

$$\lim_{i_o \to 0} H = \lim_{i_o \to 0} \frac{\left[i_a \ i_b \ i_c\right]}{i_o} = 0 \tag{8}$$

The current transfer function can be defined according to (2) and Table II as the quotient between the input line currents (dependent variable) and the output current (independent variable). By observing (7a), matrix H has that form, and thus matches the converter's current transfer function characteristics. Consequently, matrix H is the converter's transfer function, and relates the input and output electric variables as shown in (9).

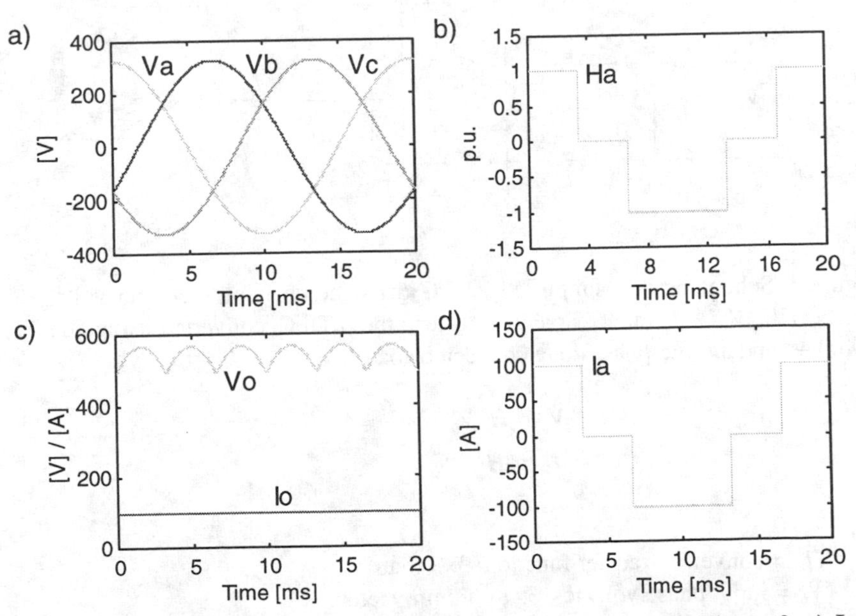

Figure 3. Electric variables of three-phase rectifier shown in Figure 2. a) Input phase voltages (v_a, v_b, v_c), b) rectifier transfer function element H_a (phase a), c) output dc voltage v_o and load current i_o, d) input line current i_a.

$$v_o = \begin{bmatrix} v_a & v_b & v_c \end{bmatrix} \cdot H$$
$$\begin{bmatrix} i_a & i_b & i_c \end{bmatrix} = H^T \cdot i_o \tag{9}$$

Figure 3 shows these relations graphically for the system shown in Figure 2, using a diode bridge as a rectifier. Figure 3a) shows the independent input phase voltages of the ac mains (220 V_{rms}). Figure 3b) shows the phase a transfer function component H_a. It should be noticed that this term is simply the normalized line current (7a). Figure 3c) shows the dependent output voltage v_o obtained using (9), and the independent load current i_o. Finally, Figure 3d) shows the dependent input line current i_a obtained using (9).

The DFC is used as a case study in this chapter. Figure 4 shows a simplified converter-load system used for modeling purposes. Using the transfer function [20], and the electrical variables classifications given in Table II, the converter's input and output voltages and currents relationships can be written as shown in (10) considering Figure 4.

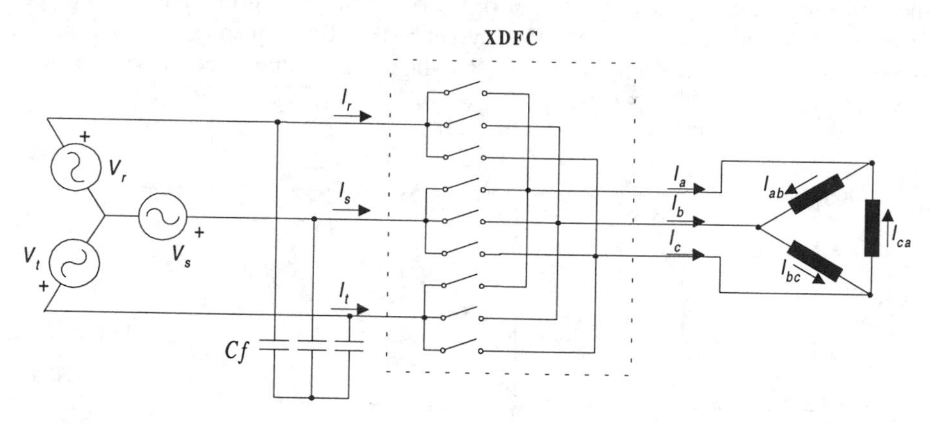

Figure 4. Schematic of a simplified XDFC drive, comprizing the input voltage source (V_r, V_s, V_t), input capacitive filter, the XDFC converter using ideal switches, and a three-phase delta connected load.

$$V_o = H \cdot V_{in}$$
$$I_{in} = H^T \cdot I_o \tag{10}$$

where,

H = converter transfer function, 3x3 matrix;
V_{in} = input phase voltages, 3x1 column vector;
V_o = output line voltages, 3x1 column vector;
I_o = output phase currents, 3x1 column vector;
I_{in} = input line currents, 3x1 column vector.

Expanding (10) yields the following two equations.

$$\begin{bmatrix} V_{ab} \\ V_{bc} \\ V_{ca} \end{bmatrix} = \begin{bmatrix} h_{11} & h_{12} & h_{13} \\ h_{21} & h_{22} & h_{23} \\ h_{31} & h_{32} & h_{33} \end{bmatrix} \cdot \begin{bmatrix} V_r \\ V_s \\ V_t \end{bmatrix} \tag{11}$$

$$\begin{bmatrix} I_r \\ I_s \\ I_t \end{bmatrix} = \begin{bmatrix} h_{11} & h_{21} & h_{31} \\ h_{12} & h_{22} & h_{32} \\ h_{13} & h_{23} & h_{33} \end{bmatrix} \cdot \begin{bmatrix} I_{ab} \\ I_{bc} \\ I_{ca} \end{bmatrix} \tag{12}$$

Elements h_{ij} of transfer function H can assume values only in {-1,0,1} to assure that Kirchoff's voltage law is satisfied in (10).

Matrix H can be written as a function of the converter's switches as

$$H = \begin{bmatrix} S1-S2 & S4-S5 & S7-S8 \\ S2-S3 & S5-S6 & S8-S9 \\ S3-S1 & S6-S4 & S9-S7 \end{bmatrix} \tag{13}$$

This equation is deduced by referring to Figure 1 and Table I. Each element h_{ij} in H can be determined by observing how the input phase voltage V_i is reflected to the corresponding output line voltage V_j. For example, element h_{11} reflects input phase voltage V_r **positively** to the output line voltage V_{ab} through switch S1, and reflects phase voltage V_r **negatively** to the output line voltage V_{ab} through switch S2, and does not affect V_{ab} at all with switch S3. Thus, element h_{11} is defined by (14),

$$h_{11} = S1\text{-}S2. \tag{14}$$

The remaining elements h_{ij} of matrix H can be determined in the same way. Transfer function H relates the input line currents of the converter with the output phase currents of the load. These currents are not available if the machine's phases are delta connected. To solve this, a current transfer function is defined that relates input and output line currents by simple inspection of Figure 1. Each input line is connected to each of the three output lines, so to comply with Kirchhoff's current law, the sum of the load's line currents multiplied by the switches states must add up to the corresponding input line current. This can be stated as

$$I_{in} = H_i \cdot I_o \tag{15}$$

where,

H_i = converter current transfer function, 3x3 matrix;
I_{in} = input line currents, 3x1 column vector;
I_o = output line currents, 3x1 column vector.

Matrix H_i can also be expressed as a function of the converter's switches, as shown in (16).

$$H_i = \begin{bmatrix} S1 & S2 & S3 \\ S4 & S5 & S6 \\ S7 & S8 & S9 \end{bmatrix} \tag{16}$$

Equations (10) through (16) totally define the XDFC's operation and provide a useful tool for modeling and controlling the converter due to the minimum processing requirements of the transfer function approach.

3. Space Vector Model of the DFC

Space vectors have proven to be an extremely useful modeling technique for static power converters. Since their introduction [7], they have been employed to modulate and control rectifiers, inverters, and DFCs [21][22]. The reason for their success is that they provide the engineer a better understanding of the converter operation.

Space vectors are obtained using a three-phase to two-phase matrix transformation. In this chapter, Park's matrix is used (17).

$$P = \sqrt{\frac{2}{3}} \cdot \begin{bmatrix} 1 & -1/2 & -1/2 \\ 0 & \sqrt{3}/2 & -\sqrt{3}/2 \end{bmatrix} \tag{17}$$

Each electric state es of the DFC can be converted using (17) into a space vector sv as shown in (18).

$$\vec{sv}_{2x1} = P_{2x3} \cdot es_{3x1} \tag{18}$$

where,

$\vec{sv}$ = space vector, 2x1 column vector;
P = Park's transformation, 2x3 matrix;
es = converter electric state, 3x1 column vector.

Space vectors are bidimensional vectors, and thus can also be written in complex number notation, where element $\vec{sv}_{11}$ is the real part and element $\vec{sv}_{21}$ is the imaginary part.

Each converter electric state has two associated vectors, a voltage space vector toward the load side and a current space vector toward the input side. For voltage space vectors, es is defined by (19), and determined by (11) as a function of the input phase voltages and the converter's transfer function H.

$$es = \begin{bmatrix} V_{ab} \\ V_{bc} \\ V_{ca} \end{bmatrix} \tag{19}$$

For current space vectors, *es* is defined by (20), and determined by (15) as a function of the output line currents and the converter's current transfer function H_i.

$$es = \begin{bmatrix} I_r \\ I_s \\ I_t \end{bmatrix} \tag{20}$$

Using complex number notation, space vectors can be written as

$$s\vec{v} = \mathrm{Re}\{s\vec{v}\} + j \cdot \mathrm{Im}\{s\vec{v}\} \tag{21}$$

where its module and argument are defined by (22).

$$\begin{aligned} |s\vec{v}| &= \sqrt{\mathrm{Re}^2\{s\vec{v}\} + \mathrm{Im}^2\{s\vec{v}\}} \\ \arg(s\vec{v}) &= \tan^{-1}\left(\frac{\mathrm{Im}\{s\vec{v}\}}{\mathrm{Re}\{s\vec{v}\}}\right) \end{aligned} \tag{22}$$

Table III shows the voltage and current space vectors for each of the 27 XDFC's electric states. For the sake of simplicity, sinusoidal input voltages and sinusoidal output currents have been considered. These are shown in (23).

$$\begin{aligned} V_r &= \sqrt{2}V \cdot \cos(w_i t) & I_a &= \sqrt{2}I \cdot \cos(w_i t) \\ V_s &= \sqrt{2}V \cdot \cos(w_i t - 2\pi/3) & I_b &= \sqrt{2}I \cdot \cos(w_i t - 2\pi/3) \\ V_t &= \sqrt{2}V \cdot \cos(w_i t + 2\pi/3) & I_c &= \sqrt{2}I \cdot \cos(w_i t + 2\pi/3) \end{aligned} \tag{23}$$

Figure 5 depicts voltage and current space vectors in the two-phase (α–β) plane. Space vectors 1 to 18 are stationary, i.e., they do not rotate; however, their phase changes in ±180° as their module varies sinusoidally in time as a function of time and the input frequency for voltage space vectors, and as a function of time and the output frequency for current space vectors. Space vectors 19 to 24 have a fixed module and a varying phase; consequently, they rotate as a function of time and the input frequency for voltage space vectors, and as a function of time and the output frequency for current space vectors. Space vectors 25, 26, and 27 are named null space vectors, as they produce zero output voltages and zero input currents.

TABLE III
DFC VOLTAGE AND CURRENT SPACE VECTORS

N°	$\lvert V_o \rvert$	arg(V_o)	$\lvert I_i \rvert$	arg(I_i)
1	$\sqrt{6}\cdot V\cdot\cos(\omega_i t+\pi/6)$	30°	$\sqrt{2}\cdot I\cdot\cos(\omega_o t)$	-30°
2	$\sqrt{6}\cdot V\cdot\cos(\omega_i t-\pi/2)$	30°	$\sqrt{2}\cdot I\cdot\cos(\omega_o t)$	90°
3	$\sqrt{6}\cdot V\cdot\cos(\omega_i t+5\pi/6)$	30°	$\sqrt{2}\cdot I\cdot\cos(\omega_o t)$	-150°
4	$\sqrt{6}\cdot V\cdot\cos(\omega_i t+\pi/6)$	90°	$\sqrt{2}\cdot I\cdot\cos(\omega_o t+2\pi/3)$	150°
5	$\sqrt{6}\cdot V\cdot\cos(\omega_i t-\pi/2)$	90°	$\sqrt{2}\cdot I\cdot\cos(\omega_o t+2\pi/3)$	-90°
6	$\sqrt{6}\cdot V\cdot\cos(\omega_i t+5\pi/6)$	90°	$\sqrt{2}\cdot I\cdot\cos(\omega_o t+2\pi/3)$	30°
7	$\sqrt{6}\cdot V\cdot\cos(\omega_i t+\pi/6)$	150°	$\sqrt{2}\cdot I\cdot\cos(\omega_o t-2\pi/3)$	-30°
8	$\sqrt{6}\cdot V\cdot\cos(\omega_i t-\pi/2)$	150°	$\sqrt{2}\cdot I\cdot\cos(\omega_o t-2\pi/3)$	90°
9	$\sqrt{6}\cdot V\cdot\cos(\omega_i t+5\pi/6)$	150°	$\sqrt{2}\cdot I\cdot\cos(\omega_o t-2\pi/3)$	-150°
10	$\sqrt{6}\cdot V\cdot\cos(\omega_i t+\pi/6)$	-150°	$\sqrt{2}\cdot I\cdot\cos(\omega_o t)$	150°
11	$\sqrt{6}\cdot V\cdot\cos(\omega_i t-\pi/2)$	-150°	$\sqrt{2}\cdot I\cdot\cos(\omega_o t)$	-90°
12	$\sqrt{6}\cdot V\cdot\cos(\omega_i t+5\pi/6)$	-150°	$\sqrt{2}\cdot I\cdot\cos(\omega_o t)$	30°
13	$\sqrt{6}\cdot V\cdot\cos(\omega_i t+\pi/6)$	-90°	$\sqrt{2}\cdot I\cdot\cos(\omega_o t+2\pi/3)$	-30°
14	$\sqrt{6}\cdot V\cdot\cos(\omega_i t-\pi/2)$	-90°	$\sqrt{2}\cdot I\cdot\cos(\omega_o t+2\pi/3)$	90°
15	$\sqrt{6}\cdot V\cdot\cos(\omega_i t+5\pi/6)$	-90°	$\sqrt{2}\cdot I\cdot\cos(\omega_o t+2\pi/3)$	-150°
16	$\sqrt{6}\cdot V\cdot\cos(\omega_i t+\pi/6)$	-30°	$\sqrt{2}\cdot I\cdot\cos(\omega_o t-2\pi/3)$	150°
17	$\sqrt{6}\cdot V\cdot\cos(\omega_i t-\pi/2)$	-30°	$\sqrt{2}\cdot I\cdot\cos(\omega_o t-2\pi/3)$	-90°
18	$\sqrt{6}\cdot V\cdot\cos(\omega_i t+5\pi/6)$	-30°	$\sqrt{2}\cdot I\cdot\cos(\omega_o t-2\pi/3)$	30°
19	$3\cdot V/\sqrt{2}$	$\omega_i t+\pi/6$	$\sqrt{3/2}\cdot I$	$\omega_o t$
20	$3\cdot V/\sqrt{2}$	$\omega_i t-\pi/2$	$\sqrt{3/2}\cdot I$	$\omega_o t+2\pi/3$
21	$3\cdot V/\sqrt{2}$	$\omega_i t+5\pi/6$	$\sqrt{3/2}\cdot I$	$\omega_o t-2\pi/3$
22	$3\cdot V/\sqrt{2}$	$-\omega_i t-5\pi/6$	$\sqrt{3/2}\cdot I$	$-\omega_o t-2\pi/3$
23	$3\cdot V/\sqrt{2}$	$-\omega_i t-\pi/6$	$\sqrt{3/2}\cdot I$	$-\omega_o t+2\pi/3$
24	$3\cdot V/\sqrt{2}$	$-\omega_i t+\pi/2$	$\sqrt{3/2}\cdot I$	$-\omega_o t$
25	0	-	0	-
26	0	-	0	-
27	0	-	0	-

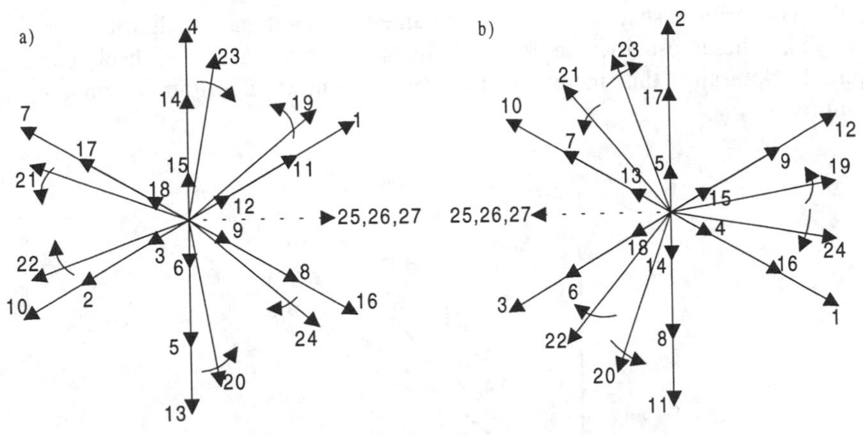

Figure 5. Space vector representation of the DFC states.
a) Voltage space vectors and b) current space vectors at $\omega t = 0°$.

4. Expert Knowledge-Based SVM

A modulation technique with low commutation frequency (lower than 1 kHz) enables the converter to work with higher power ratings. It results in increased converter efficiency and reduced switches' stress. Therefore, converters may operate with power levels previously suitable only for six-pulse converters.

A modulation technique that falls within these characteristics is the predictive current loop SVM. This modulation technique has been used for VSI converters and DFCs [13][21][22]. The technique is capable of controlling the output line currents keeping their distortion beneath a desired value, usually 4% to 6%. This is done while operating with a maximum commutation frequency of 600 Hz throughout the whole output frequency range (usually under 120 Hz).

Although this technique offers a remarkable converter performance, it poses some problems [21][22]. The technique forces the load's line current space vector I_l to stay within the error zone in the α–β plane of the current space vector reference I_{ref}. Whenever I_l falls out of the accepted error zone surrounding I_{ref}, the controller predicts the current trend for every converter state. It then selects the state that brings the current back to the accepted error zone for the longest time. This operation is depicted in Figure 6. Naturally, the current prediction is limited by the output frequency at which the converter is operating. Usually, for high output frequencies, the algorithm must be modified to avoid the unacceptable time delay produced by the required processing. Specifically, the current trend is predicted at a different time, not only when the controller detects the current error. The algorithm presents another drawback, which is due to the simplified load model used to actually predict the current trend for the different converter states. In order to assure a proper prediction for the converter's state selection, a parameter identification algorithm must be employed, where, usually, the load's inductance, resistance, and back e.m.f. are required. Naturally, this increases the overall control algorithm complexity and reliability.

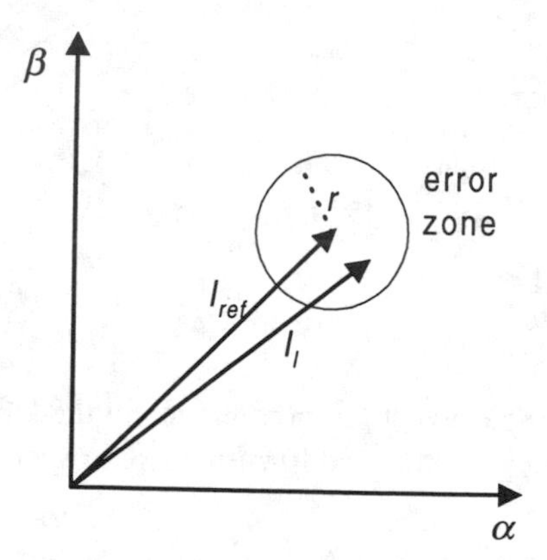

Figure 6. Predictive-current algorithm operation. Whenever the line current space vector I_l falls out of the line current reference space vector I_{ref}'s accepted error zone, the converter selects the next converter state by predicting the current trend for every converter state.

In this section a new eXpert knowledge-based SVM (XSVM) technique is presented, which is based upon an expert knowledge of the converter's operation. The technique uses a set of rules to determine the next converter state, depending on the input (measured) variables. For output and input current control, only the load's line currents and input phase voltages must be measured. The output line voltages and input line currents are also required. They are obtained by a software waveform reconstruction using the converter's transfer functions H and H_i and Equations (10) and (14). This software reconstruction reduces and simplifies the measuring and control circuitry for the XDFC. The presented XSVM technique requires a reduced processing time, being 70 times faster than the predictive current control algorithm employed in voltage source inverters. It is also independent of the load's parameters, eliminating the need for online parameter identification. This independence is achieved by the way in which the next converter state is selected, based only on how the converter's input and output variables vary in time.

The SVM presented herein is used to control both input and output currents, thus it uses two different sets of rules. These are based on a predictive current controlled converter operation, and, therefore, present performances similar to that technique. Basically, the XDFC with XSVM simultaneously controls the input and output currents. The load current distortion is kept under 6%, and the input current distortion is diminished. The input-output control slightly increases the converter's commutation frequency. However, it keeps the maximum below 850 Hz. The modulation presented allows the XDFC to operate with a unity ac-ac voltage gain. This fact is an important achievement that enables XDFC drives to operate with the system's nominal voltage level, hence eliminating the need for coupling transformers used to counteract the converter's voltage loss of other DFCs.

5. XDFC Control

5.1 XDFC Control Strategy and Operation

The converter employs a fuzzy logic controller and the expert knowledge based SVM introduced in Section 4. Software waveform reconstruction is performed by using transfer functions H and H_i, which model the converter operation. According to Table I matrix H can take 25 different forms from the 27 different electric states. These are transformed into space vectors using Park's matrix (Table III), which are required by the XSVM used for this converter.

The modeling approach chosen, based on the converter transfer function H, sets two different control objectives. The first one is to control the output line currents of the XDFC, which is the prime objective as they are the load's currents. This control is realized with H using the voltage space vectors. The second objective is to reduce the input line current distortion and, thus, increase the input power factor. This control is also realized with H (or H_i), but this time using the current space vectors in the XSVM. Clearly, there are two completely different control goals for the XDFC, and both must be fulfilled using the same means, i.e., the converter's transfer function H. To solve this dilemma, a fuzzy logic controller is used. The fuzzy controller determines which XDFC side has higher priority, either the output-load side or input-utility side. Then, it hands over command of the converter to the XSVM algorithm of the chosen port while the next converter state is selected.

Figure 7 shows a block diagram of the converter's control strategy. The input phase voltages and output line currents are measured and used by a block named Main Controller. This block measures, at every sampling instant, the output and input errors, e_o and e_i, respectively. If either accepted error is exceeded, then the fuzzy controller determines which port has higher priority. Then, the selected XSVM modulator determines the next converter state. For this, it requires the output and input voltage and current space vectors, obtained by the Main Controller block. Finally, from the selected state, the gating signals to the switches are generated.

5.2 Fuzzy Logic Controller

Fuzzy logic is a logical system that seeks to emulate human thinking and natural language [15][16]. Fuzzy control, which has emerged as one of the most active branches of fuzzy logic due to its intrinsic characteristics, provides a means of converting a control strategy comprised of a set of linguistic rules based on expert knowledge into an automatic control strategy. This control has proven extremely useful in various industrial applications [17-19], where, usually, control by conventional methods produces inferior results, especially when information being processed is inexact or uncertain.

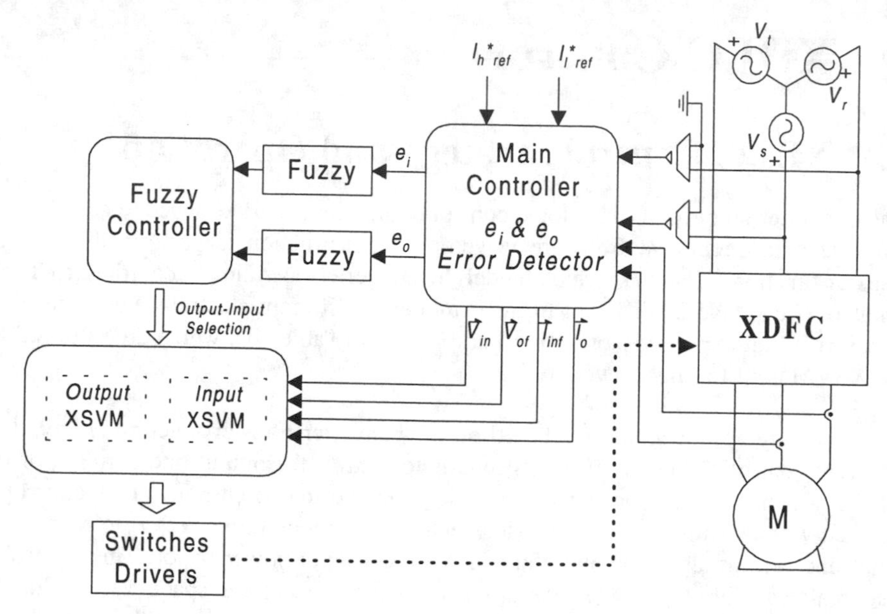

Figure 7: Control algorithm block diagram for XDFC drive.

Fuzzy logic has a unique and distinct feature of allowing partial membership, that is, a given element can be a partial member of more than one fuzzy set, with various membership degrees. The degree of membership varies from 0 (nonmember) to 1 (full member). In conventional or *crisp* sets, an element can either be a member or not of a certain set. Figure 8 shows the differences between a fuzzy set and a crisp set of a vehicle speed control system. In Figure 8a) a vehicle doing 73 km/hr is *cruising*, even though the speed limit for *fast* is 75 km/hr. In Figure 8b), using fuzzy sets, the same vehicle is a partial member of both *cruise* and *fast*, being closer to being a full member of *fast* than of *cruise*.

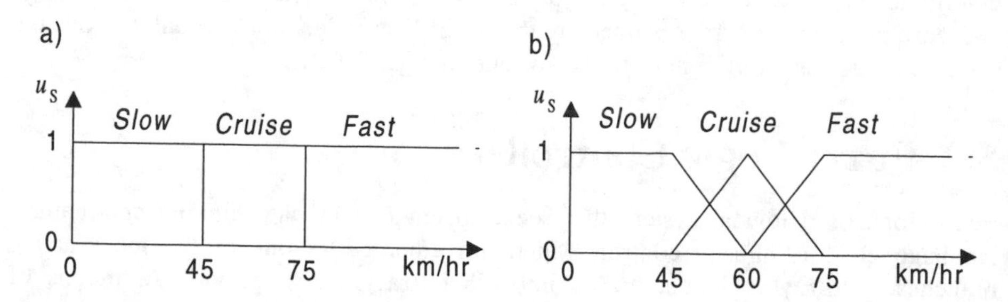

Figure 8: Representation of vehicle speed using a) crisp sets, and b) fuzzy sets theory.

In this particular application, control of the XDFC, fuzzy logic is used to determine which converter port has higher priority and, thus, should be controlled. Once the decision has been made, the fuzzy controller passes the converter's command to the XSVM controlling either the output or input terminals while the next converter state is being selected. To accomplish this, the fuzzy controller uses two fuzzy variables. Namely, the output line current error and the input harmonic current error. Whenever these variables trespass the corresponding accepted errors, the fuzzy controller is engaged. The fuzzified variables are then processed using the set of linguistic rules developed based on expert knowledge of the XDFC. From these rules the final decision is taken, specifying which converter port has higher priority and, thus, should be controlled in order to comply with both control objectives.

It seems clear now that the fuzzy controller is critical for the converter operation. It is basically the brains of it. This specific controller was fuzzified due to the intrinsic operation that this logic offers for controlling processes. Specifically, fuzzy control in this particular case realizes a linear interpolation between the two possible control actions it has, controlling either output or input converter ports, so the overall action is a smooth transition between both converter ports. On the contrary, in case this controller was not fuzzified, a threshold decision maker would be required to actually select the converter side with higher priority. This would produce a nonlinear transition between both possible control actions, creating a step transition instead of a linear one. As a result, the converter's commutation frequency would double. The global effect produced by the fuzzy controller in the XDFC operation is that the converter's commutation frequency is only lightly increased, being able to maintain it beneath 850 Hz controlling both output and input currents. This represents a significant result, as the maximum commutation frequency reaches almost 600 Hz when controlling only the output currents.

The fuzzy controller employs two fuzzy variables and one control variable [15][16]. The fuzzy variables are the fuzzy output or current error e_o, and the fuzzy input or harmonic error e_i. The output and input errors are defined as

$$e_o = \left|\vec{I}_{ref} - \vec{I}_l\right| \tag{24}$$

$$e_i = \left|\vec{I}_h\right| \tag{25}$$

Where I_{ref} is the reference current space vector, I_l is the load current space vector, and I_h is the filtered input line current space vector, or the harmonics current space vector.

As shown in Figure 9a), the universe of discourse of the fuzzy variables is divided into three fuzzy sets, namely, null error (*N*), small error (*S*), and big error (*B*). A triangular form membership distribution was chosen for linear interpolation.

The control variable *c* is the converter port. As the ports are crisp, *c* does not require a membership distribution.

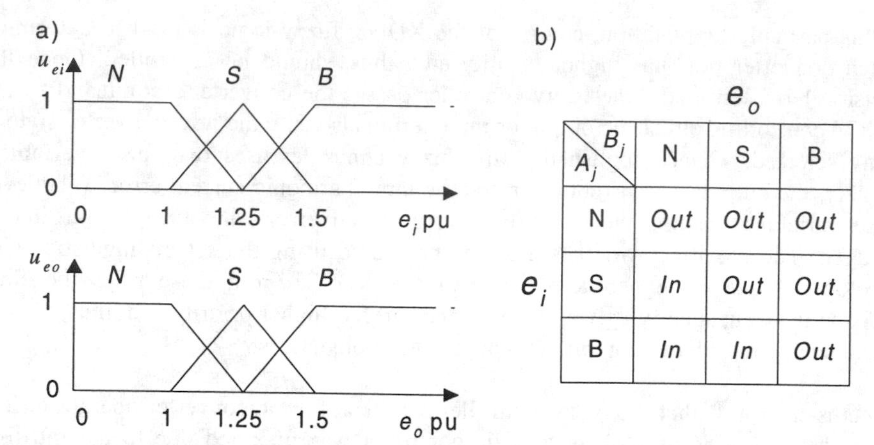

Figure 9: a) Membership functions for fuzzy variables used; namely, input error e_i, and output error e_o. b) Set of fuzzy rules for e_i and e_o, where *Out* and *In* refer to the converter port to be controlled, output and input, respectively.

For both fuzzy variables the number of fuzzy segments was chosen to have maximum control with a minimum number of rules (Figure 9b). Each rule can be stated as

$$R_j\text{: if } e_i \text{ is } A_j \text{ and } e_o \text{ is } B_j \text{ then } c \text{ is } C_j,$$

where A_j and B_j represent fuzzy segments *N*, *S*, or *B* associated to fuzzy variables e_i and e_o, respectively. As an example, consider the following values for e_i and e_o:

$$e_i = 1.2$$
$$e_o = 0.7$$

Now, using the membership functions of both fuzzy variables (Figure 9a), their degree of membership for the different fuzzy sets can be determined. For e_I, these are the following:

$$\mu_N(e_i) = 0.2$$
$$\mu_S(e_i) = 0.8$$

The output error e_o is a full member of *N* and, thus, its degree of membership is unity.

$$\mu_N(e_i) = 1$$

Then, using the fuzzy rules given in Figure 9b) the following fuzzy rules can be written for this particular case:

$$R_1\text{: if } e_i \text{ is } N \text{ and } e_o \text{ is } N \text{ then } c \text{ is } Out.$$
$$R_2\text{: if } e_i \text{ is } S \text{ and } e_o \text{ is } N \text{ then } c \text{ is } In.$$

Now, a method to determine which rule applies is required to actually make the control action, that is to decide which converter port is to be controlled. In this chapter, the fuzzy interface method employed for this purpose is the minimum operation rule used as a fuzzy implementation function. As has already been shown, the membership distribution functions of the fuzzy sets associated to each fuzzy variable and control variable, i.e., A_j, B_j, and C_j, are respectively given by μ_{Aj}, μ_{Bj}, and μ_{Cj}. Then, the firing strength of the jth rule is represented by

$$\alpha_j = \min\left(\mu_{Aj}(e_i), \mu_{Bj}(e_o)\right) \tag{26}$$

where the firing strength α_j is a measure of the contribution of jth rule to the fuzzy control action. In the example considered, the firing strengths of both active rules are given by

$$\alpha_1 = min\left(\mu_N(e_i), \mu_N(e_o)\right) = min(0.2,1) = 0.2$$

$$\alpha_2 = min\left(\mu_S(e_i), \mu_N(e_o)\right) = min(0.8,1) = 0.8$$

With fuzzy reasoning of the first type [18][19], Mamdani's minimum operation rule as a fuzzy implication function, the jth rule leads to the following control decision:

$$\mu_{Cj} = min\left(\alpha j, \mu_{Cj}(c)\right) \tag{27}$$

Therefore, the output's membership function μ_C of the output c is pointwise given by

$$\mu_C(c) = \max_{j=1}^{9}\left(\mu_{Cj}(c)\right) \tag{28}$$

Since the output c is crisp, the maximum criterion is used for defuzzification. This criterion uses, as control, output the point where the possibility distribution of the control action reaches a maximum value. With this criterion, the output for the example under study would be the maximum of the two active rules, which is rule R_2, of firing strength α_2. Therefore, the converter port to be controlled would be the *Input* port.

5.3 Load's Line Current Control

This controller is in charge of the converter's output line currents. The sole objective of this controller is to keep the load's current space vector within the accepted error zone of the reference current vector. This is the same control objective of the predictive current control; the difference lies in the way the objective is accomplished. Figure 6 shows the reference current vector and load's current vector in the α–β plane.

The controller will act upon reception of the order from the fuzzy controller, once this controller has determined that the output port has higher priority. It will then pass the command of the converter to the load's line current controller. Granted this, the controller will select the next converter state, which will be the one that will bring the current vector back to the reference current vector's error zone, and do so for the longest amount of time. The actual converter state selection is realized using the XSVM.

The output port control requires obtaining the input phase voltages, output line voltages, and the output line currents. To fulfill these requirements, only the input phase voltages and output line currents need to be physically measured by proper equipment, that is transformers and current sensors. The output line voltages are obtained by a software waveform reconstruction. This operation can be done using the converter's transfer function H, as shown in (29).

$$\begin{bmatrix} V_{ab} \\ V_{bc} \\ V_{ca} \end{bmatrix} = H \cdot \begin{bmatrix} V_r \\ V_s \\ V_t \end{bmatrix} \tag{29}$$

Sinusoidal three-phase systems produce sinusoidal two-phase systems when transformed by Park's matrix, thus producing a rotating space vector of constant magnitude. This is not the case for the output line voltages reconstructed with (29), as the line voltages are pulses varying their average value in order to follow a sinusoidal reference. So, in order to obtain the desired line voltage's space vector, the fundamental component of the line voltages is required. These are simply obtained by filtering the respective waveforms. To implement the filters, a digital approach is chosen, as it lacks all the problems associated to analog filters, specifically the parameters variation and the tuning of it. The digital filter is realized by software, and can be precisely designed to produce the required filtering characteristics.

The digital filter used for this converter is an IIR digital low pass filter [23]. It is used to obtain the fundamental frequency component V_{lf} out of the line voltages obtained with (29). The filter design parameters are shown in Table IV.

TABLE IV
Digital Filters Data

Filters	Type	Order	Cutoff f [Hz]	Pass/stop ripple [dB]
V_o, I_{in}	elliptic, low pass	4th	150	0.01 - 20
I_{in}	elliptic, band pass	14th	200 - 1050	0.01 - 20

After filtering the line voltages, these are transformed with Park's matrix into voltage space vectors together with the three phase load line currents I_l which are transformed with (17) into a space vector as in (30).

$$\vec{V}_{lf\,2x1} = P_{2x3} \cdot V_{lf\,3x1}$$
$$\vec{I}_{l2x1} = P_{2x3} \cdot I_{l3x1} \tag{30}$$

With these space vectors, the XSVM can be realized. Basically, the converter state is selected applying the following set of rules.

1. At every sampling interval, calculate the current error e_i, the phase error e_θ, the input phase voltage vector's phase α, and the load current vector module m_i. When indicated by the fuzzy controller, select the converter state as follows.

2. Determine the converter's output line voltage vector zone in the complex plane using the filtered line voltage vector V_{lf} defined in (30), (Figure 10 a).

3. Assign space vectors V_x and V_y according to the zone determined in 2 (Figure 10 b), and the time zone given by α. V_x is the space vector leading the line voltage vector, and V_y is the one lagging it. V_x and V_y are in $\{V_1, V_2, \ldots, V_{18}\}$ as shown in Table I. The input phase voltage vector's phase α determines six time zones for each vector cycle [330°-30°, 30°-90°,…,270°-330°]. Each time zone denotes the voltage space vectors with bigger modules in each zone in the complex plane (Figure 10a). For example, if the voltage vector zone is I, then V_x will be in $\{V_1, V_2, V_3, V_{10}, V_{11}, V_{12}\}$, depending on the time zone given by α.

4. If M>0.9, then
   a) If $e_\theta > 0$ then the next state is $St = V_y$.
   b) If $e_\theta < 0$ then the next state is $St = V_x$.

5. If M<.09, then
   a) If $m_i > m_{ref}$ and $e_\theta > \theta_{min}$ then the next state is $St = V_{25}, V_{26}$ *or* V_{27}. The null state is chosen to minimize switch commutations based on the actual converter state,
   b) else i) If $e_\theta > 0$ then the next state is $St = V_y$.
   ii) If $e_\theta < 0$ then the next state is $St = V_x$.

where,

$$e_i = \left|\vec{I}_l - \vec{I}_{ref}\right|$$
$$e_\theta = \arg\left(\vec{I}_l\right) - \arg\left(\vec{I}_{ref}\right)$$

m_{ref} = reference current vector's module;
θ_{min} = minimum lagging angle for null state connection;
M = modulation index;
St = converter's state.

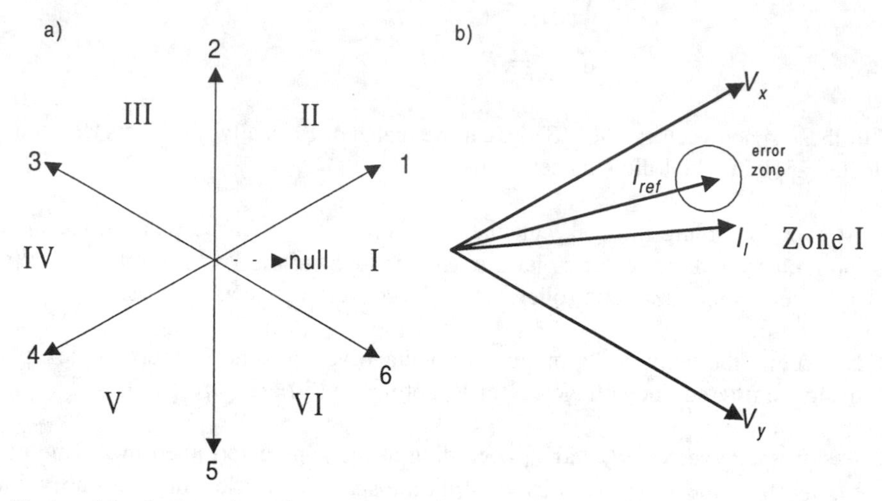

Figure 10. a) Complex plane zones defined by the converter. b) Space vectors V_x and V_y during commutation instant, i.e., I_l is out of I_{ref}'s error zone.

5.4 Input's Line Current Control

This controller is in charge of the converter's input line currents. It has a single objective which is to reduce the input current's harmonic distortion. To accomplish this goal it must keep the magnitude of the input harmonic currents restrained to a specified maximum. This is done by appropriately choosing the next converter state using the XSVM technique presented in Section 4.

The controller acts, just as the load's line current controller, upon reception of the order coming from the fuzzy controller. Once that this controller has decided that the input port has higher priority, it will pass command of the converter to the input's line current control. After the decision has been made, the input controller will select the converter state that will reduce the converter's input current distortion, which also improves the converter's power factor.

The input current control requires measurement of the input and output line currents, and the input voltages in order to select the next converter state according to the control's objective. The output line currents and input phase voltages are already measured by the load's line current control, thus leaving only the input line currents I_{in} unknown. These are obtained by software waveform reconstruction, using the converter's current transfer function H_i for this purpose. The actual operation is shown in (31).

$$\begin{bmatrix} I_r \\ I_s \\ I_t \end{bmatrix} = H_i \cdot \begin{bmatrix} I_a \\ I_b \\ I_c \end{bmatrix} \tag{31}$$

The converter input currents are also pulse-like waveforms, just like the output line voltages of the converter, therefore, they require filtering in order to produce a space vector with constant magnitude when transformed using Park's matrix. Taking advantage of the design characteristics of digital filters, two digital IIR filters are employed, a low pass filter and a band pass filter tuned to pass from the 3rd to 19th harmonic currents. With them the input line currents I_{in} are filtered, obtaining the filtered input line currents Ii_{nf} and I_h of the converter. The filtered waveforms are then used to obtain two current space vectors (17) and (32).

$$\begin{aligned} \vec{I}_{inf\,2x1} &= P_{2x3} \cdot I_{inf\,3x1} \\ \vec{I}_{h2x1} &= P_{2x3} \cdot I_{h3x1} \end{aligned} \tag{32}$$

The filter's design parameters are shown in Table IV. Finally, the converter state is selected applying the set of rules from the expert knowledge based SVM technique. These are shown below.

1. At every sampling interval, calculate the harmonic error e_i, the phase error e_γ, the input phase voltage vector's (V_{in}) phase α, and the output current vector's phase β. When indicated by the fuzzy controller, select the converter state following the next steps.

2. Assign V_x and V_y considering the input line current vector I_i zone in the complex plane (Figure 10), and the time zone of the output line current vector I_{lo}. V_x is the space vector leading the line current vector, and V_y is the one lagging it, with V_x and V_y in $\{V_1, V_2, \ldots, V_{18}\}$ as shown in Table I. The output line current vector's phase β determines six time zones for each vector cycle [0°-60°, 60°-120°,…,300°-360°]. Each time zone denotes the current space vectors with bigger modules in each zone in the complex plane (Figure 10b). For example, in case the load current vector zone is I, then V_x will be in $\{V_3, V_6, V_9, V_{12}, V_{15}, V_{18}\}$ (Figure 5), depending on the time zone given by β.
   a) If $e_\gamma > 0$ then the next state is $St = V_y$.
   b) If $e_\gamma < 0$ then the next state is $St = V_x$.

where,

$$e_i = \left|\vec{I}_h\right|$$

$$e_\gamma = \arg\left(\vec{I}_{inf}\right) - \arg\left(\vec{V}_{in}\right)$$

St = converter's state.

6. Results

The previous sections of this chapter have described the modeling, operation, and control of the presented XDFC. However, the advantages of the proposed converter have yet to be shown. In this section, results of the converter operation are given. For this purpose computer simulations were performed to validate the control algorithm presented.

The XDFC is evaluated in an Adjustable Speed Drive (ASD) like the one shown in Figure 4. The converter is controlled to produce a constant *V/f* characteristic up to 50 Hz, and constant power in the field weakening region. The load's current distortion is set to a maximum of 6%, and the input current harmonic distortion is reduced. The load considered is a 20 kVA squirrel cage induction machine. The test circuit parameters are shown in Table V.

TABLE V
TEST CIRCUIT PARAMETERS

Parameters	Values
Input rms phase voltage	120 V
Input frequency	50 Hz
Power rating	20 KVA
Squirrel cage induction machine phase parameters	X=2.4 Ω/phase R=6 Ω /phase
Input capacitive filter (Wye connected, ungrounded)	86 μF/phase

Computer simulations of the XDFC were performed using Matlab under Windows environment. Although Matlab is a computer language and not a circuit simulator, it offers multiple advantages due to its graphics interface and matrix-like functions which are proper for circuit representations. The converter's transfer function can be extensively employed when simulating under these conditions, being a powerful tool for modeling static power converters.

Example– Computer simulation using Matlab

As an example for simulating under Matlab environment, the rectifying system depicted in Figure 2 will be analyzed, but with a resistive-inductive (*R-L*) load instead of the current source. In order to simulate any circuit, all its equations must be written in the time domain. The equations for the circuit in Figure 2 are the following.

$$v_o(t) = \begin{bmatrix} v_a(t) & v_b(t) & v_c(t) \end{bmatrix} \cdot \begin{bmatrix} H_a(t) & H_b(t) & H_c(t) \end{bmatrix}^T \quad (33)$$

$$v_o(t) = R \cdot i_o(t) + L \cdot \frac{di_o(t)}{dt} \quad (34)$$

$$\begin{bmatrix} i_a(t) & i_b(t) & i_c(t) \end{bmatrix} = \begin{bmatrix} H_a(t) & H_b(t) & H_c(t) \end{bmatrix} \cdot i_o(t) \quad (35)$$

Now, if a time step of Δt is used to discretize the time variable t, Equation (34) can be rewritten as shown in (36) using a first order approximation for the current derivative.

$$v_o(t) = R \cdot i_o(t) + L \cdot \frac{i_o(t) - i_o(t - \Delta t)}{\Delta t} \tag{36}$$

The load current i_o may be determined at time instant t.

$$i_o(t) = \frac{\left(v_o(t) + L \cdot \frac{i_o(t - \Delta t)}{\Delta t} \right)}{\left(R + L/_{\Delta t} \right)} \tag{37}$$

Finally, to simulate the circuit the following steps are required:

1. Create voltages v_a, v_b, v_c as time vectors of length n, where n is given by (38).

$$n = \frac{Total_{time}}{\Delta t} \tag{38}$$

2. Create transfer functions (Figure 3b) H_a, H_b, H_c as time vectors of length n.

3. Run the following loop n times.
   a) Determine $v_o(t)$ using (33).
   b) Determine $i_o(t)$ using (37).
   c) Determine $i_a(t)$, $i_b(t)$, $i_c(t)$ using (35).

Figure 11a) shows the input phase voltage v_a and input line current i_a, and Figure 11b) shows the output voltage v_o and load current i_o. The simulations results were obtained using the following load parameters.

Initial $i_o = 0$ A;
Load Resistance = 5 Ω;
Load Inductance = 10 mH.

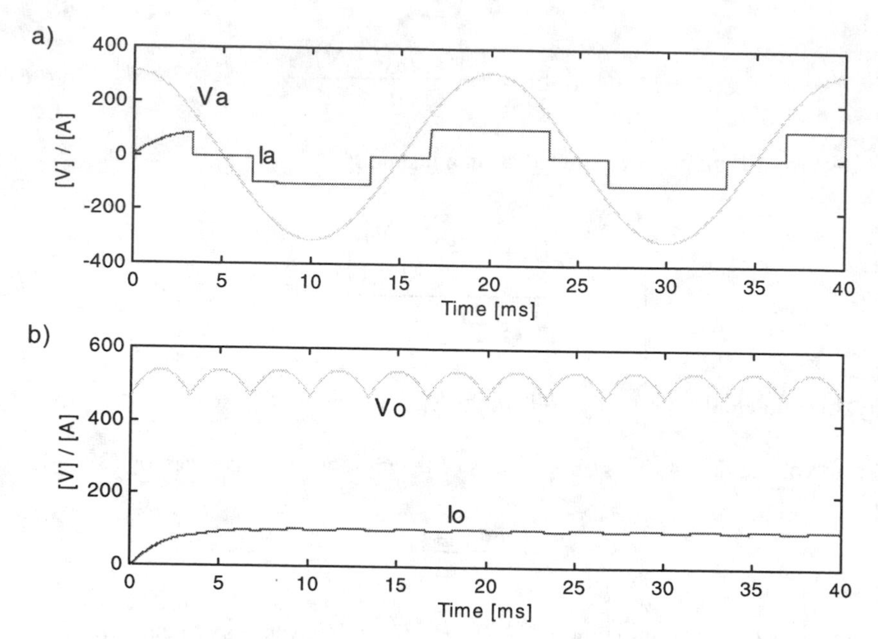

Figure 11. Simulation results of rectifying-system shown in Figure 2 using an *R-L* load instead of a current source.

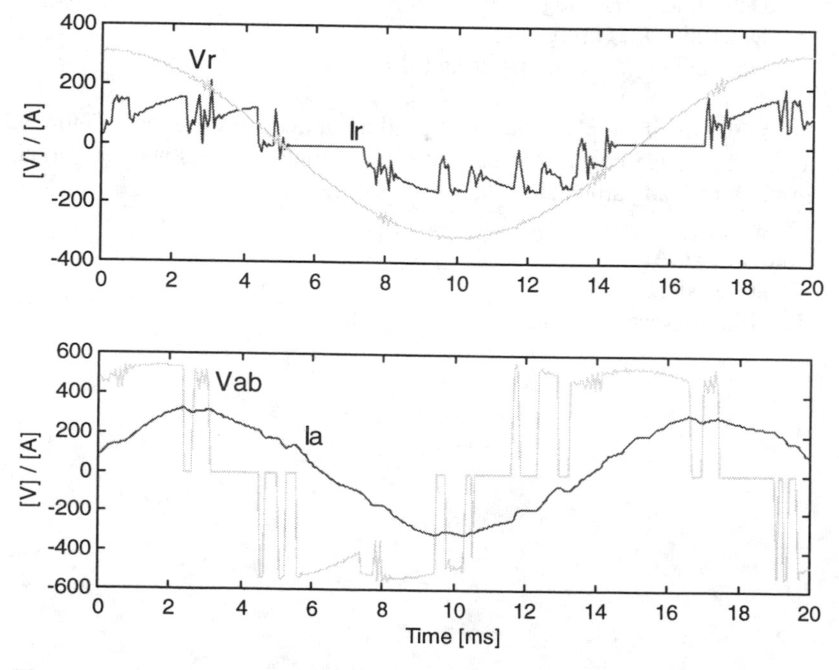

Figure 12. Input phase voltage V_r, input line current I_r, output line voltage V_{ab}, and output line current I_a of the XDFC operating at 70 Hz.

The simulations results obtained for the XDFC adjustable speed drive considered are shown in Figure 12. The converter is operating at 70 Hz output frequency. The output line voltages and line currents show this. The converter's low commutation frequency can be appreciated by looking at the output line voltage. This waveform has a reduced number of pulses, implying low commutation frequency on the converter. The load's line currents have a distortion lower than the 6% limit imposed, which is why they clearly resemble sinusoidal waveforms. The input current control action is shown in the figure. This control produces the series of chops in each current pulse border. Regarding the input displacement factor, the phase voltage and line current are nearly in phase.

7. Evaluation

This section presents an evaluation of the proposed XDFC working in an ASD. In order to compare the proposed control algorithm, the Synchronous DFC (SDFC) presented in [13] is used as reference. That converter also achieved a unity ac-ac voltage gain, and kept the converter's commutation frequency below 550 Hz. Thus, it also belongs to the high voltage gain and low commutation frequency/losses trend previously described in Section 1.

The SDFC uses the fictitious link concept to model and control the converter [6]. Under this approach, the converter's transfer function H is split into a rectifying transfer function HR, and an inverting transfer function HI. Therefore, its operation can be described as a fictitious rectifier producing a fictitious dc link, which is then inverted at the desired output frequency and amplitude by a fictitious inverter. Matrix H is redefined as the dyadic product of column vectors HI and HR as follows:

$$H = HI \cdot HR^T \tag{39}$$

where,

H = DFC transfer function, 3x3 matrix;
HR = Fictitious rectifier transfer function, 3x1 column vector;
HI = Fictitious inverter transfer function, 3x1 column vector.

With this decomposition of matrix H, the converter's input-output relations (9) can be rewritten in the following way:

$$\begin{aligned} V_o &= HI \cdot \left(HR^T \cdot V_{in}\right) \\ V_o &= HI \cdot V_{dc} \\ I_{in} &= HR \cdot \left(HI^T \cdot I_o\right) \\ I_{in} &= HR \cdot I_{dc} \end{aligned} \tag{40}$$

where,

V_{dc} = dc link voltage produced by the fictitious rectifier;
I_{dc} = dc link current drawn by the fictitious inverter.

Transfer functions *HR* and *HI* elements can take values from {-1,0,1}, being forced to add to zero in order to comply with Kirchhoff's voltage law. These transfer functions can then take seven possible combinations or Electric States, which are illustrated in Table VI.

TABLE VI
TRANSFER FUNCTION STATES FOR 3-PHASE CONVERTERS

H	S_1	S_2	S_3	S_4	S_5	S_6	S_7
H_1	1	0	-1	-1	0	1	0
H_2	0	1	1	0	-1	-1	0
H_3	-1	-1	0	1	1	0	0

The SDFC models the fictitious rectifier as a diode bridge, and the fictitious inverter as a space vector modulated VSI. Consequently, its waveforms resemble the conventional drives. The SDFC is controlled by a predictive current loop which selects the next inverter state (thus, DFC state) by predicting the current trend produced by all the converter states, and chooses the one that satisfies the control objectives of this technique. These are to keep the load's line currents space vector within the reference currents space vector V_r. Figure 13 shows the input and output waveforms of the SDFC operating at 70 Hz in an ASD with the parameters shown in Table V.

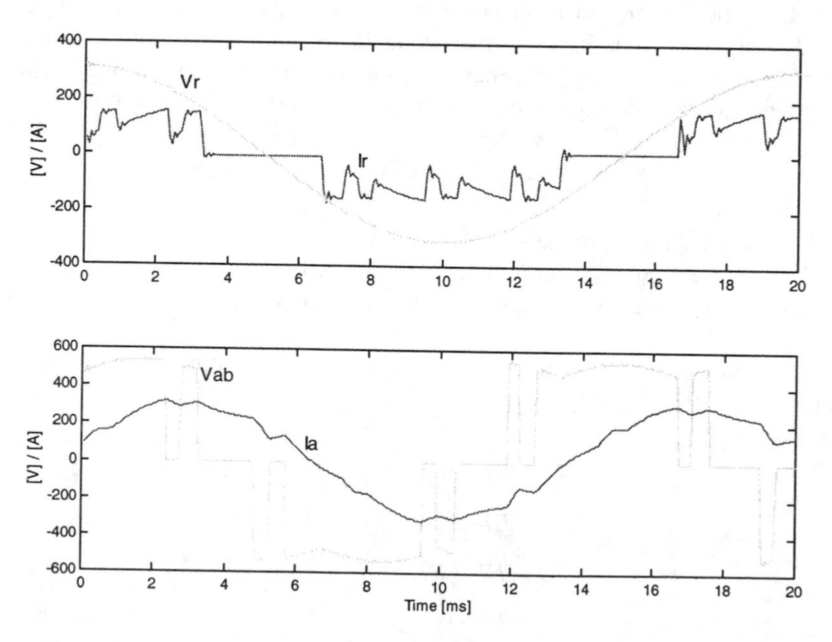

Figure 13. Input phase voltage V_r, input line current I_r, output line voltage V_{ab}, and output line current I_a of the SDFC operating at 70 Hz.

The evaluation considers the ac-ac voltage gain (Gv), the input power factor (pf), the input current total harmonic distortion (THD), and the commutation frequency of both converters operating in an ASD. These results are shown in Figure 14. In order to assess the impact of the operation of multiple converters connected to a common feeder, the total input current waveforms and corresponding THD as a function of the number of converters connected was also determined. These results are important as converters are usually imbedded in environments with multiple nonlinear (converters) and linear loads. So the interaction between them should be considered.

Regarding the converters' ac-ac voltage gain Gv, both DFCs achieve a unity voltage gain. The converters' input power factor shows that the presented XDFC converter is superior to the SDFC. The input current distortion THD is also reduced in the XDFC when compared to the SDFC. The presented DFC offers better results; it has considerably reduced the input current distortion compared to the SDFC. This reduction is a consequence of the input current control introduced as second control objective of the XDFC. Finally, the converters' commutation frequency is also shown.

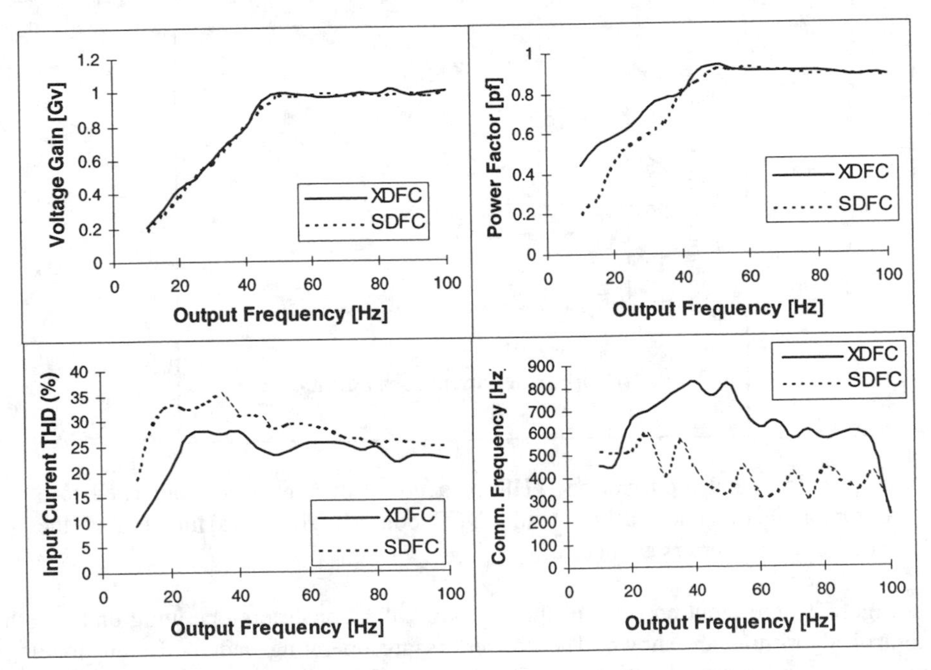

Figure 14. Evaluation performed between the XDFC and SDFC. Figures show voltage gain Gv, input power factor *pf*, input current THD and commutation frequency.

The SDFC has a lower commutation frequency than the presented XDFC. This slight increase in the converter's commutation frequency is a result of the realized input current control. Nevertheless, the XDFC keeps the commutation frequency beneath 850 Hz, maintaining the advantages of a low commutation frequency converter, i.e., reduced power losses and possible higher power rating. It should be mentioned that the increase in commutation frequency is a slight variation thanks to the fuzzy controller for the converter control. Without this controller, the commutation frequency doubles the one of the converter with only output current control (SDFC).

The multiple converters evaluation is shown in Figure 15, where the total input current distortion is depicted as a function of the number of converters connected to a common feeder. Two groups of converters were considered, one of SDFC and one of XDFC, each group separately connected to individual common feeders. The presented XDFC shows an asynchronous operation, producing an increasing harmonic cancellation of the common feeder's input current. Consequently, the THD value diminishes as the number of converters increases. On the contrary, the SDFC does not present a THD reduction when the number of connected converters increases. This is so because the harmonic currents drawn from different converters ride in phase, as each SDFC is synchronized by a fictitious diode bridge rectifier [6], whose input currents have unity displacement factor.

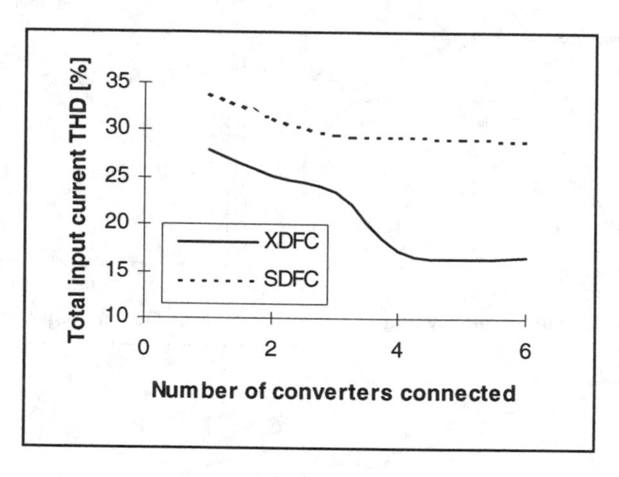

Figure 15. Total input current THD of a group of converters connected to a common feeder under XDFC and SDFC control. This as a function of the number of converters connected.

In Figure 16 the input currents of these groups of 6 converters operating under both control algorithms are shown. The converters are operating under different loading conditions and output frequencies. Clearly, the SDFC input current shows that no harmonic cancellation has taken place, as the waveform resembles a typical six-pulse converter line current. This is not the case for the XDFC. This converter offers a much better performance, as the input currents are subject to harmonic current cancellation and, therefore, present a low current distortion, with sinusoidal waveform resemblance.

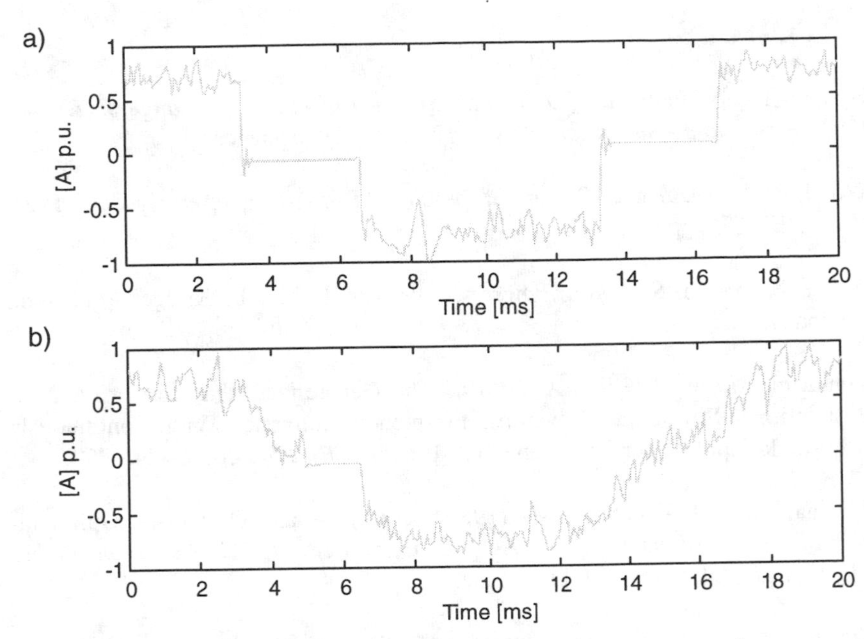

Figure 16. Input current of a group of 6 converters connected to a common feeder, operating with different load conditions and various operating speeds; a) shows the SDFC converters, and b) the XDFC converters.

8. Conclusion

The XDFC main feature is to achieve a unity ac-ac voltage gain. Consequently, it eliminates the need for coupling transformers, enabling the use of motors rated at the system's nominal voltage. It operates with a low commutation frequency, below 850 Hz throughout the output frequency range. The converter showed the feasibility of simultaneously controlling both input and output currents. This was accomplished using a fuzzy controller that determined which converter side had higher priority, thus requiring immediate control actions. The converter's load current distortion was kept below a desired percentage, and the input current distortion was diminished. The converter produced an asynchronous operation, which proved to increase the performance toward the utility side when multiple XDFC drives are connected to a common feeder. The XDFC used an XSVM technique. This technique uses a set of rules to determine the next converter state, effectively controlling the converter and maintaining a low commutation frequency. When compared to similar performance techniques, it offers a reduced processing time, being 70 times faster than the predictive current control. It is also independent of the load's parameters, and thus eliminates the need for on-line parameters identification. Finally, the converter transfer function was used to realize a waveform software reconstruction, which reduced the measuring requirements of the XDFC.

References

[1] Gyugyi, L. and Pelly, B. (1976), *Static Power Frequency Changers: Theory, Performance and Application,* New York, Wiley-Interscience.

[2] Wood, P. (1979), General Theory of Switching Power Converters, in Conf. Proc. *IEEE PESC'79*, pp. 3-10.

[3] Wood, P. (1981), *Switching Power Converters*, Robert E. Krieger Publishing Company.

[4] Venturini, M. and Alesina, A. (1980), The Generalized Transformer: A New Bidirectional Sinusoidal Waveform Frequency Converter With Continuously Adjustable Input Power Factor, in Conf. Proc. *IEEE PESC'80*, pp. 242-252.

[5] Alesina, A. and Venturini, M. (1989), Analysis and Design of Optimum-Amplitude Nine-Switch Direct AC-AC Converters, *IEEE Trans. on Power Electronics*, Vol. 4, pp. 101-112.

[6] Ziogas, P., Khan, S., and Rashid, M. (1985), Some Improved Forced Commutated Cycloconverter Structures, *IEEE Trans. on Industry Applications*, Vol. 21, pp. 1242-1253.

[7] Van der Broeck, H., Skudenly, H., and Viktor, G. (1988), Analysis and Realization of a Pulsewidth Modulator Based on Voltage Space Vectors, *IEEE Trans. on Industry Applications*, Vol. 24, pp. 142-150.

[8] Huber, L. and Borojevic, D. (1991), Space Vector Modulation with Unity Input Power Factor for Forced Commutated Cycloconverters, in Conf. Proc. *IEEE IAS'91*, pp. 1032-1041.

[9] Casadei, D., Grandi, G., Serra, G., and Tani, A. (1993), Space Vector Control of Matrix Converter with Unity Input Power Factor and Sinusoidal Input/Output Waveforms, in Conf. Proc. *EPE'93*, Vol. 7, pp. 170-175.

[10] Huber, L. and Borojevic, D. (1995), Space Vector Modulated Three-Phase to Three-Phase Matrix Converter with Input Power Factor Correction, *IEEE Trans. on Industry Applications*, Vol. 31, pp. 1234-1246.

[11] Zhang, L., Watthanasarn, C., and Sheperd, W. (1996), Analysis and Implementation of a Space Vector Modulation Algorithm for Direct Matrix Converters, *EPE Journal*, Vol. 6, pp. 7-15.

[12] Wiechmann, E., Espinoza, J., Salazar, L., and Rodriguez, J. (1993), A Direct Frequency Converter Controlled by Space Vectors, in Conf. Rec. *IEEE PESC'93*, pp. 314-320.

[13] Wiechmann, E., Garcia, A., Salazar, L., and Rodriguez, J. (1997), High Performance Direct Frequency Converters Controlled by Predictive Current Loop, *IEEE Trans. on Power Electronics*, Vol. 12, pp. 547-557.

[14] Wiechmann, E., Burgos, R., Salazar, L., and Rodriguez, J. (1997), Fuzzy Logic Controlled Direct Frequency Converters Modulated by An Expert Knowledge-Based Space Vector Technique, in Conf. Proc. *IEEE IAS'97*, pp. 1437-1446.

[15] Lee, C. (1990), Fuzzy Logic in Control Systems:Fuzzy Logic Controller-Part I, *IEEE Trans. on Systems, Man, and Cybernetics*, Vol. 20, pp. 404-418.

[16] Lee, C. (1990), Fuzzy Logic in Control Systems: Fuzzy Logic Controller-Part II, *IEEE Trans. on Systems, Man, and Cybernetics*, Vol. 20, pp. 419-435.

[17] Sousa, G. and Bose, B. (1994), A Fuzzy Set Theory Based Control of a Phase-Controlled Converter DC Machine Drive, *IEEE Trans. on Industry Applications*, Vol. 30, pp. 34-44.

[18] So, W., Tse, C., and Lee, C. (1996), Development of a Fuzzy Logic Controller for DC/DC Converters: Design, Computer Simulation, and Experimental Evaluation, *IEEE Trans. on Power Electronics*, Vol. 11, pp. 24-32.

[19] Mir, S., Zinger, D., and Elbuluk, M. (1994), Fuzzy Controller for Inverter Fed Induction Machines, *IEEE Trans. on Industry Applications*, Vol. 30, pp. 78-84.

[20] Wiechmann, E., Ziogas, P., and Stefanovic, V. (1987), Generalized Functional Model for Three-Phase PWM Inverter/Rectifier Converters, *IEEE Trans. on Industry Applications*, Vol. 23, pp. 236-246.

[21] Holtz, J. (1992), Pulsewidth Modulation -A Survey, *IEEE Trans. on Industrial Electronics*, Vol. 39, pp. 410-420.

[22] Divan, D., Lipo, T., and Habetler, T. (1990), *PWM Techniques for Voltage Source Inverters*, Tutorial Notes *IEEE PESC'90.*

[23] Parks, T. and Burrus, C. (1994), *Digital Filter Design*, John Wiley & Sons, Inc.

Chapter 4:

Design of an Electro-Hydraulic System Using Neuro-Fuzzy Techniques

DESIGN OF AN ELECTRO-HYDRAULIC SYSTEM USING NEURO-FUZZY TECHNIQUES

P.J. Costa Branco and **J.A. Dente**
Mechatronics Laboratory
Department of Electrical and Computer Engineering
Instituto Superior Técnico, Lisbon
Portugal

Increasing demands in performance and quality make drive systems fundamental parts in the progressive automation of industrial processes. Their conventional models become inappropriate and have limited scope if one requires a precise and fast performance. So, it is important to incorporate learning capabilities into drive systems in such a way that they improve their accuracy in real time, becoming more autonomous agents with some "degree of intelligence."

To investigate this challenge, this chapter presents the development of a learning control system that uses neuro-fuzzy techniques in the design of a tracking controller to an experimental electro-hydraulic actuator. We begin the chapter by presenting the neuro-fuzzy modeling process of the actuator. This part surveys the learning algorithm, describes the laboratorial system, and presents the modeling steps as the choice of actuator representative variables, the acquisition of training and testing data sets, and the acquisition of the neuro-fuzzy inverse-model of the actuator.

In the second part of the chapter, we use the extracted neuro-fuzzy model and its learning capabilities to design the actuator position controller based on the feedback-error-learning technique. Through a set of experimental results, we show the generalization properties of the controller, its learning capability in actualizing in real time the initial neuro-fuzzy inverse-model, and its compensation action improving the electro-hydraulics' tracking performance.

1. Introduction

Recent integration of new technologies involving new materials, power electronics, microelectronics, and information sciences made relevant new demands in

0-8493-9804-5/99/$0.00+$.50

performance and optimization procedures for drive systems. To handle command and control problems, the dynamic behavior of a drive must be modeled taking into account the electromagnetic and mechanical phenomena. However, if one requires a precise and fast performance, the control laws become more complex and nonlinear and the classical models become inappropriate and of limited scope.

The existing models are not sufficiently accurate, the parameters are poorly known, and, also, because physical effects like thermal behavior, magnetic saturation, friction, viscosity, are in general time-variants, they are difficult to develop with the necessary simplicity and accuracy. So, it is important to develop drive systems that incorporate learning capabilities in a way that their control systems automatically improve accuracy in real time and become more autonomous.

To investigate the possibilities of incorporating learning capabilities into drive systems, we present the implementation of a control system that uses neuro-fuzzy modeling and learning procedures to design a tracking controller to an electro-hydraulic actuator. The learning capability of the neuro-fuzzy models is employed to permit the controller to achieve actuator inverse dynamics and thus compensate the possible unstructured uncertainties to improve performance in trajectory following.

In the first part of this chapter, we present the actuator modeling using the neuro-fuzzy methodology. In this way, the information about its dynamic behavior is expressed in a multimodel structure by a rule set composing the neuro-fuzzy model. Each region of actuator's operating domain is characterized by a rule subset describing its local behavior. The neuro-fuzzy model permits the actuator's information, codified into it, can be generalized, and use its neural-based-learning capabilities in a manner to permit modifying and/or adding knowledge to the model when necessary.

Today, conventional fuzzy controllers are publicized by industry as being "intelligent." Although, to define some "intelligence" degree, it is essential to have learning mechanisms that they do not have. Initially, some approaches have been proposed to improve the performance of conventional controllers using fuzzy logic. The first used fuzzy logic to tune gain parameters of PID controllers [34], [35], or substitute PID controllers by their fuzzy approximation [23], [36].

Some papers in the literature address control systems using learning mechanisms based on neural networks [7], [8], [12], and others introduced the idea of fuzzy learning controllers using a self-organizing approach [38], [39], or, more recently, by neuro-fuzzy structures [16], [17], [20], [37].

The second part of the chapter presents the implementation of the learning control system to the electro-hydraulic actuator combining its neuro-fuzzy inverse-model with a conventional proportional controller. This scheme results in the indirect compensation control scheme named feedback-error-learning proposed by Kawato in

[5], [15], and initially explored by the authors in an unsupervised way in [18]. The controller was implemented on a Personal Computer (PC) with a 80386 CPU and an interface with A/D (analog to digital) and D/A (digital to analog) converters. All programming was done in C language, including the neuro-fuzzy algorithm and actuator signal's acquisition and conditioning.

The implemented system is constituted by realtime learning and control cycles. During these cycles, the inverse-model of the actuator uses its neural-based-learning capabilities to extract rules not incorporated into the initial model, and even change itself to characterize a possible new actuator's dynamic.

We show experimental results concerning the position control of the electro-hydraulic actuator. At each control cycle, the incorporated learning mechanism extracts its inverse-model and generates a compensation signal to the actuator. The results show that the controller is capable of generalizing its acquired knowledge for new trajectories; it can acquire and introduce new system's information in real time using the sensor signals; and it can compensate possible nonlinearities in the system to progressively reduce its trajectory errors.

The structure of the chapter is as follows. In Section 2, the fuzzy system employed is characterized by its fuzzy logic operations. In Section 3, we review fuzzy modeling processes in the literature. Section 4 describes the neuro-fuzzy modeling algorithm. Section 5 presents the experimental system and the technique used to obtain a good training data set from the electro-hydraulic system. In Section 6, we extract the inverse-model of the actuator using the modeling algorithm presented in Section 4 and the training set of Section 5. Section 7 describes the neuro-fuzzy control system using the feedback-error-learning algorithms and presents some experimental tests.

2. The Fuzzy Logic System

Fuzzy sets establish a mechanism for representing linguistic concepts like big, little, small and, thus, they provide new directions in the application of pattern recognition based on fuzzy logic to automatically model drive systems [31], [32]. These computational models are able to recognize, represent, manipulate, interpret, and use fuzzy uncertainties through a fuzzy system.

A fuzzy logic system consists of three main blocks: fuzzification, inference mechanism, and defuzzification. The following subsections briefly explain each block, and characterize them with regard to the type of fuzzy system we used.

2.1 Fuzzification

Fuzzification is a mapping from the observed numerical input space to the fuzzy sets defined in the corresponding universes of discourse. The *fuzzifier* maps a numerical value denoted by $\boldsymbol{x}' = (x_1', x_2', \ldots, x_m')$ into fuzzy sets represented by membership functions in **U**. These functions are Gaussian, denoted by $\mu_{A_j^i}(x_j')$ as we expressed in Equation (1).

$$\mu_{A_j^i}(x_j') = a_j^i \exp\left[-\frac{1}{2}\left(\frac{x_j' - b_j^i}{c_j^i}\right)^2\right] \tag{1}$$

In this equation, $1 \le j \le m$ refers to the variable (j) from m considered input variables; $1 \le i \le n_j$ considers the i membership function among all n_j membership functions considered for variable (j); a_j^i defines the maximum of each Gaussian function, here $a_j^i = 1.0$; b_j^i is the center of the Gaussian function; and c_j^i defines its shape width.

2.2 Inference Mechanism

Inference mechanism is the fuzzy logic reasoning process that determines the outputs corresponding to fuzzified inputs.

The fuzzy rule-base is composed by IF-THEN rules like

$$R^{(l)}\text{: IF } (x_1 \textit{ is } A_1^{(l)} \text{ and } x_2 \textit{ is } A_2^{(l)} \text{ and} \ldots\ x_m \textit{ is } A_m^{(l)}) \text{ THEN } (y \text{ is } \omega^{(l)}),$$

where: $R^{(l)}$ is the lth rule with $1 \le l \le c$ determining the total number of rules; $x_1, x_2, \ldots x_m$ and y are, respectively, the input and output system variables; $A_j^{(l)}$ are the antecedent linguistic terms (or fuzzy sets) in rule (l) with $1 \le j \le m$ being the number of antecedent variables; and $\omega^{(l)}$ is the rule conclusion being, for that type of rules, a real number usually called *fuzzy singleton*. The conclusion, a numerical value and not a fuzzy set, can be considered as a *pre-defuzzified* output that helps to accelerate the inference process.

Each IF-THEN rule defines a fuzzy implication between condition and conclusion rule parts and denoted by expression $A_1^{(l)} \times \ldots \times A_m^{(l)} \to \omega^{(l)}$. The implication operator used in this work is the *product-operator*, as shown in expression (2). The right-hand term $\mu_{A_1^{(l)} \times \ldots \times A_m^{(l)}}(\boldsymbol{x}')$ represents the condition degree and is defined in Equation (3).

$$\mu_{A_1^{(l)} \times \ldots \times A_m^{(l)} \to \omega^{(l)}}(\boldsymbol{x}', y') = \mu_{A_1^{(l)} \times \ldots \times A_m^{(l)}}(x_1', \ldots, x_m') . \mu_{\omega^{(l)}}(y') \tag{2}$$

The symbol " * " in Equation (3) is the t-norm corresponding to the conjunction *and* in the rules. The most commonly used t-norms between linguistic expressions u *and* v are: fuzzy intersection defined by operation $min(u,v)$, algebraic product uv, and the bounded sum $max(0, u+v-1)$. This work uses algebraic product as the t-norm operator.

$$\mu_{A_1^{(l)} \times \ldots \times A_m^{(l)}}(\boldsymbol{x}') = \mu_{A_1^{(l)}}(x_1') * \ldots * \mu_{A_m^{(l)}}(x_m') \tag{3}$$

Since the rule conclusion $\omega^{(l)}$ is considered a fuzzy singleton, the value of its membership degree $\mu_{\omega^{(l)}}(y')$ in expression (2) stays 1.0. So, the final expression for fuzzy implication degree (2) results in multiplication of each condition membership degree (3) and equal to expression (4).

$$\mu_{A_1^{(l)} \times \ldots \times A_m^{(l)} \to \omega^{(l)}}(\boldsymbol{x}') = \mu_{A_1^{(l)} \times \ldots \times A_m^{(l)}}(\boldsymbol{x}').(1.0) = \prod_{j=1}^{m} \mu_{A_j^{(l)}}(x_j') \tag{4}$$

For this type of fuzzy system, the *product inference* in Equation (3) expresses the activation degree of each identified rule by measured condition variables, and equals the expression for implication degree in (4).

The reasoning process combines all rule contributions $\omega^{(l)}$ using the centroid defuzzification formula in a weighted form, as indicated by inference function (5). This equation maps input process states (x_j') to the value resulting from inference function $Y(\boldsymbol{x}')$. If we fix the structure made by the Gaussian membership functions, the parameters of the fuzzy logic system to be learned will be the rule conclusion value $\omega^{(l)}$.

$$Y(\boldsymbol{x}') = \frac{\sum_{l=1}^{c} \left(\prod_{j=1}^{m} \mu_{A_j^{(l)}}(x_j') \right) \omega^{(l)}}{\sum_{l=1}^{c} \left(\prod_{j=1}^{m} \mu_{A_j^{(l)}}(x_j') \right)} \tag{5}$$

2.3 Defuzzification

Basically, defuzzification maps output fuzzy sets defined over an output universe of discourse to crisp outputs. It is employed because in many practical applications a crisp output is required. A defuzzification strategy is aimed at producing the nonfuzzy output that best represents the possibility distribution of an inferred fuzzy output. At present, the commonly used strategies are described as the following.

1) The Max Criterion Method

The max criterion method produces the point at which the possibility distribution of the fuzzy output reaches a maximum value.

2) The Mean of Maximum Method

The mean of maximum generates an output which represents the mean value of all local inferred fuzzy outputs whose membership functions reach the maximum. In the case of a discrete universe, the inferred fuzzy output may be expressed as

$$z_0 = \sum_{j=1}^{l} \frac{w_j}{l}$$

where w_j is the support value at which the membership function reaches the maximum value $\mu_z(w_j)$ and l is the number of such support values.

3) The Center of Area Method

The center of area generates the center of gravity of the possibility distribution of the inferred fuzzy output. In the case of a discrete universe, this method yields

$$z_0 = \frac{\sum_{j=1}^{n} \mu_z(w_j) w_j}{\sum_{j=1}^{n} \mu_z(w_j)}$$

where n is the number of quantization levels of the output.

3. Fuzzy Modeling

Basic principles of fuzzy models, also known in literature by *fuzzy modeling*, were first introduced by Zadeh in [2] and [13]. First applications in modeling systems using fuzzy-logic consisted initially in duplication of expert experience to process control [22]. Although, this *qualitative information* can present limitations as the acquired knowledge usually presents errors and even some gaps.

Another source of information is *quantitative information.* It is acquired by acquisition of numerical data from most representative system variables, and can be used together with the anterior qualitative information to complete it or even produce new information [3].

The acquisition of models using fuzzy logic is usually divided into two types as shown in Figure 1: a linguistic approach composed by relational and natural models and a hybrid approach concerning the neuro-fuzzy models.

The main difference between these approaches is related to the knowledge representation in the model. While linguistic approach describes the system behavior using rules of IF-THEN using only fuzzy sets (linguistic variables), the hybrid approach uses linguistic variables in the condition rule part (IF) and uses a numerical value in the conclusion part (THEN) which is considered as a function of input variables [3], [4].

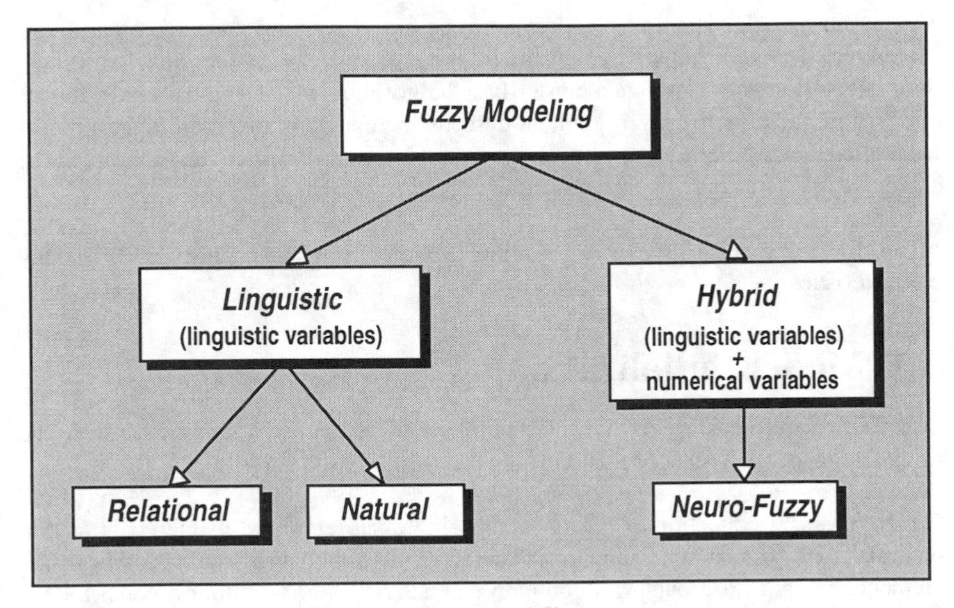

Figure 1. Fuzzy modeling types.

Linguistic modeling can be divided into two types: relational modeling and *natural* modeling. Relational modeling [25-28] establishes a set of all possible rules based on an attributed linguistic partition for each input-output variable. It computes for each rule the respective true value of how much that rule contributes to describe system behavior. The set of all rules composes, in a computational way, a multidimensional matrix called relational matrix. Using the theory of relational equations [29], [30], each matrix element can be computed as being the rule membership degree in the extracted system's model .

The second type of linguistic modeling is denoted by *natural* modeling. It does not use relational equations to obtain the model. The rules are codified from information supplied by the process operator and/or from knowledge obtained from the literature. The first application examples of this type of modeling were the fuzzy controllers in [22] and [23].

Fuzzy modeling based on hybrid approach permits employing learning techniques used by neural-networks in the identification of each rule [16], [17], [20]. The parameter set composing rule condition part are the membership functions width and their position in the respective universe of discourse. In the conclusion part, the parameters are the function terms that compute the rule answer.

4. The Learning Mechanism

The learning mechanism uses two data sets: one for the training stage and other to test the extracted model. Initially, using the training set, we extract the model rules and their conclusion value through a cluster-based algorithm [19]. Then, the model has its conclusion values tuned by a gradient-descent method [24] to produce the process neuro-fuzzy model. Since the test set has examples not presented during the training stage, we use it to verify the generalization model performance.

In the following subsections, we recapitulate the learning mechanism and its main characteristics.

4.1 Model Initialization

The first modeling stage of the electro-hydraulic actuator is concerned with the initialization of each rule conclusion using the cluster-based algorithm.

Cluster means a collection of *objects* composing a subset where its elements form a natural group among all exemplars. Therefore, it establishes a subset where the elements compose a group with common characteristics constituting a pattern. This concept applied to the fuzzy partition of system's operating domain divides it into clusters, each one interpreted as a rule $R^{(l)}$ describing, in our case, the actuator's local behavior.

The cluster concept when used with fuzzy logic [33] associates to each data point a value among zero and one representing its membership degree in the rule. This allows each sample data to belong to multiple rules with different degrees.

In Figure 2, we illustrate the cluster concept applied to a fuzzy system. Suppose, for simplicity, a system with two inputs denoted by x_1 and x_2, and one output y. As shown in the figure, each domain variable x_1 and x_2 is equally partitioned by symmetric triangular fuzzy sets characterizing each linguistic term, for example, with *PM- Positive Medium*, *NB- Negative Big*, *ZE- Zero*, and other fuzzy sets.

A data set composed by system examples is acquired to be used in the training stage. Figure 2(a) displays the examples covering the domain, each one formed by a data sample like $(x_1^{'}, x_2^{'}) \rightarrow y^{'}$. The examples are grouped in clusters for each respective rule $R^{(l)}$. In the figure, we exemplify the rule acquisition expressed in statement (6).

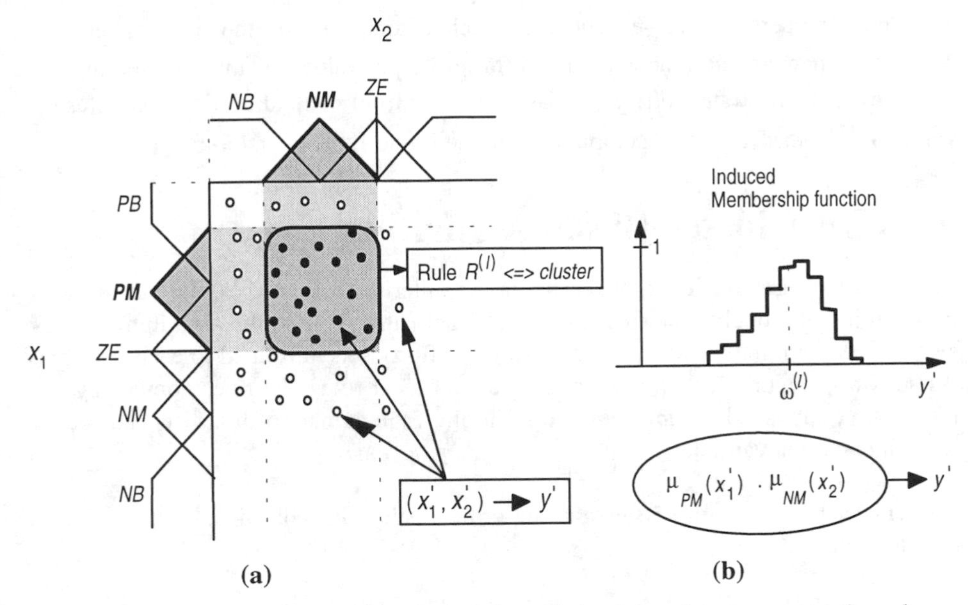

Figure 2. (a) Set of examples selected from the training data to extract the rule with antecedents defined by fuzzy sets *PM* and *NM*. (b) Membership function induced by weighted output values y' into the specified rule region, and the computed conclusion value $\omega^{(l)}$.

$$\textbf{IF } [(x_1 \text{ is } \textit{Positive-Medium}) \textbf{ and } (x_2 \text{ is } \textit{Negative-Medium})] \textbf{THEN } [y \text{ is } \omega^{(l)}] \quad (6)$$

The condition rule part is characterized by fuzzy sets *PM (Positive-Medium)* and *NM (Negative-Medium)*. The conclusion part, characterized by a numerical value $\omega^{(l)}$, is extracted based on the examples contained into the domain region covered by the two fuzzy sets *PM* and *NM*. This set of examples is represented in Figure 2 by filled circles into the rule region $R^{(l)}$.

Using the fuzzy cluster concept, it attributes to each example a certain degree of how much it belongs to that cluster or, in other words, how much each example contributes to the extraction of conclusion value $\omega^{(l)}$ of that rule $R^{(l)}$.

Suppose an example $(x_1', x_2') \rightarrow y'$ inside the rule region. Its contribution degree is computed by the product of each condition membership degree in fuzzy sets *PM* and *NM* of specified rule region, as expressed in (7) and displayed in Figure 2(b). The computed contribution degree then weights the corresponding output value y'.

$$\mu_{PM}(x_1').\mu_{NM}(x_2') \quad (7)$$

The anterior operations are executed for each example inside the rule region, and compose a membership function defined for all output values y' into the rule region, as Figure 2(b) illustrates to rule (6). Using the centroid method, the final conclusion value $\omega^{(l)}$ for that rule (l) is computed from the induced membership function.

4.2 The Cluster-Based Algorithm

The algorithm uses the ideas introduced in the anterior section to extract each rule to build an initial model to the electro-hydraulic actuator. At first, the algorithm divides system's domain into a set of clusters using the fuzzy sets attributed to each variable. As shown in Figure 2, each cluster represents a local rule. The rules composing the model are established *a priori* by multiplication of the number of fuzzy sets attributed to each condition variable.

The cluster-based algorithm steps are described below in more detail, and a simple example illustrates it.

Starting with rule one ($l = 1$) and the kth training example, the cluster-based algorithm summarizes the following steps to extract its conclusion value $\omega^{(1)}$:

Step 1) Establish the variable set better characterizing the actuator's behavior;

Step 2) Set the limits of each universe of discourse and the number of fuzzy sets for the selected input-output variables in step 1. The algorithm uses symmetric Gaussian membership functions uniformly distributed by each universe of discourse;

Step 3) The algorithm begins with the extraction of the first rule (l = 1). From the training set, we take the kth numerical example $(x_1'(k), x_2'(k), \ldots, x_m'(k)) \rightarrow y'(k)$, and calculate, for all condition variables, their respective membership degrees in the fuzzy sets composing the rule as expressed in (8).

$$\mu_{A_1^{(l)}}(x_1'(k)), \mu_{A_2^{(l)}}(x_2'(k)), \ldots, \mu_{A_m^{(l)}}(x_m'(k)) \tag{8}$$

Step 4) Calculate the membership degree of corresponding output value $y'(k)$ in rule (l), or its membership degree in cluster (l), as indicated in (9) by the term $S1^{(l)}(k)$.

$$S1^{(l)}(k) = \mu_{A_1^{(l)}}(x_1'(k)) . \mu_{A_2^{(l)}}(x_2'(k)) . \cdots . \mu_{A_m^{(l)}}(x_m'(k)) \tag{9}$$

Step 5) The output value $y'(k)$ is weighted by its membership degree $S1^{(l)}(k)$ in rule (l), as described in Equation (10) by $S2^{(l)}(k)$.

$$S2^{(l)}(k) = y^{'}(k).S1^{(l)}(k) \tag{10}$$

Step 6) In this step, the algorithm adds recursively the value $S2^{(l)}(k)$ and the membership degree $S1^{(l)}(k)$ as indicated in (11). The variable *Numerator* adds to rule (l) all weighted contributions made by the n data values $y^{'}(k)$ in the training set. The variable *Denominator* sums each membership degree in order to normalize the conclusion value $\omega^{(l)}$.

$$\begin{cases} Numerator \rightarrow Numerator + (S2^{(l)}(k)) \\ Denominator \rightarrow Denominator + (S1^{(l)}(k)) \end{cases} \tag{11}$$

Get the next example. If there are no more examples, go to step 7 and compute the conclusion value $\omega^{(l)}$. If not, go to step 3 and pick up the next example as indicated in (12).

$$k \rightarrow k+1 \tag{12}$$

Step 7) If the training set has finished $(k = n)$, compute the conclusion value $\omega^{(l)}$ for rule (l) using equation (13).

$$\omega^{(l)} = \frac{Numerator}{Denominator} \tag{13}$$

Step 8) The algorithm now goes to next rule (14), begins again with the first training example (15), and returns to step 3. If there are no more rules $(l = c)$, the algorithm stops.

$$l \rightarrow l+1 \tag{14}$$

$$k \rightarrow 1 \tag{15}$$

4.3 Illustrative Example

This example illustrates the anterior steps for one training period. It uses the two examples shown in (16) to demonstrate the computation of $\omega^{(l)}$ for a certain specified rule (l).

$$\begin{cases} (x_1^{'}(1), x_2^{'}(1)) \rightarrow y^{'}(1) \quad (k=1) \\ (x_1^{'}(2), x_2^{'}(2)) \rightarrow y^{'}(2) \quad (k=2) \end{cases} \tag{16}$$

This example considers a system with two antecedent variables denoted by x_1 and x_2, and one output variable, y. The variables are partitioned by symmetric triangular membership functions. The use of a triangular partition instead of a Gaussian one helps us to better visualize the algorithm steps. We attributed 7 fuzzy sets to variable x_1 (Figure 3a), 5 fuzzy sets to x_2 (Figure 3b), and 5 fuzzy sets to y (Figure 3c).

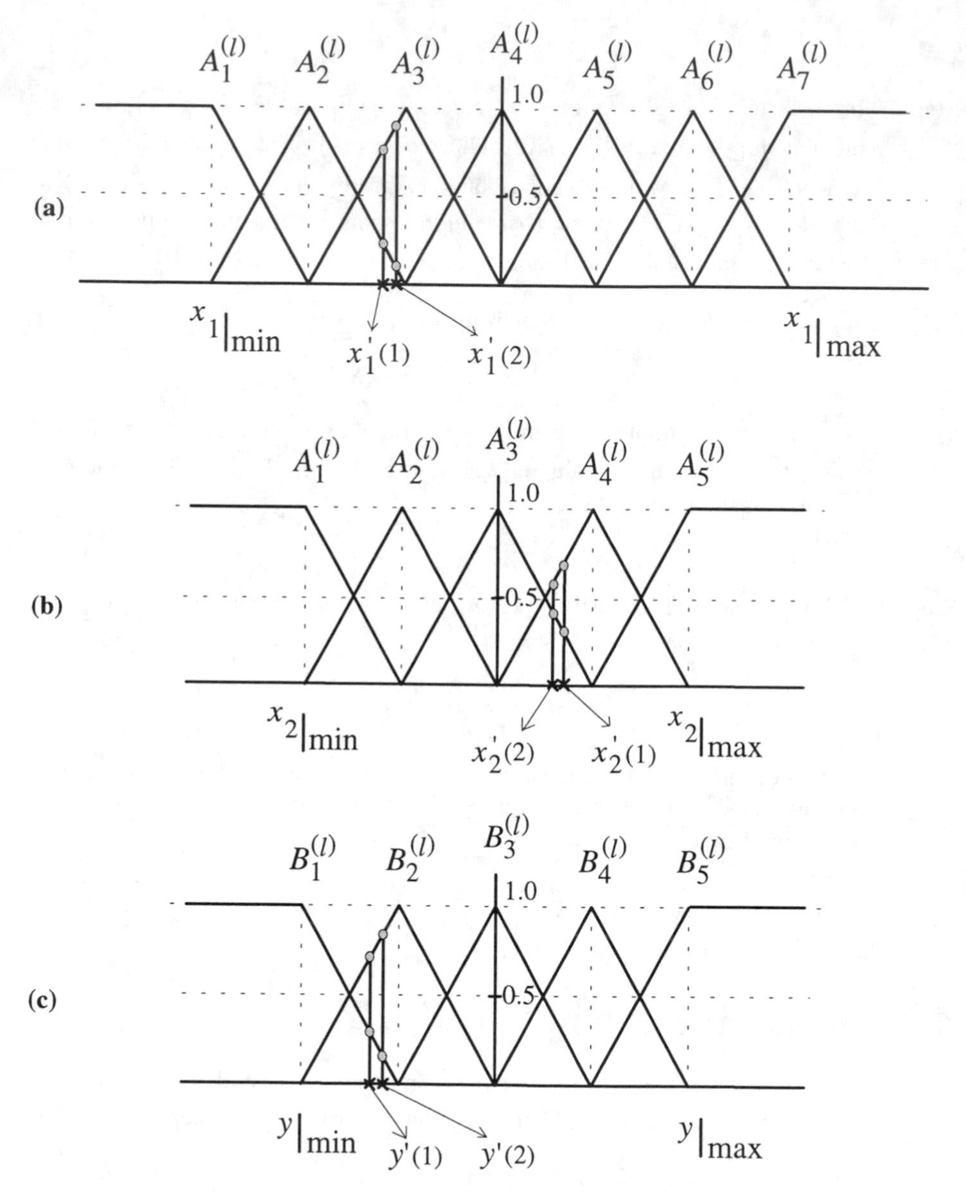

Figure 3: (a) Partition of variable x_1 with 7 fuzzy sets. (b) Partition of variable x_2 with 5 fuzzy sets. (c) Partition of variable y with 5 fuzzy sets.

Suppose in this example that we want to extract the consequent value $\omega^{(1)}$ for rule $(l = 1)$ described in (17).

$$R^{(1)}(k)\text{: } \textbf{IF } (x_1 \text{ } is \text{ } A_3^{(1)} \textbf{ and } x_2 \text{ } is \text{ } A_4^{(1)}) \textbf{ THEN } \omega^{(1)} = ? \qquad (17)$$

Each variable in the two training samples in (16) has a membership degree in each antecedent fuzzy set $A_3^{(1)}$ and $A_4^{(1)}$. In expressions (18) to (20), we show the corresponding degrees attributed to values $x_1^{'}$, $x_2^{'}$, and $y^{'}$, for examples in (16). The triangular partition causes all numerical values to always have two non-zero membership degrees and a null degree in the other fuzzy sets, as illustrated in Figure 3 for each training example. The difference using Gaussian functions is that each variable would have a number of degrees equal to the attributed fuzzy sets.

$$\begin{cases} x_1^{'}(1) \rightarrow \left[0,\ \mu_{A_2^{(1)}}(x_1^{'}(1)),\ \mu_{A_3^{(1)}}(x_1^{'}(1)),\ 0,\ 0,\ 0,\ 0 \right] \\ x_1^{'}(2) \rightarrow \left[0,\ \mu_{A_2^{(1)}}(x_2^{'}(2)),\ \mu_{A_3^{(1)}}(x_2^{'}(2)),\ 0,\ 0,\ 0,\ 0 \right] \end{cases} \tag{18}$$

$$\begin{cases} x_2^{'}(1) \rightarrow \left[0,\ 0,\ \mu_{A_3^{(1)}}(x_2^{'}(1)),\ \mu_{A_4^{(1)}}(x_2^{'}(1)),\ 0 \right] \\ x_2^{'}(2) \rightarrow \left[0,\ 0,\ \mu_{A_3^{(1)}}(x_2^{'}(2)),\ \mu_{A_4^{(1)}}(x_2^{'}(2)),\ 0 \right] \end{cases} \tag{19}$$

$$\begin{cases} y^{'}(1) \rightarrow \left[\mu_{B_1^{(1)}}(y^{'}(1)),\ \mu_{B_2^{(1)}}(y^{'}(1)),\ 0,\ 0,\ 0 \right] \\ y^{'}(2) \rightarrow \left[\mu_{B_1^{(1)}}(y^{'}(2)),\ \mu_{B_2^{(1)}}(y^{'}(2)),\ 0,\ 0,\ 0 \right] \end{cases} \tag{20}$$

The algorithm extracts the value of $\omega^{(1)}$ using steps 5 to 8. It considers the fuzzy sets $A_3^{(l)}$ and $A_4^{(l)}$ of condition part in rule $R^{(1)}$. Therefore, the conclusion value is computed by Equation (21) combining each output value $y^{'}(k)$, weighted and normalized by their contribution degrees to the specified rule.

$$\omega^{(1)} = \frac{\left[\mu_{A_3^{(l)}}(x_1^{'}(1)).\mu_{A_4^{(l)}}(x_2^{'}(1)) \right].y^{'}(1) + \left[\mu_{A_3^{(l)}}(x_1^{'}(2)).\mu_{A_4^{(l)}}(x_2^{'}(2)) \right].y^{'}(2)}{\left[\mu_{A_3^{(l)}}(x_1^{'}(1)).\mu_{A_4^{(l)}}(x_2^{'}(1)) \right] + \left[\mu_{A_3^{(l)}}(x_1^{'}(2)).\mu_{A_4^{(l)}}(x_2^{'}(2)) \right]} \tag{21}$$

4.4 The Neuro-Fuzzy Algorithm

The neuro-fuzzy algorithm developed by Wang [24] uses the hybrid model developed by Takagi-Sugeno in [3]. In this type of model, condition part uses linguistic variables and the conclusion part is represented by a numerical value which is considered a function of system's condition expressed in the variables $x_1, x_2, \ldots, x_m$ (22). These models are suitable for neural-based-learning techniques as gradient methods to extract the rules [6] and generate models with a reduced number of rules.

$$\omega^{(l)} = g(x_1, x_2, \ldots, x_m) \tag{22}$$

The neuro-fuzzy algorithm uses membership functions of Gaussian type. With Gaussian fuzzy sets, the algorithm is capable of utilizing all information contained in the training set to calculate each rule conclusion, which is different when using triangular partitions.

Figure 4 illustrates the neuro-fuzzy scheme for an example with two input variables (x_1, x_2) and one output variable (y). In the first stage of the neuro-fuzzy scheme, the two inputs are codified into linguistic values by the set of Gaussian membership functions attributed to each variable. The second stage calculates to each rule $R^{(l)}$ its respective activation degree. Last, the inference mechanism weights each rule conclusion $\omega^{(l)}$, initialized by the cluster-based algorithm, using the activation degree computed in the second stage. The error signal between the model inferred value Y and the respective measured value (or teaching value) $y^{'}$, is used by the gradient-descent method to adjust each rule conclusion. The algorithm changes the values of $\omega^{(l)}$ to minimize an objective function E usually expressed by the mean quadratic error (23). In this equation, the value $y^{'}(k)$ is the desired output value related with the condition vector $\boldsymbol{x}^{'}(k) = (x_1^{'}, x_2^{'}, \cdots, x_m^{'})_k$. The element $Y(\boldsymbol{x}^{'}(k))$ is the inferred response to the same condition vector $\boldsymbol{x}^{'}(k)$ and computed by Equation (24).

$$E = \frac{1}{2}\left[Y(\boldsymbol{x}^{'}(k)) - y^{'}(k)\right]^2 \tag{23}$$

$$Y(\boldsymbol{x}^{'}(k)) = \frac{\sum_{l=1}^{c}\left(\prod_{j=1}^{m} \mu_{A_j^{(l)}}(x_j^{'}(k))\right)\omega^{(l)}(k)}{\sum_{l=1}^{c}\left(\prod_{j=1}^{m} \mu_{A_j^{(l)}}(x_j^{'}(k))\right)} \tag{24}$$

Equation (25) establishes adjustment of each conclusion $\omega^{(l)}$ by the gradient-descent method. The symbol α is the learning rate parameter, and t indicates the number of learning iterations executed by the algorithm.

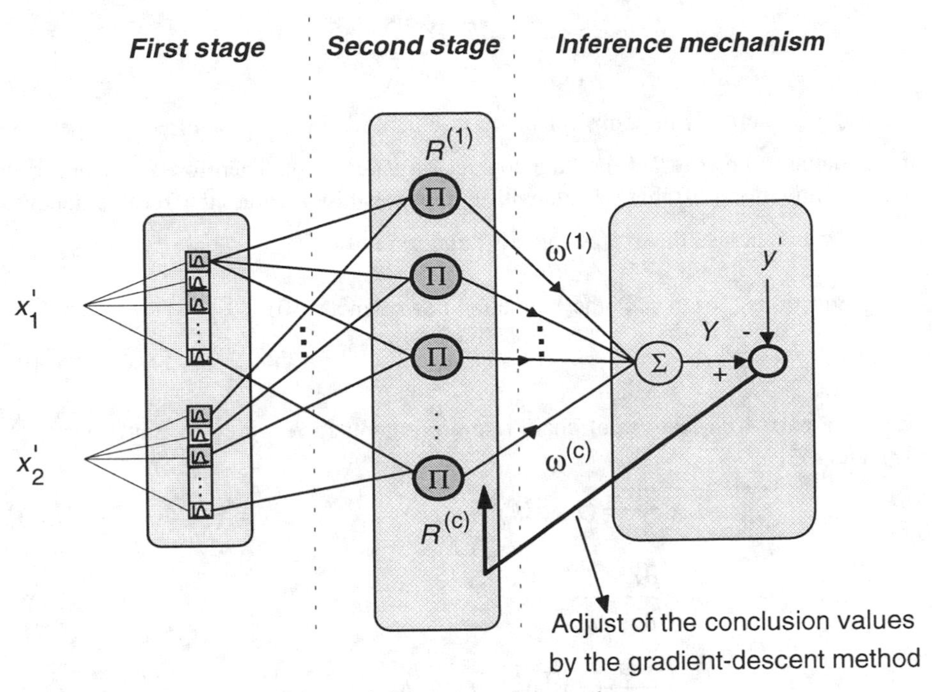

Figure 4: The neuro-fuzzy scheme.

$$\omega^{(l)}(t+1) = \omega^{(l)}(t) - \alpha \frac{\partial E}{\partial \omega^{(l)}} \tag{25}$$

The inference function (24) depends on $\omega^{(l)}$ only through its numerator. The expression composing the numerator is now denoted by a and is shown in (26).

$$a = \sum_{l=1}^{c} \left(\prod_{j=1}^{m} \mu_{A_j^{(l)}}(x_j'(k)) \right) . \omega^{(l)}(t) \tag{26}$$

The denominator of function (24) is dependent on a term $d^{(l)}$, defined in (27), and denoted by b in (28).

$$d^{(l)} = \prod_{j=1}^{m} \mu_{A_j^{(l)}}(x_j'(k)) \tag{27}$$

$$b = \sum_{l=1}^{c} (d^{(l)}) \tag{28}$$

To calculate the adjustment of each conclusion value $\omega^{(l)}$, it is necessary to compute the variation of the objective function E, ∂E, in relation to the variation that occurred in $\omega^{(l)}$ in the anterior instant, $\partial \omega^{(l)}$. Therefore, using the chain rule to calculate $\partial E / \partial \omega^{(l)}$ results in expression (29).

$$\frac{\partial E}{\partial \omega^{(l)}} = \frac{\partial E}{\partial Y}\frac{\partial Y}{\partial a}\frac{\partial a}{\partial \omega^{(l)}} \tag{29}$$

The use of the chain rule looks for the term contained in E that is directly dependent on the value to be adjusted, i.e., the conclusion value $\omega^{(l)}$. Therefore, we can verify by chain Equation (29) that it starts with E dependent of Y value, the Y value depends on term a and, at last, the expression a is a function of $\omega^{(l)}$.

Using Equations (26) to (28), the Y function is written as (30).

$$Y(\boldsymbol{x}'(k)) = \frac{a}{b} \tag{30}$$

The three partial derivatives of chain rule are computed resulting in Equations (31), (32), and (33).

$$\frac{\partial E}{\partial Y} = (Y(\boldsymbol{x}'(k)) - y'(k)) \tag{31}$$

$$\frac{\partial Y}{\partial a} = \frac{1}{b} \tag{32}$$

$$\frac{\partial a}{\partial \omega^{(l)}} = \prod_{j=1}^{m} \mu_{A_j^{(l)}}(x'_j(k)) = d^{(l)} \tag{33}$$

Substituting the three derivatives in chain Equation (29), the final partial derivative of E in relation to $\omega^{(l)}$ results in expression (34).

$$\frac{\partial E}{\partial \omega^{(l)}} = \frac{(Y(\boldsymbol{x}'(k)) - y'(k))d^{(l)}}{b} \tag{34}$$

The replacement of derivative $\partial E / \partial \omega^{(l)}$ in Equation (25) gives the final result presented in (35). In this equation, $d^{(l)}$ represents the activation degree of rule (l) by condition $\boldsymbol{x}'(k)$. The expression $\sum_{l=1}^{c}(d^{(l)})$ is the normalization factor of value $d^{(l)}$. Using these two considerations, the adjustment to be made in $\omega^{(l)}$ can be interpreted as being proportional to the error between the neuro-fuzzy model response and the supervising value, but weighted by the contribution of rule (l), denoted by $d^{(l)}$, to the final neuro-fuzzy inference.

$$\begin{aligned} \omega^{(l)}(t+1) &= \omega^{(l)}(t) - \alpha \frac{(Y(\boldsymbol{x}'(k)) - y'(k))d^{(l)}}{b} \\ &= \omega^{(l)}(t) - \alpha \frac{(Y(\boldsymbol{x}'(k)) - y'(k))d^{(l)}}{\sum_{l=1}^{c}(d^{(l)})} \end{aligned} \tag{35}$$

5. The Experimental System

The experimental control system is composed by a permanent-magnet (P.M.) synchronous motor driving a hydraulic pump that sends fluid to move a linear piston. Figure 5 shows a diagram of the system incorporating two control loops. The interior loop, in gray, is responsible for the motor speed control. The loop is composed of an electrical drive with a PI controller to command the motor speed. The exterior loop, in black, controls the piston position using a proportional controller that gives the motor speed reference to the electrical drive.

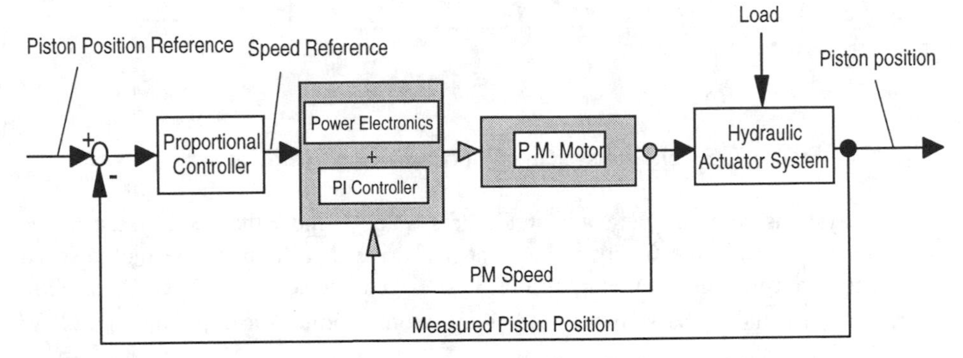

Figure 5: Diagram of the experimental electro-hydraulic drive system.

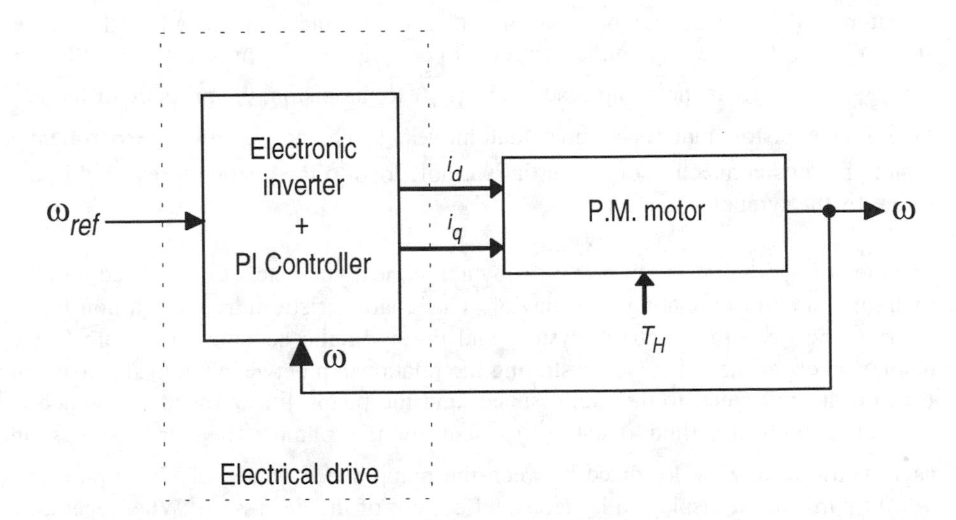

Figure 6: First subsystem composed by the electrical drive and the P.M. motor.

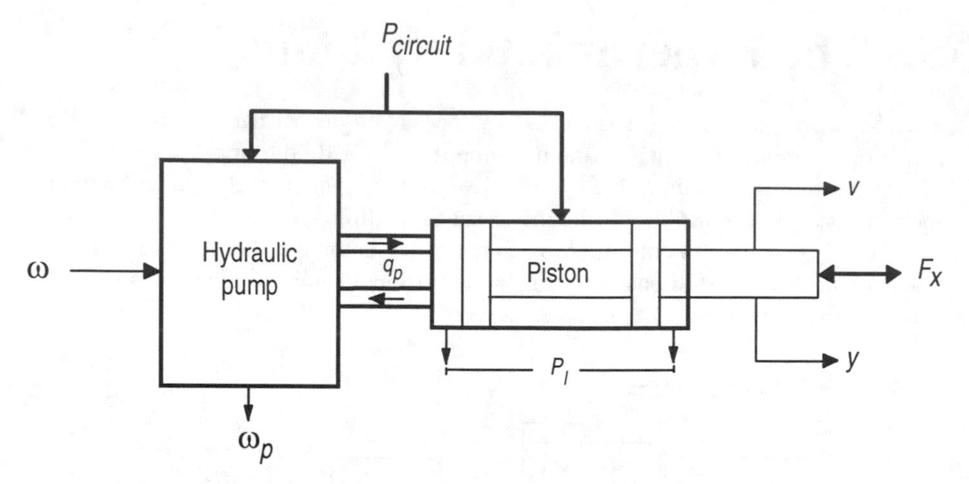

Figure 7: Second subsystem composed by the hydraulic system.

Two subsystems compose the actuator. Figures 6 and 7 show these subsystems. The first subsystem in Figure 6 shows the electrical drive that controls the motor speed (ω). The electronic inverter employs IGBTs to generate currents i_1, i_2, i_3, in *Park* coordinates i_d and i_q as shown in the figure, commanding the P.M. motor (220V/ 1.2Nm/ ±3000 rpm). The speed controller is composed of a PI regulator. The motor load is denoted by T_H, and it comes from the hydraulic pump connected to the motor.

In Figure 7, we show the second subsystem that composes the electro-hydraulic actuator. The hydraulic pump is assumed to rotate at the same speed as the motor ($\omega = \omega_p$), with the hydraulic circuit operating at a pressure of 40 bar ($P_{circuit} = 40bar$). As the pump sends fluid (q_p) to the piston, the pressure difference (P_l) in the piston induces a force that moves it. The implemented experimental system permits connection of an inertial variable load to the piston represented in the figure by the symbol F_x.

The electro-hydraulic system is marked by a nonlinear characteristic localized into the hydraulic circuit dominating its behavior. This characteristic introduces a non-linear interface between the electrical system and the hydraulic actuator. In Figure 8, we display an experimental curve illustrating the relationship between pump speed signal (ω), considered equal to the motor speed, and the piston linear speed (v) which is associated with the fluid quantity q_p sent by the pump. The curve shows an asymmetric dead-zone localized between the pump speed values of -700 r.p.m. and +900 r.p.m., and it displays a hysteresis effect out of the dead-zone. When operating into the dead-zone, the two actuator subsystems stay disconnected and the piston stops as the fluid stream q_p debited by the pump is near zero. Out of the dead-zone, the inclination of the two lines shows that the pump debits slightly more hydraulic fluid when rotating in one direction than rotating to the other.

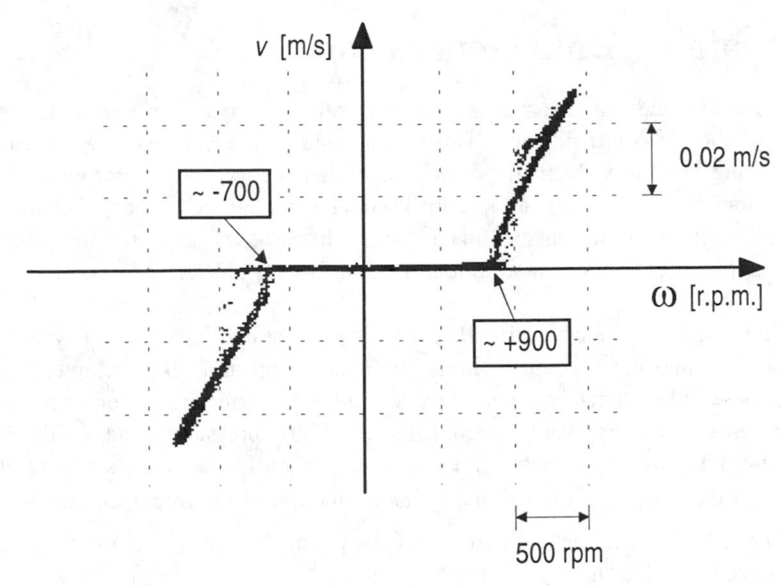

Figure 8: Experimental curve showing the nonlinear characteristic present in the hydraulic circuit.

In Figure 9, we illustrate the piston asymmetric behavior when operating in open-loop (without the proportional controller) for a sinusoidal reference to the motor speed (Figure 9a). We can observe in Figure 9b that the piston moves more to one direction than to the other. Therefore, after some sinusoidal periods, the piston halts at the end of its course of 0.20m. This behavior is mainly caused by the nonlinear characteristic with the asymmetry of the dead-zone, sending more fluid for one pump speed direction than to the other.

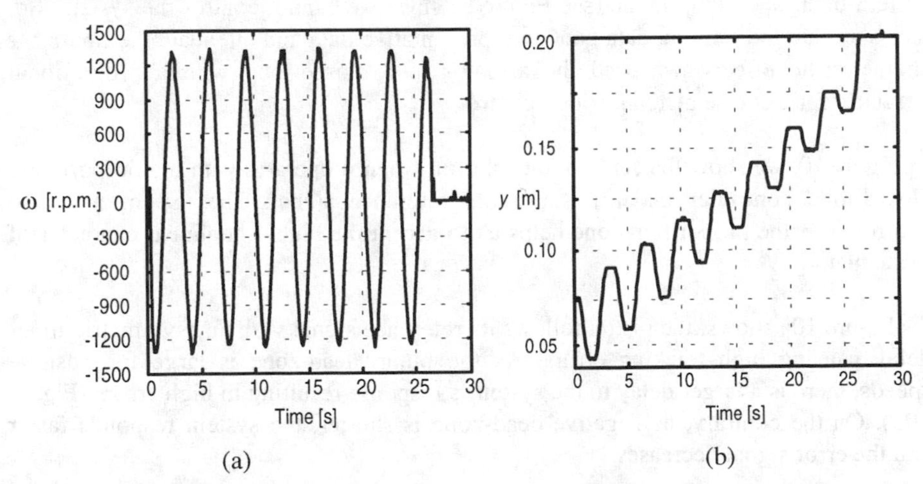

Figure 9: Actuator's response for a sinusoidal reference signal with amplitude and frequency constants, operating in an open-loop mode. (a) Motor speed signal ω. (b) Piston position signal y.

5.1 Training Data Generation

To obtain some relevant information for the training process, we used theoretical knowledge about system physics. This knowledge is present when we model the actuator using electromechanical power conversion theory and hydrodynamic laws. As the system contains a great number of variables that can be chosen to characterize its dynamic, it is important to make some hypotheses and simplifications to concentrate our attention to a small but representative variable set.

As shown before, the electro-hydraulic actuator is separated into two subsystems: the electrical drive and the hydraulic circuit with the pump and piston elements. If we consider these subsystems as "black-boxes" and make some considerations, as, for example, not consider relevant the contribution of the pressure signal in the circuit ($P_{circuit}$) because it remains approximately constant during actuator's operation, we can interpret the piston position signal (y) as a function of the reference signal (y_{ref}), the motor speed (ω), and the linear speed of the piston (v). Thus, the direct model can be represented by relation (36).

$$y = f(y_{ref}, \omega, v) \tag{36}$$

To extract function $f(.)$, it is necessary to use some numerical data available from the system. For this, two different sets of experimental values are added to the modeling process, one set for training and the other for testing.

As described, the actuator has an asymmetric behavior dominated by the presence of a nonlinear characteristic. If we need to acquire some training data that characterizes a significant part of the electro-hydraulic system's operating domain, we cannot use the system in an open-loop mode (see Figure 9) since we cannot control the system. So, to assure that the training data contain representative data and attenuate the nonlinear characteristic effects, we used the actuator in a closed-loop with a proportional regulator for a coarse piston position control.

In Figure 10, we show the actuator's evolution when it operates with the proportional closed-loop controller under a sinusoidal reference signal. The use of a coarse controller as the proportional one helps us to accent the highly nonlinear character of the actuator.

As Figure 10a shows, the piston follows its reference signal with an asymmetric time-delay causing high tracking errors. As the pump dead-zone is large for positive speeds, there is a larger delay in the system's response resulting in high errors (Figure 10b). On the contrary, as negative dead-zone is shorter, the system responds faster and the error signal decreases.

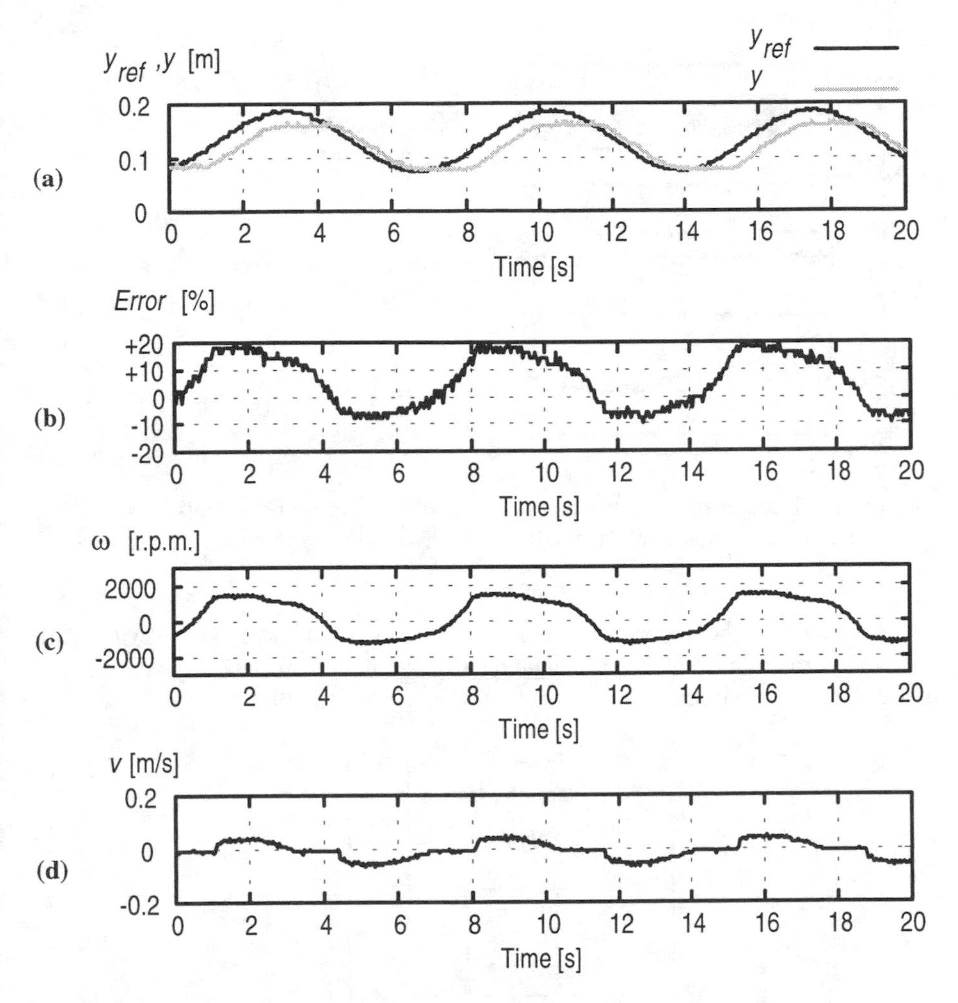

Figure 10: Electro-hydraulic system behavior when operating with the closed-loop proportional controller. (a) Reference signal evolution (y_{ref}) and the piston position signal (y). (b) Error signal evolution displayed in a percentage scale. (c) Evolution of the pump speed signal (ω). (d) Evolution of the piston speed signal (v).

If we link the pump speed signal displayed in Figure 10c with the respective piston speed signal in Figure 10d, we can note that there is a set of operating regions where, although the pump rotates, the piston does not move. Figure 11 shows a zoom of this behavior. For the pump speed signal, we mark the speed interval corresponding to the dead-zone. Below, we mark the corresponding regions where the piston speed is zero. When the pump operates into the dead-zone, the hydraulic circuit is decoupled from the electrical part. The pump, although rotating, does not debit fluid into the hydraulic system and so there is no pressure difference on the piston to move it.

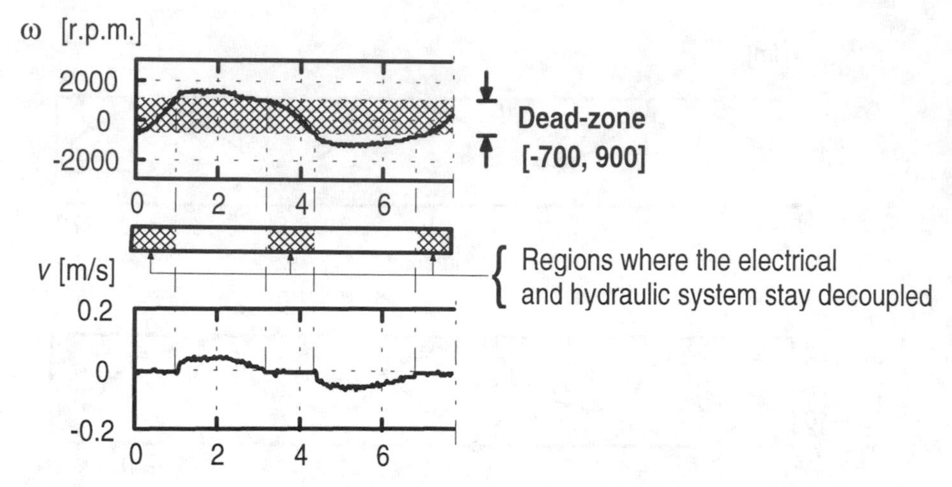

Figure 11: Picture detail of the pump speed and piston speed signals. It shows the effect of the dead-zone decoupling the hydraulic part from the electrical one.

To complement the theoretical knowledge about the experimental system with more objective information, some experimental data is acquired. This data set is used in the training stage and is composed of the system's behavior examples.

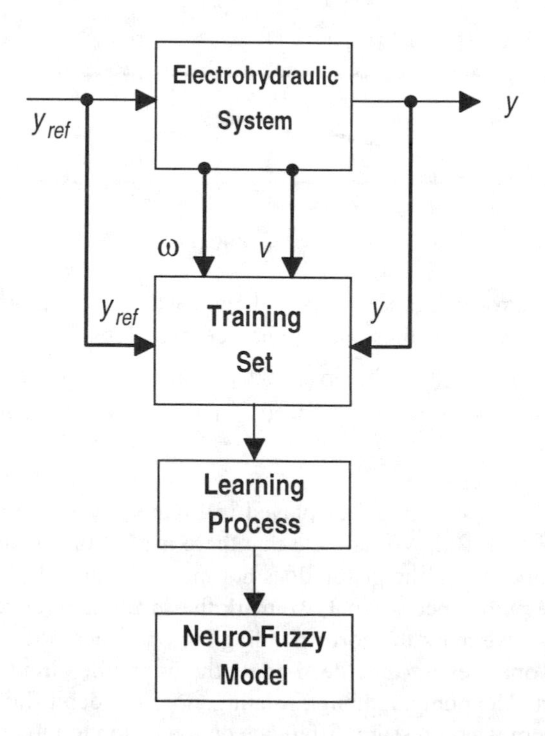

Figure 12: Diagram scheme representing the modeling stages.

Usually, to construct a training set, a Pseudo-Random Binary Signal (PRBS) is injected into the system in the manner that collected data spans during system's operating domain, although, this signal is not good to excite drive systems as pointed out in [7]. So, a better technique is to use an excitation signal of sinusoidal type composed of different magnitudes and frequencies, but within drive's response limits.

For the electro-hydraulic actuator, we used a sinusoidal signal as the reference for piston position with its amplitude ranging from 0 to 0.2m (the piston course limits) and frequencies among 0 and 1Hz because, for higher frequency values, the actuator begins filtering the reference signal.

The modeling process is described by a diagram in Figure 12. Initially, a data set with four system signals (y_{ref}, ω, v, y) is acquired using the anterior training procedures. Figure 13 displays the acquired training set composed of the sinusoidal reference signal y_{ref} with respective position y, the hydraulic pump speed signal ω, and the piston speed signal, v.

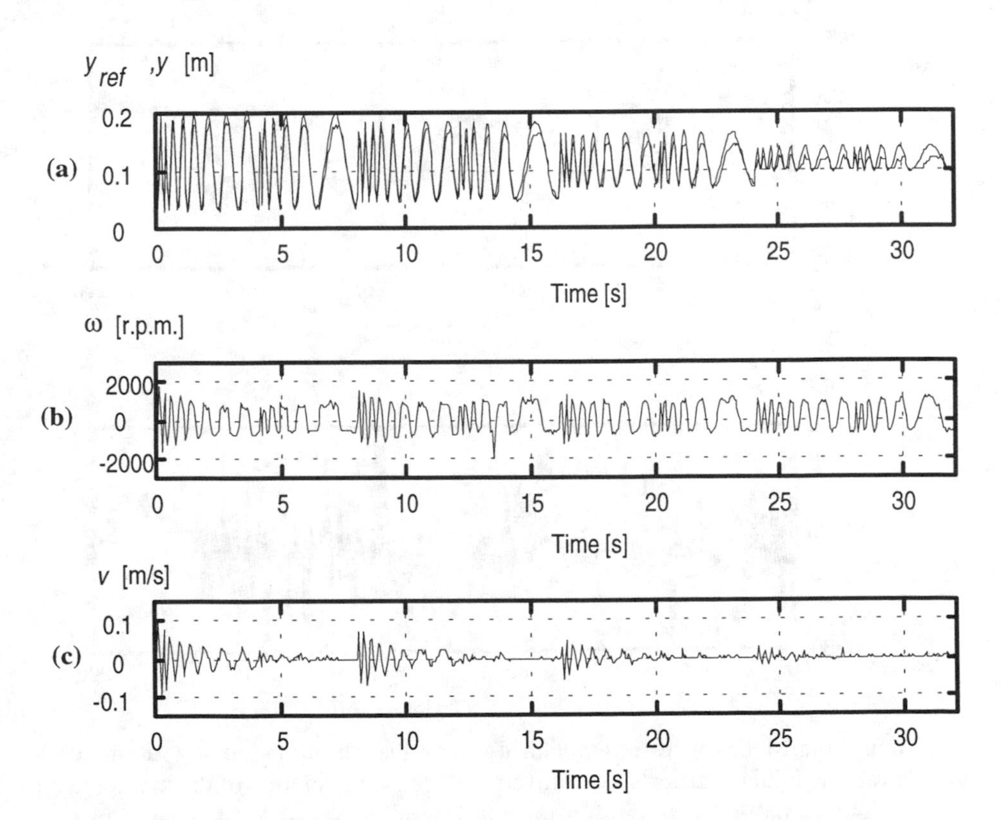

Figure 13: The acquired training data set. (a) Reference and position signal (y_{ref} and y). (b) Hydraulic pump speed (ω). (c) Piston speed (v).

6. Neuro-Fuzzy Modeling of the Electro-Hydraulic Actuator

In this section, the actuator is modeled using the neuro-fuzzy algorithm based on training data set of Figure 13. The experiment consists of obtaining the inverse model of the actuator represented by relation $\omega = h(y_{ref}, y, v)$.

The fuzzy model is composed of 7 membership functions attributed to the reference signal y_{ref}, 11 membership functions to the piston position signal y, and 7 membership functions attributed to the piston speed v. The functions are of Gaussian type, as explained before, with their shape b_j^i settled in 60% of each partition interval for each variable (j).

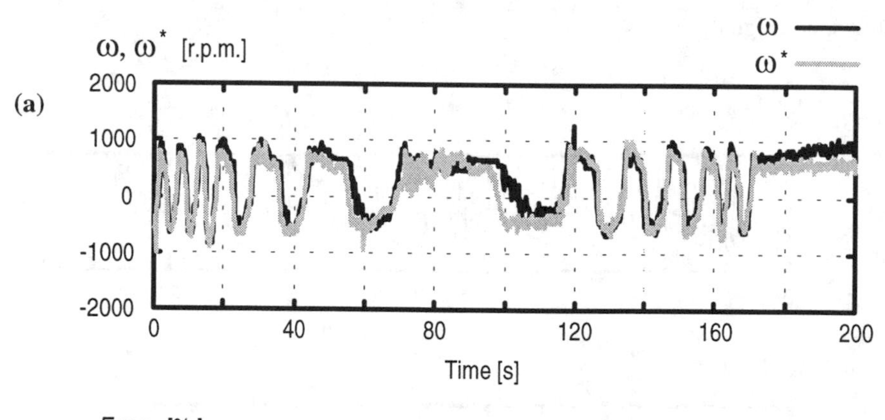

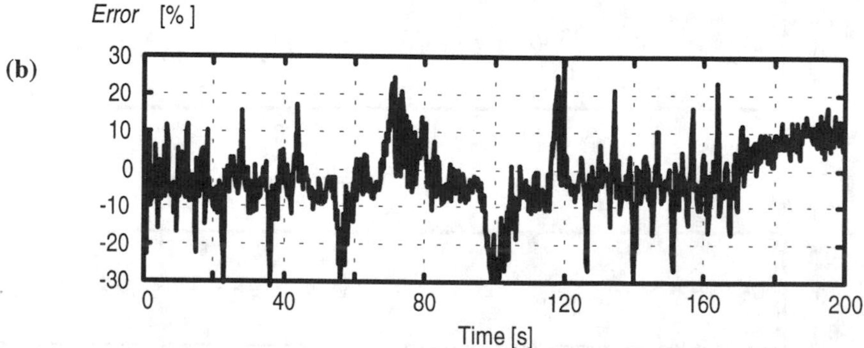

Figure 14: Modeling results obtained using the cluster-based algorithm to extract the initial actuator's fuzzy inverse-model. (a) Evolution of the measured (ω) and the inferred pump speed signal (ω^*). (b) Error signal evolution.

The first step of modeling process uses the cluster-based algorithm to extract the initial fuzzy model. To verify the generalization capability of the learned model, we use a test data set with actuator's examples not presented to the learning algorithm during the training stage. Figure 14 displays the generalization results obtained after extracting the fuzzy inverse-model.

Figure 14a shows the inferred pump speed (ω^*) from the fuzzy model and the measured one (ω). Through the error signal displayed in Figure 14b, we can observe that there are high errors for some operating regions. These are caused mainly by

- those domain regions where a small number or even no examples were acquired, because there was not enough information to extract a representative rule set for those regions;
- when the actuator operates into the dead-zone, it cannot be defined an inverse functional relation and the model generates high prediction errors;
- other errors appear as a consequence of noise presence in acquired signals y and v, which can deviate the inferred pump speed values from their correct predictions within a certain degree.

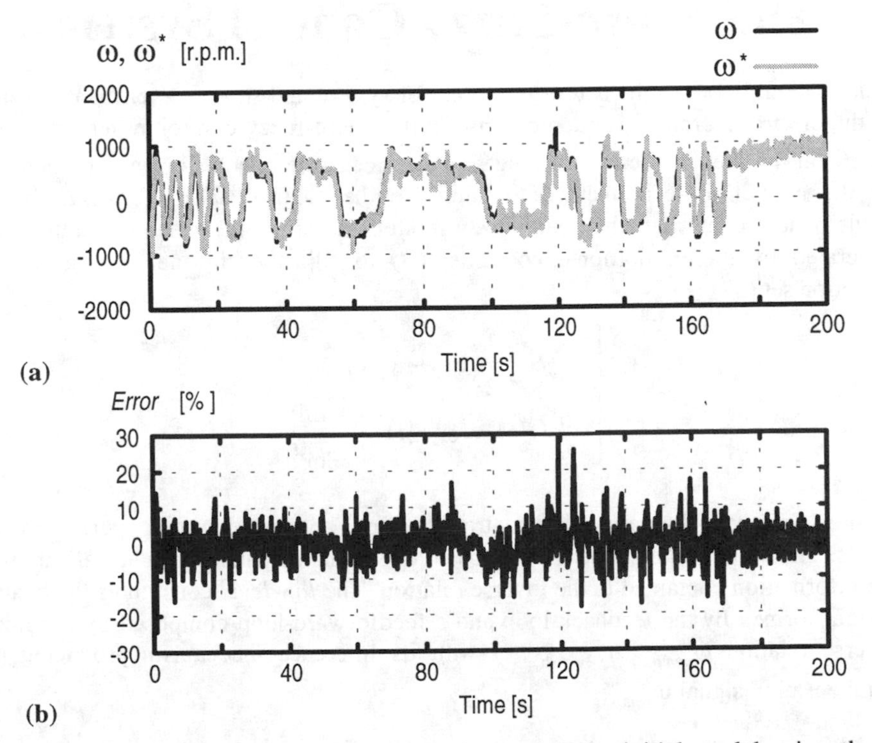

Figure 15: Modeling results obtained after tuning the initial model using the neuro-fuzzy algorithm. (a) Evolution of the measured (ω) and the inferred signal (ω^*). (b) Error signal evolution.

In the next experiment, we consider the anterior initial model and the use of the gradient-descent method explained in Section 4.4 to fine adjust it. For the learning process, the parameters used by the algorithm were: a number of 50 iterations $(K = 50)$, a learning rate of 0.8 $(\alpha = 0.8)$, and the same fuzzy model structure used in the cluster-based algorithm. The results obtained are displayed in Figure 15. They show the good tuning made by the neuro-fuzzy algorithm that reduces the error signal to about 10%.

It is important to note that the number of iterations and the learning rate were chosen in a manner that prohibits the model from incorporating noise dynamics by over fitting. Another aspect of domain regions where the number of acquired examples is small, the neuro-fuzzy algorithm continues to present high inference errors because it had little or no information to do a good tuning and extract representative rules. These points reveal the necessity of acquiring real-time information from the process. In this way, the learning mechanism can collect more information to correct and/or incorporate other rules into the model and reduce its prediction errors.

7. The Neuro-Fuzzy Control System

This section describes the neuro-fuzzy control system and shows experimental results of the electro-hydraulic position control. In the neuro-fuzzy control system, which is based on the feedback-error-learning scheme, each rule conclusion $\omega^{(l)}$ is modified by the gradient-descent method to minimize the mean quadratic error *E*. In the implemented controller, the neuro-fuzzy model minimizes the mean quadratic error generated by the proportional controller (P) to adjust each rule as indicated in Equation set (37).

$$\begin{cases} E = \frac{1}{2}\left(\mathrm{P}(y - y_{ref})\right)^2 \\ \omega^{(l)}(t+1) = \omega^{(l)}(t) - \alpha \frac{\partial E}{\partial \omega^{(l)}} \end{cases} \quad (37)$$

Figure 16 shows a diagram of our control scheme. The control system operates in two levels. The *high level* contains the responsible learning mechanism by actualization of the information contained in the inverse relation. The *low level* constitutes the control system formed by the feedback-loop and a feedforward-loop composed by the fuzzy inverse relation $\omega_{comp} = h(y_{ref}, v, y)$ with its inference mechanism producing the compensation signal ω_{comp}.

At each control iteration, the learning system collects the present values of the reference signal (y_{ref}), piston speed (v), and the current piston position (y), through the available sensor set. These signals express actuator's operating condition and make each model rule active to some degree (see expression (3)). The inference mechanism uses the model rules with corresponding activation degrees and computes the compensation signal (ω_{comp}) to be added to the proportional controller command (ω_p), as illustrated in Figure 16. The final signal, denoted by ω_{ref} and equal to the sum of ω_{comp} and ω_p ($\omega_{ref} = \omega_p + \omega_{comp}$), is sent to the electro-hydraulic actuator as its command signal.

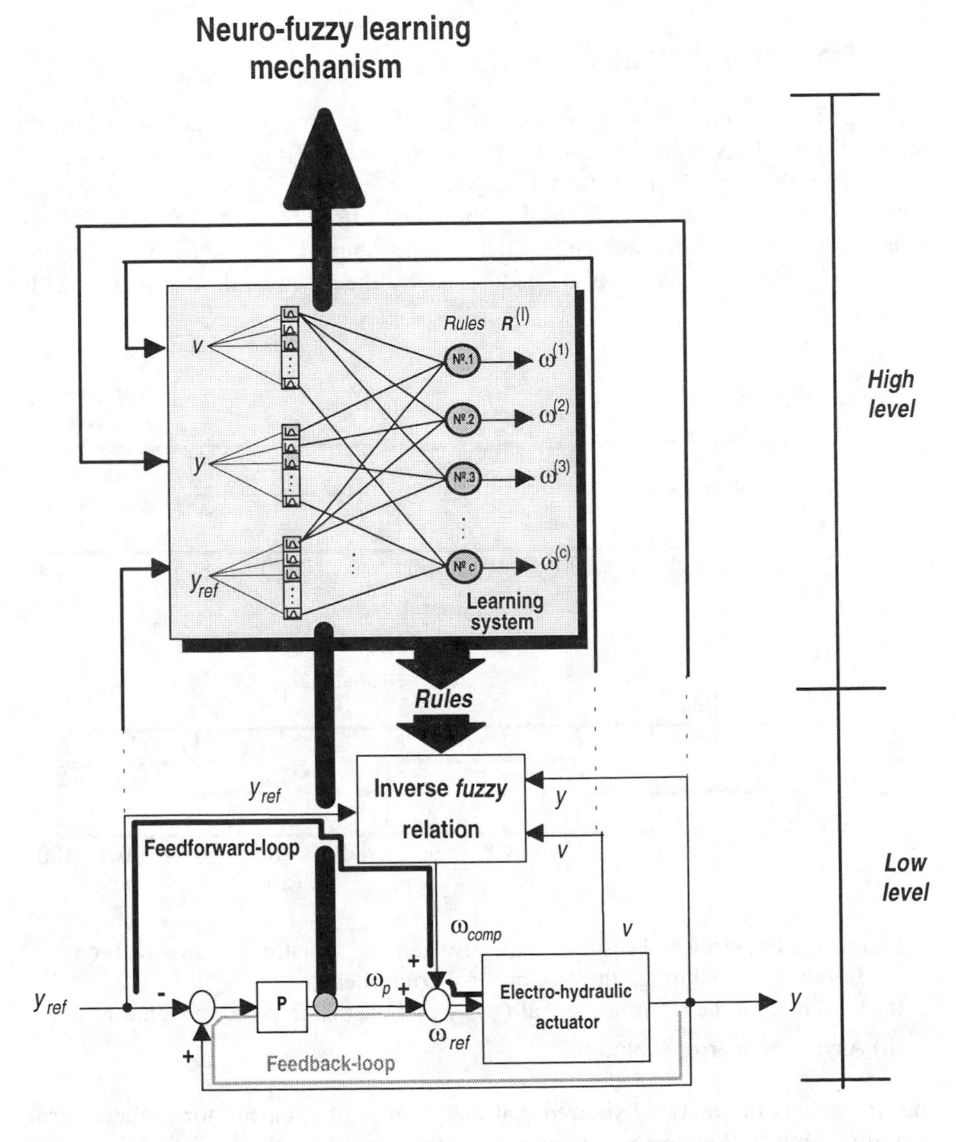

Figure 16: Diagram of the implemented neuro-fuzzy control system.

To conclude the control cycle, the error signal generated by the proportional controller after the application of computed compensation signal is used to adjust each rule. The inverse relation is then adjusted based on the performance attained by the compensation made with the anterior rule set and verified through the magnitude of the proportional controller signal. Each rule is then adjusted proportionally to its anterior activation degree, interpreted as a measure of how much the rule contributed to the actuator's actual performance.

7.1 Experimental Results

The experimental results use a square wave as the reference position signal to the piston. Figure 17 shows the results of the first test. In this test, the actuator is controlled only through the feedback-loop with the proportional controller without any compensation signal. The results show an offset error signal between the reference position and that attained by the piston (Figure 17a). The asymmetric error evolution shown in Figure 17b is conditioned by the asymmetric dead-zone in the hydraulic circuit.

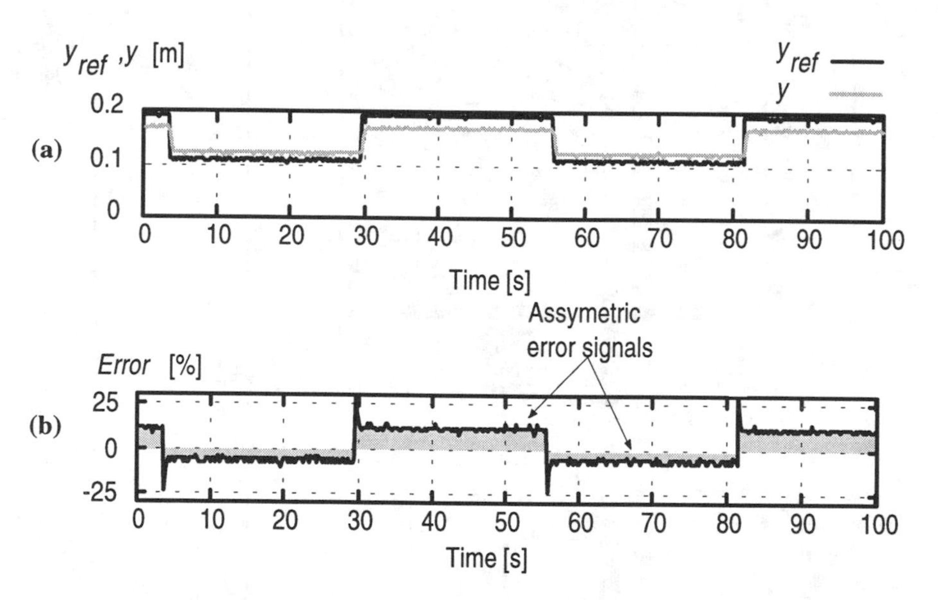

Figure 17: Experimental results obtained when the actuator operates with only the feedback-loop through the proportional controller.
(a) Evolution of the reference signal (y_{ref}) and the piston position signal (y).
(b) Asymmetric error evolution.

The first results in Figure 17 showed that the absence of compensator in the control-loop gives high tracking errors. In the second test, we added the compensation signal generated by the neuro-fuzzy inverse-model to the command signal of the proportional controller.

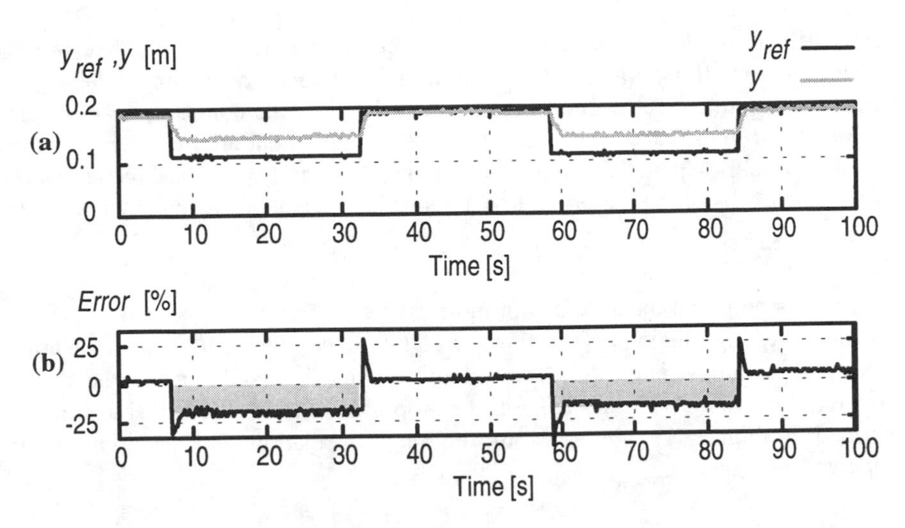

Figure 18: Experimental results when the feedforward-loop is added to the actuator system but without the neuro-fuzzy learning mechanism.
(a) Evolution of the reference signal (y_{ref}) and the piston position signal (y).
(b) Error signal evolution.

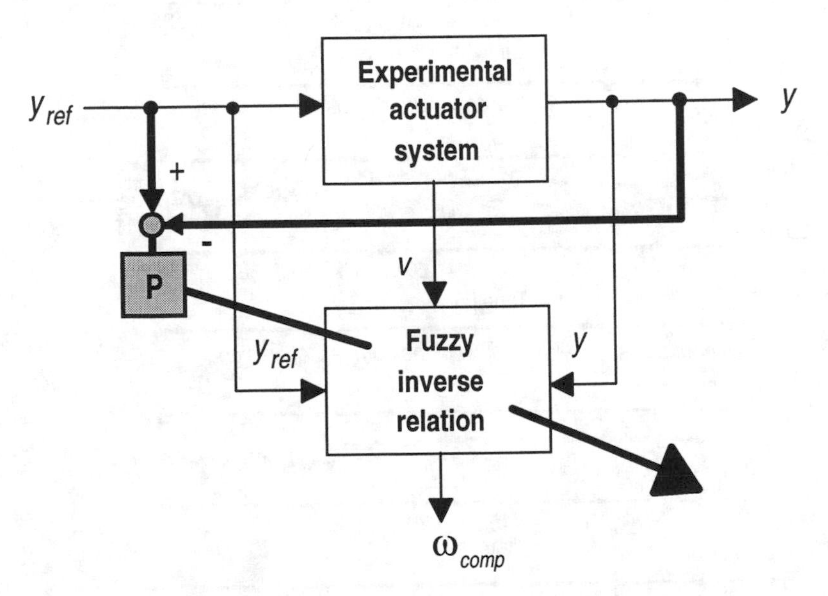

Figure 19: Diagram showing the use of the proportional controller signal to correct the inverse relation.

The results of the second test are displayed in Figures 18a and b. They use the compensation feedforward-loop with the initial extracted neuro-fuzzy model but without the learning mechanism. These results show that the compensator eliminates the error signal in the superior part of the reference signal, but produces a higher error value in the inferior part. The compensation signal generated by the inverse relation was capable of distorting the proportional controller signal (ω_p), thus increasing the error for the inferior part.

These results point out the necessity of more precise adjustment of model rules in the inferior operating region. Therefore, we introduce the neuro-fuzzy learning mechanism so the system acquires new signals in real time and corrects the rules to tune the inverse model. The system uses the proportional controller signal to adjust, as described in Figure 19, the rules of the inverse relation and then correct the compensation signal ω_{comp}.

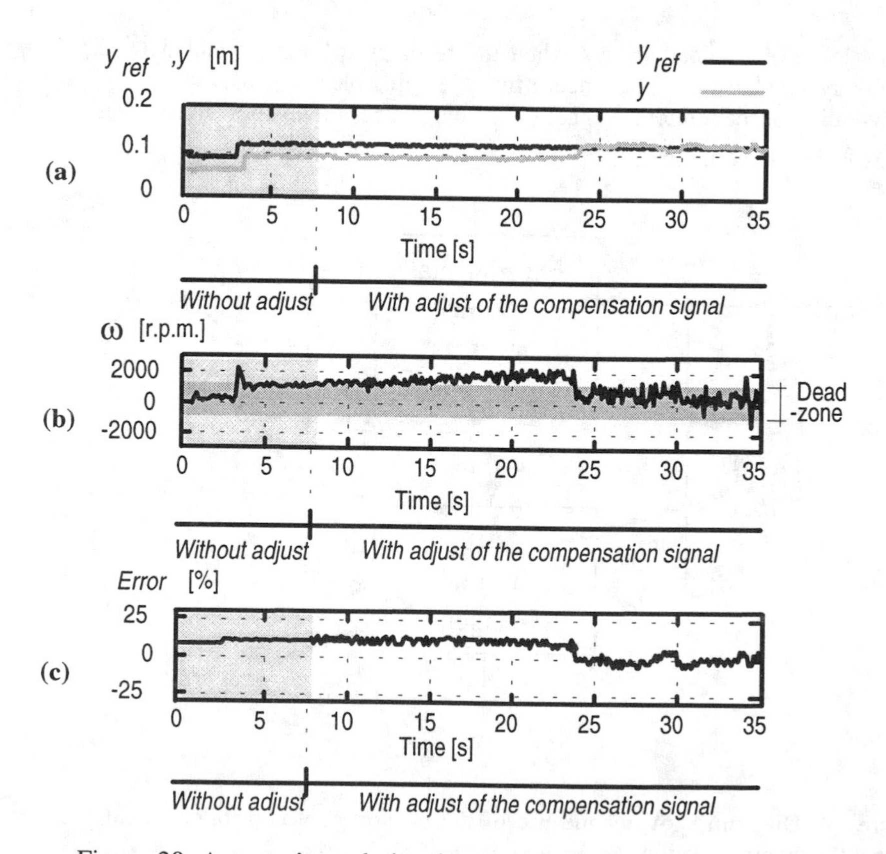

Figure 20: Actuator's evolution for a step with the learning mechanism action to adjust the compensation signal. (a) Evolution of (y_{ref}) and piston position signal (y). (b) Hydraulic pump speed (ω). (c) Error signal.

The results in Figure 20a illustrate the piston approximation to the reference signal as the rules are adjusted and the compensation signal is corrected. In this test, we used a low learning rate ($\alpha = 0.0005$) to better visualize the adjust of the compensation signal. As the learning mechanism begins to actuate, the system slowly increases the pump speed, as verified in Figure 20b, to send fluid to the hydraulic circuit and so move the piston. The pump increases its speed until its magnitude becomes sufficient to remove the actuator out of the dead-zone, adjust the model rules, and then conduct the piston to the reference position reducing the error signal as shown in Figure 20c.

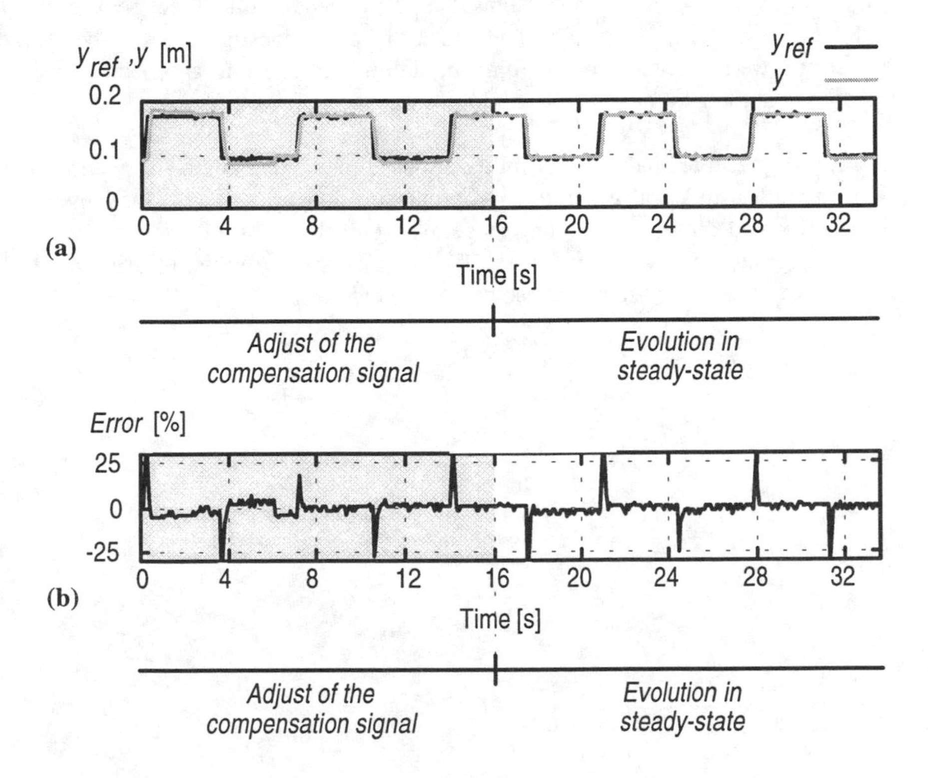

Figure 21: Actuator's evolution with the adjust, in real time, of the model rules to correct the compensation signal. (a) Evolution of the reference signal (y_{ref}) and the piston position signal (y). (b) Error signal.

In Figure 21, we present the piston evolution when the feedforward-loop and the learning mechanism are inserted into the control system. This test uses a higher learning rate ($\alpha = 0.02$) for a fast transient but without overshoots. These results show the realtime tuning until about 16 seconds where the compensation signal gradually eliminates the error offset, approximating the piston to the reference signal.

8. Conclusion

The neuro-fuzzy methodology is used to demonstrate the incorporation of learning mechanisms into control of drive systems. We believe that emerging technologies as neuro-fuzzy systems have to be used together with usual conventional controllers to produce more "intelligent" and autonomous drive systems. All the knowledge accumulated about the classical controllers and emerging techniques as fuzzy systems, neural networks, or genetic algorithms should be utilized. So, it is becoming important to investigate control designs that permit a symbiotic effect between the old and new approaches. To the concretization and investigation of the anterior objectives, we presented a neuro-fuzzy modeling and learning approach to design a position controller for an electro-hydraulic actuator.

The results presented indicate the ability of the implemented neuro-fuzzy controller in performing learning and generalization properties to quite different movements than those presented during the training stage. The compensation of nonlinearities in the electro-hydraulic system deviating the feedback controller action to drive the piston position to its reference signal is also demonstrated.

References

[1] RayChaudhuri, T., Hamey, L.G.C., and Bell, R.D. (1996), From Conventional Control to Autonomous Intelligent Methods, *IEEE Control System Magazine*, Vol. 16, No. 5.

[2] Zadeh, L. A. (1973), Outline of a New Approach to the Analysis of Complex Systems and Decision Processes, *IEEE Trans. on Systems, Man and Cybernetics*, Vol. SMC-3, No. 1, pp. 28-44.

[3] Takagi, T. and Sugeno, M. (1985), Fuzzy Identification of Systems and Its Applications to Modeling and Control, *IEEE Tran. on Systems, Man, and Cybernetics*, Vol. SMC-15, No. 1, pp. 116-132.

[4] Sugeno M. and Yasukawa, T. (1993), A Fuzzy-Logic-Based Approach to Qualitative Modeling, *IEEE Trans. on Fuzzy Systems*, Vol. 1, No. 1, pp. 7-31.

[5] Myamoto, H., Kawato, M., Setoyama, T., and Suzuki, R. (1988), Feedback-Error-Learning Neural Network for Trajectory Control of a Robotic Manipulator, *Neural Networks* 1:251-265.

[6] Wang, L. X. (1992), Back-Propagation of Fuzzy Systems as Nonlinear Dynamic System Identifiers, in *Proc. of IEEE Int. Conf. on Fuzzy Systems*, pp. 1409-1418, San Diego, CA, U.S.A.

[7] Low, T-S., Lee, T-H., and Lim, H-K. (1993), A Methodology for Neural Network Training for Control of Drives with Nonlinearities, *IEEE Trans. on Industrial Electronics*, Vol. 39, No. 2, pp. 243-249.

[8] Ohno, H., Suzuki, T., Aoki, K., Takahasi, A., and Sugimoto, G. (1994), Neural Network Control for Automatic Braking Control System, *Neural Networks*, Vol. 7, No. 8, pp. 1303-1312.

[9] Miller III, W.T., Hewes, R.P., Glanz, F.H., and Kraft III, L.G. (1990), Realtime Dynamic Control of an Industrial Manipulator Using a Neural-Network-Based Learning Controller", *IEEE Tran. on Robotics and Automation*, Vol. 6, No. 1, pp. 1-9.

[10] Bondi, P., Casalino, G., and Gambardella, L. (1988), On The Iterative Learning Control Theory for Robotic Manipulators, *IEEE Journal of Robotics and Automation*, Vol. 4, No. 1, pp. 14-22, Feb.

[11] Miller III, W.T. (1987), Sensor-Based Control of Robotic Manipulators Using a General Learning Algorithm, *IEEE Journal of Robotics and Automation*, Vol. RA-4, No. 2, pp. 157-165.

[12] Lewis, F.L. (1996), Neural Network Control of Robot Manipulators, *IEEE Expert*, pp. 65-75.

[13] Zadeh, L.A. (1965), Fuzzy Sets, *Information and Control*, Vol. 8, pp. 338-353.

[14] Sugeno, M. and Yasukawa, T. (1993), A Fuzzy-Logic-Based Approach to Qualitative Modeling, *IEEE Trans. on Fuzzy Systems*, Vol. 1, No. 1, pp. 7-31, Feb.

[15] Kawato, M., Uno, Y., Isobe, M., and Suzuki, R. (1988) Hierarchical Neural Network Model for Voluntary Movement with Application to Robotics, *IEEE Control System Magazine*, pp. 8-16, April.

[16] Lin, C.T. and Lee, C.S.G. (1991), Neural-Network-Based Fuzzy Logic Control and Decision System, *IEEE Trans. on Computers*, Vol. 40, No. 12, pp. 1320-1336.

[17] Wang, L.X. and Mendel, J.M. (1992), Fuzzy Basis Functions, Universal Approximation, and Orthogonal Least-Squares Learning, *IEEE Trans. on Neural Networks*, Vol. 3, No. 5, pp. 807-814.

[18] Costa Branco, P.J. and Dente, J.A. (1996), Inverse-Model Compensation Using Fuzzy Modeling and Fuzzy Learning Schemes. *In Intelligent Engineering Systems Through Artificial Neural Networks, Smart EngineeringSystems: Neural Networks, Fuzzy Logic and Evolutionary Programming,* Eds. C.H. Dagli, M. Akay, C.L. Philip Chen, B. Fernández, and J. Ghosh, Vol.6, pp. 237-242, ASME Press, New York, U.S.A.

[19] Wang, L.X. (1993), Training of Fuzzy-Logic Systems Using Neighborhood Clustering, *Proc. 2nd IEEE Conf. on Fuzzy Systems,* pp. 13-17, San Francisco, California, U.S.A., March.

[20] Wang, L.X. (1993), Stable Adaptive Fuzzy Control of Nonlinear Systems, *IEEE Trans. on Fuzzy Systems*, Vol. 1, No. 2, pp. 146-155.

[21] L.X. Wang, *Adaptive Fuzzy Systems and Control*, Englewood Cliffs, N.J.: Prentice-Hall, 1994.

[22] Holmblad, L.P. and Ostergaard, J.J. (1982), Control of Cement Kiln by Fuzzy Logic, M.M. Gupta and E. Sanchez, Eds. in *Approximate Reasoning in Decision Analysis*, Amsterdam: North-Holland. pp. 389-400.

[23] Mandani, E.H. (1977), Application of Fuzzy Logic to Approximate Reasoning Using Linguistic Variables, *IEEE Trans. on Computers*, Vol. C-26, pp. 1182-1191.

[24] Wang, L.X.. (1992), Back-Propagation of Fuzzy Systems as Nonlinear Dynamic System Identifiers, *Proc. IEEE Int. Conf. on Fuzzy Systems*, pp. 1409-1418, San Diego, CA.

[25] Pedrycz, W. (1984), An Identification of Fuzzy Relational Systems, *Fuzzy Sets and Systems*, Vol. 13, pp. 153-167.

[26] Czogala, E. and Pedrycz, W. (1981), On Identification in Fuzzy Systems and Its Applications in Control Problems, *Fuzzy Sets and Systems*, Vol. 6, No. 1, pp. 73-83.

[27] Costa Branco, P.J. and Dente, J.A. (1993), A New Algorithm for On-Line Relational Identification of Nonlinear Dynamic Systems, In *Proc. of Second IEEE Int. Conf. on Fuzzy Systems (IEEE-FUZZ'93)*, Vol. 2, pp. 1073-1079.

[28] Postlethwaite, B. (1991), Empirical Comparison of Methods of Fuzzy Relational Identification, *IEE Proc.-D*, Vol. 138, pp. 199-206.

[29] Higashi, M. and Klir, G.J. (1984), Identification of Fuzzy Relational Systems, *IEEE Trans. on Systems, Man, and Cybernetics*, Vol. SMC-14, No. 2, pp. 349-355.

[30] Xu, C.W. and Lu, Y.Z. (1987), Fuzzy Model Identification and Self-Learning for Dynamic Systems, *IEEE Trans. on Systems, Man, and Cybernetics*, Vol. SMC-17, No. 4.

[31] Costa Branco, P.J. and Dente, J.A. (1997), The Application of Fuzzy Logic in Automatic Modeling Electromechanical Systems, *Fuzzy Sets and Systems*, Vol. 95, No. 3, pp. 273-293.

[32] Costa Branco, P.J., Lori, N., and Dente, J.A. (1996), New Approaches on Structure Identification of Fuzzy Models: Case Study in an Electromechanical System, T. Furuhashi and Y. Uchikawa, Eds. in *Fuzzy Logic, Neural Networks, and Evolutionary Computation*, LNCS/Lecture Notes in Artificial Intelligence, Springer-Verlag, Berlin, pp. 104-143.

[33] Bezdek, J.C. and Pal, S.K., Eds. (1992), *Fuzzy Models For Pattern Recognition*, New York: IEEE Press.

[34] Tzafestas, S. and Papanikolopoulos, N.P. (1990), Incremental Fuzzy Expert PID Control, *IEEE Trans. on Industrial Electronics*, Vol. 37, No. 5, pp. 365-371.

[35] Ollero, A. and García-Cerezo, A.J. (1989), Direct Digital Control, Auto-Tuning and Supervision Using Fuzzy Logic, *Fuzzy Sets and Systems*, Vol. 30, pp. 135-153.

[36] Czogala, E. and Rawlik, T. (1989), Modeling of a Fuzzy Controller with Application to the Control of Biological Processes, *Fuzzy Sets and Systems*, Vol. 31, pp. 13-22.

[37] Jang, J-S.R. and Sun, C-T. (1995), Neuro-Fuzzy Modeling and Control, *Proceedings of the IEEE*, Vol. 83, pp. 378-405.

[38] Procyk, T.J. and Mamdani, E.H. (1979), A Linguistic Self Organising Process Controller, *Automatica*, Vol. 15, pp. 15-30.

[39] Shao, S. (1988), Fuzzy Self-Organising Controller and Its Application for Dynamic Processes, *Fuzzy Sets and Systems*, Vol. 26, pp. 151-164.

Chapter 5:

Neural Fuzzy Based Intelligent Systems and Applications

NEURAL FUZZY BASED INTELLIGENT SYSTEMS AND APPLICATIONS

Emdad Khan
Core Technology Group, National Semiconductor
2900 Semiconductor Dr.
Santa Clara, CA 95052, U.S.A.

Neural Networks and Fuzzy Logic are the two key technologies that have recently received growing attention in solving real world, nonlinear, time variant problems. Because of their learning and/or reasoning capabilities, these techniques do not need a mathematical model of the system which may be difficult, if not impossible, to obtain for complex systems. Although these techniques have had successes in solving many real world problems, they have limitations as well. Intelligent combinations of these two technologies can exploit their advantages while eliminating their limitations. Such combinations of neural networks and fuzzy logic are called Neural Fuzzy Systems (NFS). Intelligent Systems (IS) based on neural fuzzy techniques have shown good potential to solve many complex real word problems. In this chapter, we discuss various types of Neural Fuzzy Systems, their features, and some key application areas. We use "neural nets" and "neural networks" interchangeably. By neural nets, we mean artificial neural nets which try to mimic biological neural nets.

1. Introduction

The need to solve highly nonlinear, time variant problems has been growing rapidly as many of today's applications have nonlinear and uncertain behavior which changes with time. Conventional mathematical model based techniques can effectively address linear, time invariant problems. Model based techniques can also address more complex nonlinear time variant problems, but only in a limited manner. Currently no model based method exists that can effectively address complex, nonlinear and time variant problems in a general way. These problems coupled with others (such as problems in decision making, prediction, etc.) have inspired a growing interest in intelligent techniques including Fuzzy Logic, Neural Networks, Genetic Algorithms, Expert Systems, and Probabilistic Reasoning. Intelligent Systems, in general, use various combinations of these techniques to address real world complex problems. In

0-8493-9804-5/99/$0.00+$.50

this chapter, we will be focusing on the intelligent systems based on various combinations of neural nets and fuzzy logic.

Both fuzzy logic and neural nets have been very successful in solving many nonlinear time variant problems. However, both technologies have some limitations as well which have prevented them from providing efficient solutions for a large class of nonlinear time variant problems. In fuzzy logic, it is usually difficult and time consuming to determine the correct set of rules and membership functions for a reasonably complex system. Moreover, fine tuning a fuzzy solution is even more difficult and takes longer. One cannot easily write fuzzy rules to meet a known accuracy of the solution. In neural nets, it is difficult to understand the "Black Box," i.e., how the neural net actually learns the input-output relationships and maps that to its weights. It is also difficult to determine the proper structure of a neural net that will effectively address the current problem. An appropriate combination of these two technologies (NFS) can effectively solve the problems of fuzzy logic and neural nets and, thus, can more effectively address the real world complex problems.

NFS have numerous applications including controls (automotive, appliances), fast charging of various kinds of batteries, pattern recognition (speech and handwriting recognition), language processing (translation, understanding), decision making, forecasting, planning and acting (e.g., in Intelligent Agents).

In this chapter, we discuss various types of Neural Fuzzy Systems, their features, and some key applications. First, we talk about the advantages and disadvantages of the neural net and fuzzy logic in Section 2. We have omitted the basics of neural net and fuzzy logic for simplicity. A good review of fuzzy logic and neural nets can be found in [7, 14, 21]. In Section 3 we discuss the capabilities of the NFS. Various types of NFS are discussed in Section 4. Detailed descriptions of a few NFS are given in Section 5. In Section 6, we discuss some key applications. The conclusion is given in Section 7.

2. Advantages and Disadvantages of Fuzzy Logic and Neural Nets

2.1 Advantages of Fuzzy Logic

Fuzzy logic converts complex problems into simpler problems using approximate reasoning. The system is described by fuzzy rules and membership functions using human type language and linguistic variables. Thus, one can effectively use his/her knowledge to describe the system's behavior.

A fuzzy logic description can effectively model the uncertainty and nonlinearity of a system. It is extremely difficult, if not impossible, to develop a mathematical model of a complex system to reflect nonlinearity, uncertainty, and variation over time. Fuzzy logic avoids the complex mathematical modeling.

Fuzzy logic is easy to implement using both software on existing microprocessors or dedicated hardware. Fuzzy logic based solutions are cost effective for a wide range of applications (such as home appliances) when compared to traditional methods.

2.2 Disadvantages of Fuzzy Logic

Fuzzy logic has been proven successful in solving problems in which conventional, mathematical model based approaches are either difficult to develop or inefficient and costly. Although easy to design, fuzzy logic brings with it some critical problems.

As the system complexity increases, it becomes more challenging to determine the correct set of rules and membership functions to describe system behavior. A significant time investment is needed to correctly tune membership functions and adjust rules to obtain a good solution. For complex systems, more rules are needed, and it becomes increasingly difficult to relate these rules. The capability to relate the rules typically diminishes when the number of rules exceeds approximately 15. A hierarchical rule base can be used but even then the problem remains, as relating rules at different hierarchies is difficult. For many systems, it is impossible to find a sufficient working set of rules and membership functions.

In addition, the use of fixed geometric-shaped membership functions in fuzzy logic limits system knowledge more in the rule base than in the membership function base. This results in requiring more system memory and processing time.

Fuzzy logic uses heuristic algorithms for defuzzification, rule evaluation, and antecedent processing. Heuristic algorithms can cause problems mainly because heuristics do not guarantee satisfactory solutions that operate under all possible conditions. Moreover, the generalization capability of fuzzy logic is poor compared to neural nets. The generalization capability is important in order to handle unforeseen circumstances.

Once the rules are determined, they remain fixed in the fuzzy logic controller, which is unable to learn (except in adaptive fuzzy systems, which allow some limited flexibility).

Conventional fuzzy logic cannot generate rules (users cannot write rules) that will meet a pre-specified accuracy. Accuracy is improved only by trial and error.

Conventional fuzzy logic does not incorporate previous state information (very important for pattern recognition, like speech) in the rule base. A recurrent fuzzy logic (described later) incorporates the past information and hence is more effective for context sensitive information systems.

2.3 Advantages of Neural Nets

Recently, neural nets have seen revived interest from many areas of industry and education. Neural net research began in the early 1940s; however, it remained dormant until the '80s. Neural nets try to mimic the human brain's learning mechanism. Like fuzzy logic, neural net based solutions do not use mathematical modeling of the system.

Neural nets learn system behavior by using system input-output data. Neural nets have good generalization capabilities. The learning and generalization capabilities of neural nets enable it to more effectively address nonlinear, time variant problems, even under noisy conditions. Thus, neural nets can solve many problems that are either unsolved or inefficiently solved by existing techniques, including fuzzy logic. Finally, neural nets can develop solutions to meet a pre-specified accuracy.

2.4 Disadvantages of Neural Nets

As already mentioned, a major problem with neural nets is the "Black Box" nature, or rather, the relationships of the weight changes with the input-output behavior during training and use of trained system to generate correct outputs using the weights. Our understanding of the "Black Box" is incomplete compared to a fuzzy rule based system description.

From an implementation point of view, neural nets may not provide the most cost effective solution – neural net implementation is typically more costly than other technologies, in particular fuzzy logic (embedded control is a good example). A software solution generally takes a long time to process and a dedicated hardware implementation is more common for fuzzy logic than neural nets, due to cost.

It is difficult, if not impossible, to determine the proper size and structure of a neural net to solve a given problem. Also, neural nets do not scale well. Manipulating learning parameters for learning and convergence becomes increasingly difficult.

Artificial neural nets are still far away from biological neural nets, but what we know today about artificial neural nets is sufficient to solve many problems that were previously unsolvable or inefficiently solvable at best.

3. Capabilities of Neural Fuzzy Systems (NFS)

A marriage between fuzzy logic and neural nets can alleviate the problems associated with each of these technologies. Neural net technology can be used to learn system behavior based on system input-output data. This learned knowledge can be used to generate fuzzy logic rules and membership functions, significantly reducing the development time. This provides a more cost effective solution as fuzzy implementation is typically a less expensive alternative than neural nets for embedded control applications. This combination also helps solve the neural net's "Black Box" problem discussed earlier. Expressing the weights of the neural net using fuzzy rules helps provide greater insights into the neural nets, thus leading to a design of better neural nets.

Neural Fuzzy Systems can generate fuzzy logic rules and membership functions for complex systems for which a conventional fuzzy approach may fail. For such systems, conventional fuzzy logic approach uses complex hierarchical rules to keep their number low so that it remains within the limits of human capabilities. However, this limits the performance and accuracy of the solution. Two examples illustrate this point. The first is a superfast battery charger and the second a washing machine.

Conventional methods for fast charging batteries use trickle charging as shown in Figure 1. The desired charge curve (derived from the battery manufacturer's data) is highly nonlinear (although it may not appear so from the figure). When exceeding a current limit set on some battery parameters, the trip point would activate and the charge rate would be adjusted to a safe level. Normally, trip points are on voltage, temperature, rate of change of voltage, rate of change of temperature, time, and accumulated charge. Normally, 3-5 trip points are used. This does not provide good, fast charging, as the resulting curve does not match the ideal charge curve well. Writing firmware to incorporate the trip points can be complicated and cumbersome as more trip points are needed to improve the solution. Theoretically, one can use numerous trip points to follow the curve but only with great expense and impracticality. Solutions using conventional fuzzy logic are possible but carry the associated problems discussed above. For example, it is difficult to determine a working set of rules and membership functions that would result in good, fast charging. A neural fuzzy approach has been found to be more effective in solving this problem [20]. Neural fuzzy approach can also account for the variations in the ideal charge curve over time.

The objective with the washing machine is to automatically determine the type of clothing and the size of the load using a minimum number of sensors and, accordingly, provide an optimal wash cycle. Washing machines typically have one pressure sensor. It detects the water pressure which is dependent on the water absorbed by the fabric. The water absorbed depends on the type and amount of fabric. Developing fuzzy logic rules and membership functions that can determine the amount of load and types of fabric based on the water pressure and how the pressure changes during the initial wash cycle is very difficult.

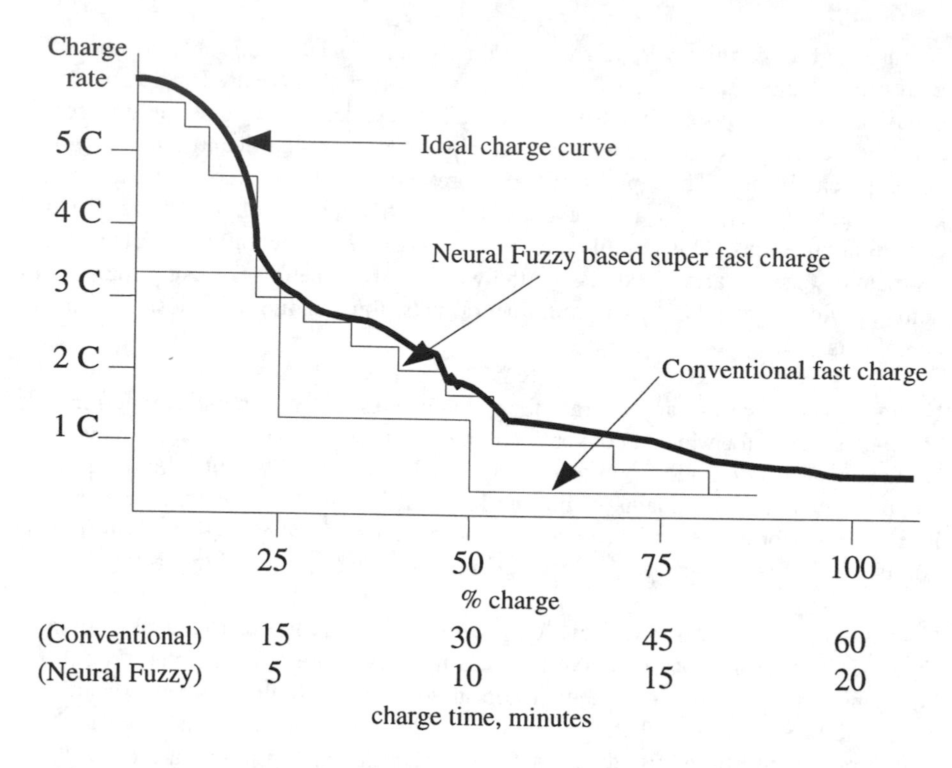

Figure 1. Fast charging of batteries using NFS and conventional methods. The problem of determining and using many trip points with conventional methods is solved by the learning capability of NFS, resulting in charging in 20 minutes compared to 60 minutes with conventional fast charging.

Washing machine manufacturer Merloni was unsuccessful with the conventional methods including fuzzy logic and neural nets. Merloni found a good solution to the problem by using a Neural Fuzzy approach [1].

The problem becomes even harder for complex, context dependent applications like speech and handwriting recognition. Writing rules for such systems that describe the context in an appropriate manner is more difficult than for the above mentioned cases. Such problems can be more effectively addressed using Recurrent Fuzzy Logic (RFL) based on Recurrent Neural Fuzzy Systems (RNFS) [11]. RFL uses the previous information as part of its antecedent, for example,

IF Input1 is Low AND Input2 is Medium AND Previous Output is High
THEN the Next Output is Low

The previous information can be extended to any number of delays (for previous inputs and outputs). RNFS can learn the system and automatically generate complex rules and associated membership functions.

By using a sufficient training set, Neural Fuzzy can learn system behavior well enough to produce working sets of rules and membership functions. The solution is generated after the net converges. The number of rules may be large and usually cannot be easily related (subset of rules can be easily related, however). This is not a significant drawback, since relating the rules is not a required function. Correct set of rules and membership functions are already generated by the system to meet desired requirements.

The Neural Fuzzy approach typically uses nonlinear membership functions. The advantage of using a nonlinear membership function is that the system knowledge can be distributed evenly between the rule base and the membership function base. This results in a reduced rule base and saves memory and overall cost.

Most important, a Neural Fuzzy System's learning and generalization capabilities allow generated rules and membership functions to provide more reliable and accurate solutions than alternative methods. In conventional approaches, one writes rules and draws membership functions, then adjusts them to improve the accuracy using trial-and-error methods. However, with the proper combination of fuzzy logic and neural networks (such as NeuFuz from National Semiconductor), it is possible to completely map (100%) the neural net knowledge to fuzzy logic. This enables users to generate fuzzy logic solutions that meet a pre-specified accuracy of outputs. This is possible because the neural net is able to learn to a pre-specified accuracy, especially for the training set (the accuracy for the test set can be controlled to be very close to the accuracy of the training set by properly manipulating the learning parameters), and learned knowledge can be fully mapped to fuzzy logic. Full mapping of the neural net to the fuzzy logic is possible when the fuzzy logic algorithms are all based on the neural net architecture. Such an elegant feature is not possible in conventional fuzzy logic, in that one cannot write fuzzy rules and membership functions to meet a pre-specified accuracy.

To guarantee full mapping of neural nets information to fuzzy logic, new fuzzy algorithms for defuzzification, rule evaluation, and antecedent processing are generally required. These algorithms are based on neural nets and thus can be made nonheuristic as opposed to conventional heuristic based algorithms. For example, Center of Gravity, COG, normally used in conventional fuzzy logic, usually works well for linear systems. With a nonlinear system, there is no guaranteed operation for a wide range of inputs or uncertainties. The rule evaluation and antecedent processing algorithms of conventional fuzzy logic are also heuristic in nature. Use of nonheuristic algorithms in NFS increases accuracy, performance, and reliability and usually reduces cost.

Another key feature of an NFS is its optimization capability. Rules and membership functions of an NFS can be optimized using neural net based efficient algorithms. The task becomes more challenging if a fuzzy logic system is used.

NFS can use fuzzy logic rules to initialize the neural nets weights. NFS can also speed up the convergence of neural nets by using variable learning rates and sigmoid functions.

4. Types of Neural Fuzzy Systems

There are various ways by which a neural net is combined (mapped) with fuzzy logic. This field is still developing and future research promises to deliver more refined and elegant approaches of fusing these technologies. In this section, we briefly discuss some key techniques of combining neural nets with fuzzy logic.

In mapping in expanded form, several parameters are used to control/adjust the shapes of the membership functions and various types of rules [8, 9, 10, 15]. In this approach information does not get lost during mapping and the relationship/mapping between the neural net and fuzzy logic is clear. Expanded mapping provides more flexibility and more effectively translates the neural nets into the fuzzy logic, but such an approach uses more neurons and takes longer to converge. In mapping in compressed form, less neurons/layers are used and this approach speeds up the convergence, but the clarity of the visualization of the mapping is lost, perhaps to a significant level. In nonheuristic mapping, nonheuristic fuzzy algorithms, developed based on the neural nets [8, 9, 10] are used. In such an approach, 100% mapping (i.e., one to one mapping between fuzzy logic and neural nets) is guaranteed, i.e., the fuzzy logic system derived from the neural net would provide the same accuracy as the trained neural net. Thus, in such systems, fuzzy logic rules can provide a pre-specified accuracy which is not possible in conventional fuzzy logic. This 100% mapping is also true in Recurrent Neural Fuzzy Systems [12, 13] discussed in Section 5.2.

In a different type of NFS, variable learning rates and sigmoid functions are used to speed up the learning of the neural net. In such a method, fuzzy logic is used to determine variable learning rates and sigmoid functions; however, the rules of fuzzy logic are usually heuristic based and can cause oscillations in the neural net.

In another kind of fuzzy logic and neural net combination, the neural net inputs, weights, and errors are treated as fuzzy numbers [6]. Using fuzzy arithmetic and extension principles, such fuzzy neural network can be trained. This approach can improve the initial fuzzy description of a system with the learning and generalization capabilities of a fuzzy neural system.

Gupta [4] used fuzzy logic based somatic and synaptic operations in the neurons. The intent was to improve the neural net's capability with the fuzzy logic descriptions.

5. Descriptions of a Few Neural Fuzzy Systems

Here we describe in detail two Neural Fuzzy Systems (NFS) – the first one uses a feed forward neural net and the second one uses a recurrent neural net. Both NFS use mapping in expanded form.

5.1 NeuFuz

5.1.1 Brief Overview

By properly combining neural nets with fuzzy logic, NeuFuz (Figure 2) attempts to solve the problems associated with both fuzzy logic and neural nets. NeuFuz learns system behavior by using system input-output data and does not use any mathematical modeling. After learning the system's behavior, NeuFuz automatically generates fuzzy logic rules and membership functions and thus solves the key problem of fuzzy logic and shortens the design cycle very significantly. The generated fuzzy logic solution (rules and membership functions) can be verified and optimized by using NeuFuz' fuzzy rule verifier and optimizer (FRVO, see Box 3 of Figure 2). The fuzzy logic solution developed by NeuFuz solves the implementation problem associated with neural nets.

Unlike conventional fuzzy logic, NeuFuz uses new defuzzification, rule inferencing, and antecedent processing algorithms which provide more reliable and accurate solutions. These new algorithms are developed based on a neural net structure. Finally, NeuFuz converts the optimized solution (rules and membership functions) into National Semiconductor's embedded controller's assembly code or into ANSI C code. This fuzzy system code then needs to be integrated with the other application code to complete the design.

Training the Neural Net

In conventional fuzzy system design, users start with some fuzzy rules and membership functions (based on their experience) and they use the development system to tune these rules and membership functions. Starting with a good set of rules and membership functions followed by proper tuning is not an easy job and takes a lot of time for a reasonably complex system. It is difficult for a human to keep track of how the rules work together when the number of rules exceeds 15. The number of antecedents and consequents in each rule complicates the system even more when these exceed 2.

NeuFuz takes a different approach to fuzzy design, mainly to eliminate the problems associated with the conventional fuzzy logic mentioned before. To take the design burden off the users, NeuFuz first learns the system behavior by using the system's input-output data. This learning is based on the learning and generalization capabilities of neural nets (Box 1 of Figure 2). The learning algorithms are described in detail in the next section. Thus, to learn how to control an application, one needs to provide input-output data for the controller to be developed. Such data can be obtained by various techniques, e.g., measurements, simulations, mathematical modeling, or experience. To adequately train the neural net, a good set of input-output data that exhibits cause-and-effect relationships of the system is required. This means that the data points should cover the possible range of operations very well. Important learning parameters are provided to properly and efficiently manage the learning process. The neural net is first initialized with some suitable values of weights that help expedite the learning and convergence. After applying a good set of input-output data for several cycles, the net converges and becomes ready to generate fuzzy rules and membership functions.

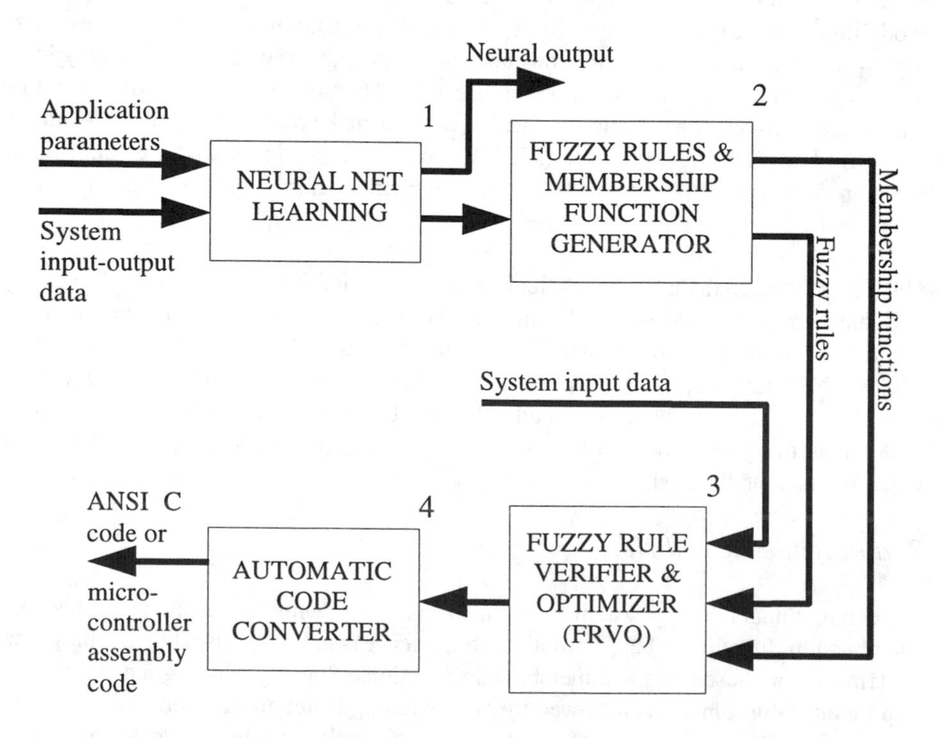

Figure 2. NeuFuz: Combining Neural Nets with Fuzzy Logic; Courtesy of National Semiconductor Corporation

Thus, users need to provide system input-output data only; no fuzzy rules or membership functions are required. However, if the user has some initial rules, they can be used to better initialize the neural net. This way the neural net may complete its learning faster.

Generating Fuzzy Rules and Membership Functions

The neural net in Box 1 of Figure 2 is properly architected so that it maps well to fuzzy logic rules and membership functions. This is shown in Figure 3 (layer 1 actually uses 4 layers, as shown in Figure 4), which is described in more detail in the next section. Neurons in layer 2 correspond to fuzzy logic rules and neurons in layer 1 correspond to the membership functions. Thus, N1 in layer 2 means

If Input 1 is Low and Input 2 is Low THEN the output is W23

where W23 is the weight between layers 2 and 3 (i.e., output layer) of the neural net.

The neuron in layer 3 does the rule evaluation and defuzzification. Thus, Box 2 of Figure 2 generates fuzzy logic rules and membership functions by directly translating Box 1 of Figure 2.

Verifying the Solution

The generated fuzzy rules and membership functions can be verified by using NeuFuz' Fuzzy Rule Verifier (Box 3 of Figure 2). A good test set should be used for the verification process. If the generated rules and membership functions do not produce satisfactory results, one can easily manipulate the appropriate parameters (e.g., more data, smaller error criterion, learning rate, etc.) so that the neural net learns more about the system behavior and finally produces a satisfactory solution.

Optimizing the Solution

The number of rules and membership functions can also be optimized using the Fuzzy Rule Verifier of NeuFuz, which is another very important feature. This reduces memory requirement and increases execution speeds – two very desirable features for many applications. Some accuracy might be lost by the optimization process and one can make some trade-offs between accuracy and cost.

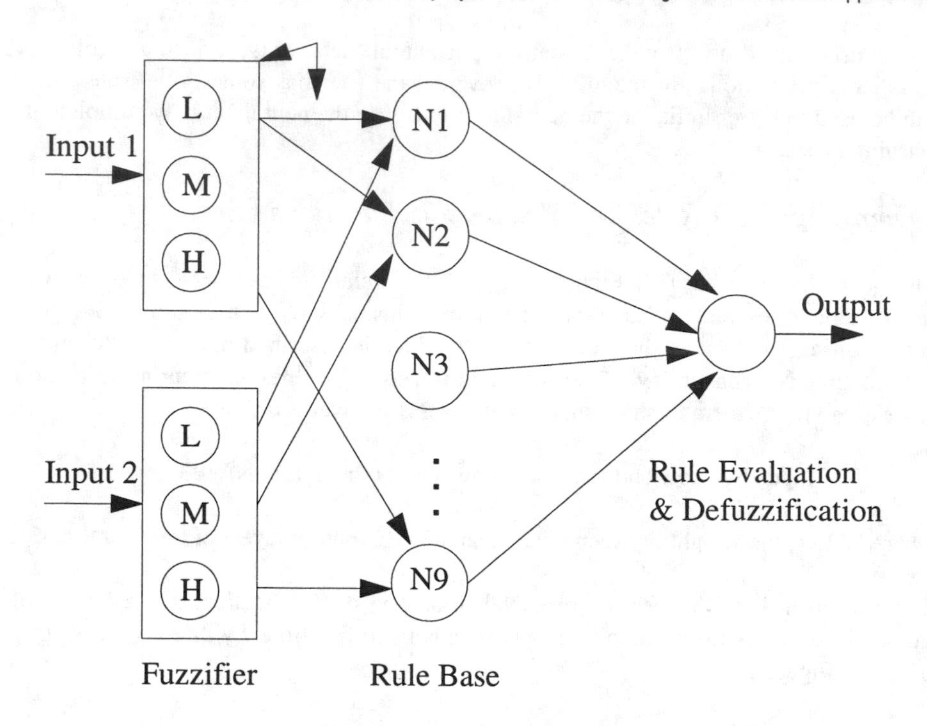

Figure 3. Learning mechanism of NeuFuz. The net is first trained with system input-output data. Learning takes place by appropriately changing the weights between the layers. After learning is completed, the final weights represent the rules and membership functions. The learned neural net, as shown above, can generate output very close to the desired output. Equivalent fuzzy design can be obtained by using generated fuzzy rules and membership functions as described in Section 5.

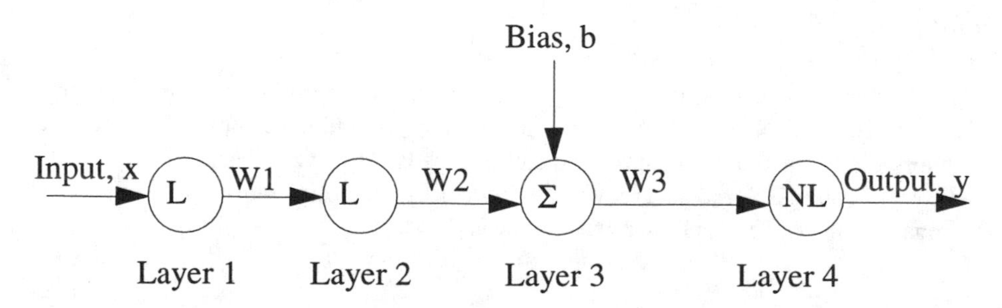

Figure 4. Neural network structure to learn membership functions.

Generating Assembly and C Code

After a satisfactory solution is obtained, NeuFuz can be used to either automatically convert the solution to an embedded processor's assembly code or generate ANSI C-code (Box 4 of Figure 2). Various options can be provided to optimize the code for accuracy, speed, or memory.

Nonlinear Membership Functions

NeuFuz uses nonlinear membership functions as opposed to membership functions of fixed geometric shapes (triangular, trapezoidal). Nonlinear membership functions can represent more system knowledge than the conventional membership functions (triangular, trapezoidal). This enables representation of a good part of the system knowledge in membership functions and, thus, reduces the number of rules (and hence cost) needed to solve the problem.

Accuracy Controlled Fuzzy Solution

The architecture of the neural net (Box 1 of Figure 2) is designed so that the neural net can be fully mapped to the fuzzy logic (Figure 3). Full mapping of the neural net solution to the fuzzy logic solution guarantees no loss of accuracy in converting a neural net based solution to a fuzzy logic based solution. Full mapping also dictates the nonheuristic algorithms used in fuzzy logic (Box 3 of Figure 2). Thus, the defuzzification, rule evaluation, and antecedent processing algorithms are derived based on the neural net architecture. Since the neural net can be trained to a pre-specified desired accuracy, the generated rules and membership functions will guarantee the same accuracy using the corresponding fuzzy logic design. In other words, the neural net will generate appropriate rules and membership functions to guarantee a pre-specified accuracy level.

Thus, unlike a conventional fuzzy system, the NeuFuz based fuzzy system can develop a solution to meet a pre-specified desired accuracy level. Pre-specified accuracy will be easily met by the training set. However, the accuracy level may not be easily met by the test set. In such a situation, the neural net can be retrained with better accuracy or part of the test set can be included in the training set (or both). This way, the performance of the test set can be improved.

Adaptability

NeuFuz can provide adaptation capabilities over time by using on-line learning capability. Thus, when implemented on embedded processors, NeuFuz can provide adaptation capability over time if on-chip learning capability is provided.

Control Parameters

NeuFuz technology provides various parameters to control and optimize the solution. Desired accuracy, learning rate, and number of membership functions are a few examples.

Understanding the Black Box

The weights of the neural nets are mapped to fuzzy logic rules and member functions. Expressing the weights of the neural net by fuzzy rules also provides better understanding of the "Black Box" and thus helps better design of the neural net itself.

5.1.2 NeuFuz Architecture

The NeuFuz architecture is shown in Figures 3 and 4. Figure 4 shows the details of layer 1 of Figure 3. Multiplicative neurons are used in the hidden layer. The output layer uses a sum neuron. The input layer, which does fuzzification and defines membership functions (layer 1 of Figure 3 and layers 1-4 of Figure 4), uses linear, nonlinear, and sum neurons. The back propagation learning algorithm is used which is properly modified to incorporate multiplicative neurons in layer 2 of Figure 3.

Fuzzification and Generating Membership Functions

The first layer of Figure 3 includes the fuzzification process, whose task is to match the values of the input variables against the labels used in the fuzzy control rule. The first layer neurons and the weights between layer 1 and 2 are also used to define the input membership functions. In fact, it is difficult to do both fuzzification and learning membership functions in one layer. Figure 4 shows a multilayer implementation for fuzzification and membership function generation. With an input level of x, the output layer 1 neuron (Figure 4) is $g1 \cdot x$ where g1 is the gain of neuron in layer 1. The input of layer 2 neuron is $g1 \cdot x \cdot W1$. Continuing this way, we have the input of layer 4 neuron, z, as

$$\begin{aligned} z &= (g1 \cdot x \cdot W1 \cdot W2 \cdot g2 + b) \cdot W3 \\ &= (a \cdot x + b) \cdot c \end{aligned} \tag{1}$$

where $a = g1 \cdot g2 \cdot W1 \cdot W2$, $c = W3$, and $g2$ = gain of layer 2 neuron.

Gains g1 and g2 can be kept constant and we can adjust only weights W1, W2, and W3 during learning. Now, if we assume the nonlinear function as an exponential function of the form $[1/(1 + e^{-z})]$, then we have the output, y, of the neuron in layer 4 as

$$y = 1 / [1 + e^{-c \cdot (a \cdot x + b)}] \tag{2}$$

By learning a, b, and c (i.e., weights W1, W2, and W3), we can easily learn an exponential membership function. The size and shape of this function is determined by weights W1, W2, W3, and bias b. By using different initial values of weights, we can generate various exponential membership functions of same type, but with different shapes, sizes, and positions. By using multiple neurons in layer 3 and 4 and using different weight values for initial W2s and W3s, we can learn any class of exponential type membership functions. These membership functions meet all the criteria to back propagate error signals. Other suitable mathematical functions could be used as well. By breaking the network in this particular way (Figure 4), we have better control of learning the membership functions. After the learning is completed, the weights remain fixed and the neural net recall operation will classify the input x in one or more fuzzy classes (each neuron in layer 4 defines a fuzzy class).

Generating Fuzzy Rules

The layer 2 neurons of Figure 3 represent the rule base. The output layer neuron provides the rule evaluation and defuzzification. Neurons in these 2 layers are linear and use a slope of unity. The weights between layers 2 and 3 (Figure 3) represent the consequent. These are singletons. After the learning is completed, layer 2 neurons along with the outputs of layer 1 neurons and the weights between layers 2 and 3 form the fuzzy rule base.

Neural Net Learning

The equivalent error, d_k^{out}, at the output layer is

$$d_k^{out} = (t_k - o_k)\, f'(net_k) \tag{3}$$

where o_k is the output of the output neuron k, t_k is the desired output of the output neuron k, and $f'(net_k)$ is the derivative which is unity for layers 2 and 3 neurons as mentioned above.

The weight modification equation for the weights between the hidden layer and output layer, W_{jk}, is

$$W_{jk}^{new} = W_{jk}^{old} + \varepsilon \cdot d_k^{out} \cdot o_j \tag{4}$$

where ε is the learning rate, W_{jk} is the weight between hidden layer neuron j and output layer neuron k, and o_j is the output from the hidden layer neuron j.

The general equation for the equivalent error in the hidden layer is

$$d_j^{hidden} = f'(net_j)\, \Sigma\, d_k^{out} \cdot W_{jk} \tag{5}$$

However, for the fuzzification layer in Figure 4, the equivalent error is different, as for the middle layer, the $netp_j$ is

$$netp_j^{hidden} = \Pi\, W_{ij} \cdot o_i \tag{6}$$

where o_i is the output of the input layer neuron i.

Thus, for the input layer (Figure 3), the equivalent error expression becomes

$$d_i^{input} = f'\,(netp_i\,)\, \Sigma\, [d_j^{hidden} \cdot W_{ij} \cdot (\Pi W_{kj} \cdot o_k)] \tag{7}$$

where both i and k are indices in the input layer and j is the index in the hidden layer; the summation is over j and product is over k.

5.1.3 Fuzzy Logic Processing

The generated fuzzy rules and membership functions are processed using the neural net based nonheuristic fuzzy logic algorithm as described below.

Antecedent Processing

We are using multiplication of the membership functions, as opposed to minimum of the membership functions, in processing the antecedents. Thus the equation to combine two antecedents is

$$\upsilon_c = \upsilon_a \cdot \upsilon_b \tag{8}$$

where υ_c is the membership function of the combinations of the membership functions υ_a and υ_b. Use of multiplication as dictated by the neural net significantly improves the results [9].

Rule Format

The rule format is

IF Input 1 is Low AND Input 2 is Low THEN the output is X

where X is a number. Thus, instead of using an output membership function, we are using singletons.

Defuzzification / Rule Evaluation

Consider the following equation for the defuzzification:

$$\text{Output} = \Sigma\, y \cdot W23 \tag{9}$$

where Output is the final defuzzified output which includes the contribution from all the rules as represented by layer 2 and y represents the output of layer 2 neurons.

Clearly, we get a defuzzification which exactly matches the neural net behavior. This defuzzification is also simpler, as it does not use the division operation used in the conventional COG (Center of Gravity) defuzzification algorithm. Another point to note is that the proposed defuzzification is actually a rule evaluation. Since the output of the rule is singleton, we actually do not need a defuzzification.

5.2 Recurrent Neural Fuzzy System (RNFS)

Figure 5 shows a recurrent neural network based neural fuzzy system. This is based on the NeuFuz architecture (Figure 3) by adding recurrent connections in the hidden layer (i.e., rule base).

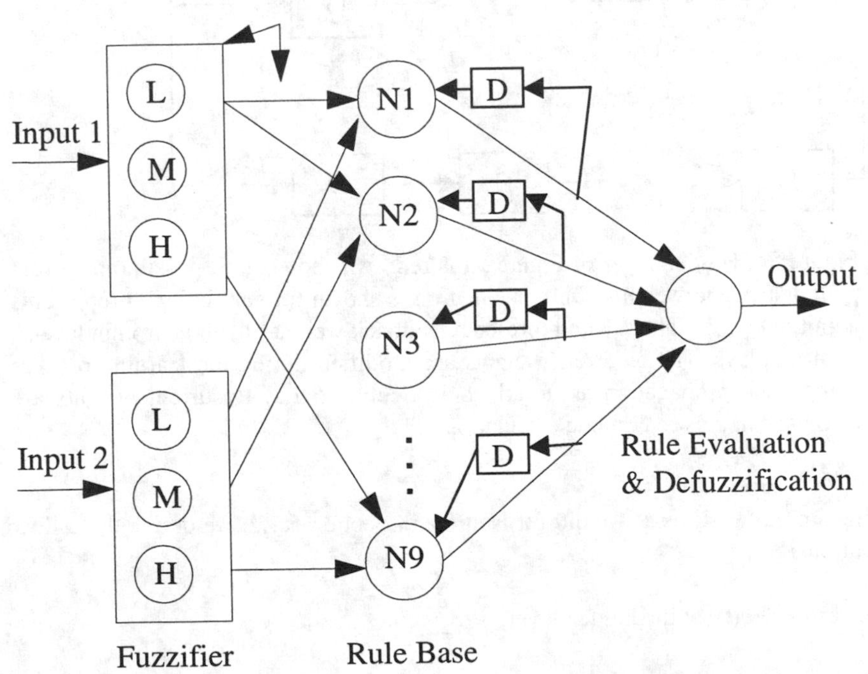

Figure 5a. Recurrent neural net using the architecture of NeuFuz shown in Figure 3. Recurrent connections are added in the hidden layers. D represents a unit delay.

5.2.1 Recurrent Neural Net

The recurrent connections in the hidden layer (Figure 5) use multiple delays. These recurrent connections are used both during learning and recall phases. In calculating the output of a neuron in layer 2, outputs of layer 1 neurons connected to layer 2 neurons are multiplied with the feedback signals of layer 2 neuron with appropriate delays. Recurrency can also be added to the input neurons to exploit more information about context.

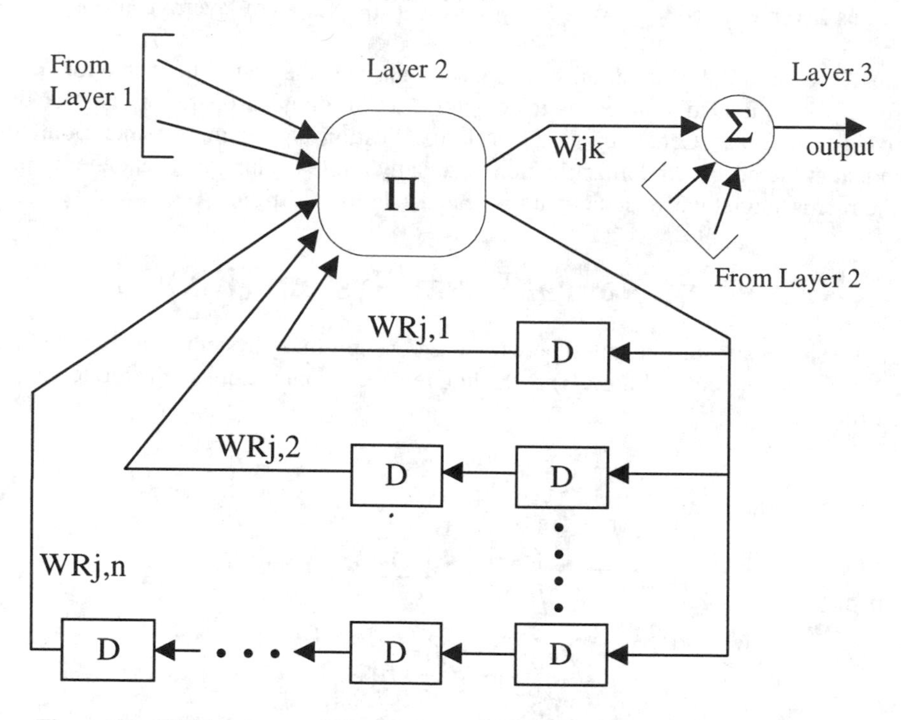

Figure 5b. Hidden layer of Figure 5a is redrawn showing all possible recurrent path delays and weights. Only one neuron is shown for simplicity. D represents a unit delay. Weight WRj,n represents feedback weight of j-th neuron in layer 2 with n delays. The recurrent weights are modified during the learning process using back propagation as described in Section 5.2.1. Recurrent weights are also used in the recall mode calculations.

Using similar analysis as we did for NeuFuz in Section 5.1.2, we derive the following equations:

The $netp_j^{hidden}(t)$, at the hidden layer, is

$$netp_j^{hidden}(t) = (\Pi\, W_{ij} \cdot o_i) \cdot (\Pi\, netp_j^{hidden}(t - m) \cdot WRj,m) \quad (10)$$

where $netp_j^{hidden}(t - m) = netp_j^{hidden}$ with delay m with respect to current time "t,"
$netp_j^{hidden}(t)$ = layer 2 (hidden layer) output at time t, and
O_i = output of the i-th neuron in layer 1.

We have used linear neurons with a slope of unity for the hidden (layer 2) and output (layer 3) layer neurons. Thus, the equivalent error at the output layer is

$$d_k^{out} = (t_k - o_k)\, f'(net_k) \tag{11}$$

where o_k is the output of the output neuron k

The general equation for the equivalent error at the hidden layer neurons using Back Propagation model is

$$d_j^{hidden} = f'(net_j)\, \Sigma d_k^{out} \cdot W_{jk} \tag{12}$$

However, for the fuzzification layer in Figure 5a, the equivalent error is different, as for the hidden layer we have product (instead of sum) neuron.

Thus, for the input layer in Figure 5a, the equivalent error expression (after incorporating the effect of the recurrent paths) becomes ([11, 12, 13])

$$d_i^{input} = f'(netp_i)\cdot \Sigma[d_j^{hidden} \cdot W_{ij} \cdot (\Pi W_{kj} \cdot o_k) \cdot \Pi netp_j^{hidden}(t - m) \cdot WRj,m] \tag{13}$$

where both i and k are indices in the input layer and j and m are the indices in the hidden layer. Also, the value of k is different from the value of i.

The weight update equation for any layer (except recurrent path in layer 2) is

$$\Delta W_{ij} = \varepsilon \cdot d_j \cdot o_i \tag{14}$$

For the feedback connections (layer 2), o_i is denoted by $o_i^{recurrent}$ and is the $netp_j^{hidden}(t\text{-}m)$ multiplied by WRj,m (which corresponds to the weight in the path with "m" delays). We add a bias term of unity to $o_i^{recurrent}$ to help convergence.

Thus, $o_i^{recurrent}$ is given by

$$o_i^{recurrent} = 1 + netp_j^{hidden}(t - m) \cdot WRj,m \tag{15}$$

and d_j is the equivalent error in the hidden layer.

Weight Update for Recurrent Path

The weight update equation for the recurrent path is somewhat different. For simplicity, consider unit delay (i.e., m = 1). For better clarification, the neuron in Figure 5b is redrawn with recurrent connection showing 2 parts: P1 and P2 (Figure 6). The output of P1 is In and the output of P2 is $netp_j^{hidden}$ as described before. Thus, we have

$$\begin{aligned} netp_j^{hidden}(t) &= In \cdot O_i^{recurrent} \\ &= In \cdot (1 + WRj,1 \cdot netp_j^{hidden}(t-1)) \end{aligned} \tag{16}$$

Let fn = $netp_j^{hidden}$. Then, the partial derivative of fn with respect to WRj,1 becomes

$$fn'(t) = In \cdot [fn(t-1) + WRj,1 \cdot fn'(t-1)] \tag{17}$$

Using standard gradient descent and chain rules, it can easily be shown that

$$\Delta WRj,1 = In \cdot d_j^{hidden} \cdot fn'(t) \tag{18}$$

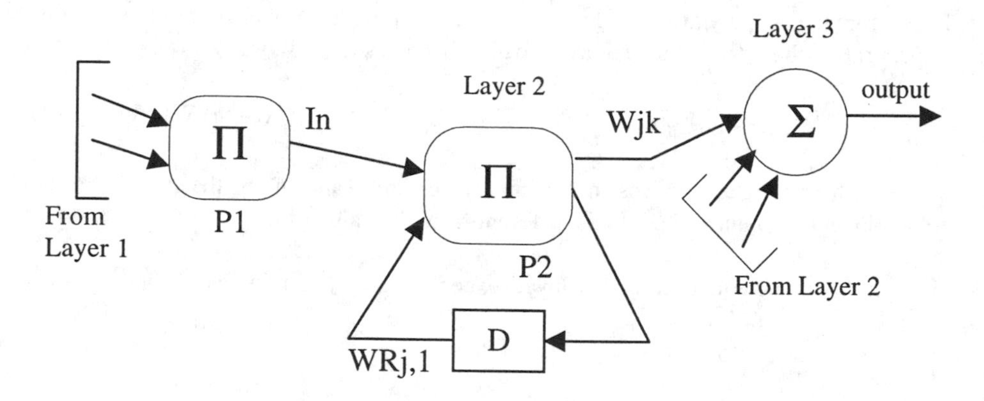

Figure 6. A neuron from layer 2 (Figure 5) is redrawn showing 2 parts: P1 and P2. The output of P1 is In and the output of P2 is $netp_j^{hidden}$ as used in Figure 5a.

5.2.2 Temporal Information and Weight Update

It is to be noted that the temporal information is accumulated by the recurrent path. Hence updates need to be done in batch mode. Thus, the above weight changes are calculated and stored for each pattern, but weight update is done only after all patterns are applied and by summing all ΔWij for each pattern. This process is then repeated until convergence is obtained. Thus, we follow schemes similar to real-time learning as reported in [7]. Accumulation of information also takes place in the "recall" mode, and it has corresponding impact on the fuzzy rules as described below. It should be noted here that the net only remembers the recent past, i.e., it gradually forgets the old context.

5.2.3 Recurrent Fuzzy Logic

Like NeuFuz, the first layer neurons (Figure 5a) include the fuzzification process whose task is to match the values of the input variables against the labels used in the fuzzy control rule. The first layer neurons and the weights between layer 1 and layer 2 are also used to define the input membership functions. The layer 2 neurons represent fuzzy rules. However, considering the recurrent connection (Figure 5a), the fuzzy rule format is accordingly changed to incorporate the recurrency. Thus, the recurrent fuzzy rule format is

IF input 1 is Low AND input 2 is Low AND previous output is Y1
THEN the new output is X

[recurrent fuzzy rule with one delay]

considering recurrent connection with unit delay. Y1 is a singleton like X. However, it can have two forms:

a) Y1 same as WRj,1 and

b) Y1 as the product of $netp_j^{hidden}(t-1)$ and WRj,1.

Clearly, the previous output information is incorporated in the rule's antecedent and is represented either by weights WRj,1 or by ($WRj,1 \cdot netp_j^{hidden}(t-1)$), although the latter makes more sense from temporal information point of view. Thus, fuzzy processing is different for these two cases, although both would essentially yield the same results.

For recurrent connection with n-delays, the recurrent fuzzy rule becomes,

IF input 1 is Low AND input 2 is Low
AND last output is Y1
AND 2nd last output is Y2
AND 3rd last output is Y3
...
...
AND n-th last output is Yn

THEN the new output is X

[recurrent fuzzy rule with n delays]

To ensure that we do not lose any information (or accuracy) in mapping neural net to fuzzy logic, a one-to-one mapping is used (for both versions of Y singletons). This approach has the key advantage of generating fuzzy logic rules and membership functions that meet a pre-specified accuracy. To ensure one-to-one mapping between the neural net and the fuzzy logic, neural network based fuzzy logic algorithms are used, as described for the NeuFuz.

Recurrency can also be used in the inputs. A typical fuzzy logic rule considering single delay would look like

IF input 1 is Low AND previous Input 1 is Medium
AND input 2 is Low AND Previous input 2 is Low
AND last output is Y1
AND 2nd last output is Y2
...
...
AND n-th last output is Yn

THEN the new output is X

[recurrent fuzzy rule with delays at inputs and outputs]

Extension to multiple delay case is straightforward.

5.2.4 Determining the Number of Time Delays

Use of too many feedback delays complicates fuzzy rules, uses more memory, and does not necessarily improve the solution. This is because the context information is useful only when the signals in the near vicinity belong to the same membership function (i.e., when autocorrelation function does not decay rapidly). With a large number of delays in the feedback path, all distant signals will not belong to the current membership class and so the recurrent information will not be useful. Typical values for number of delays are from 1 to 5. As shown in Table 1, context information with more than one delay does not help much. Also, note that the recurrency significantly improves the convergence time.

Table 1. Number of cycles used to learn a sine wave with 9 data points. 5 membership functions were used.

Accuracy	**Number of cycles to converge**				
	1D	**2D**	**3D**	**4D**	**NeuFuz (no recurrency)**
0.01	9	8	8	7	1036
0.001	172	160	160	158	5438

6. Representative Applications

Neural Fuzzy techniques can be applied to many different applications. Home appliances (vacuum cleaners, washing machines, coffee makers, cameras etc.), industrial uses (air conditioners, conveyor belts, elevators, chemical processes, etc.), automotive (antiskid braking, fuel mixture, cruise control, etc.), fast charging of batteries, and speech recognition are a few examples.

NFS (e.g., NeuFuz) learns by using the system input-output data and application parameters as shown in the first box of Figure 2. After the learning is completed, fuzzy rules and membership functions are generated, verified, optimized, and converted to assembly code.

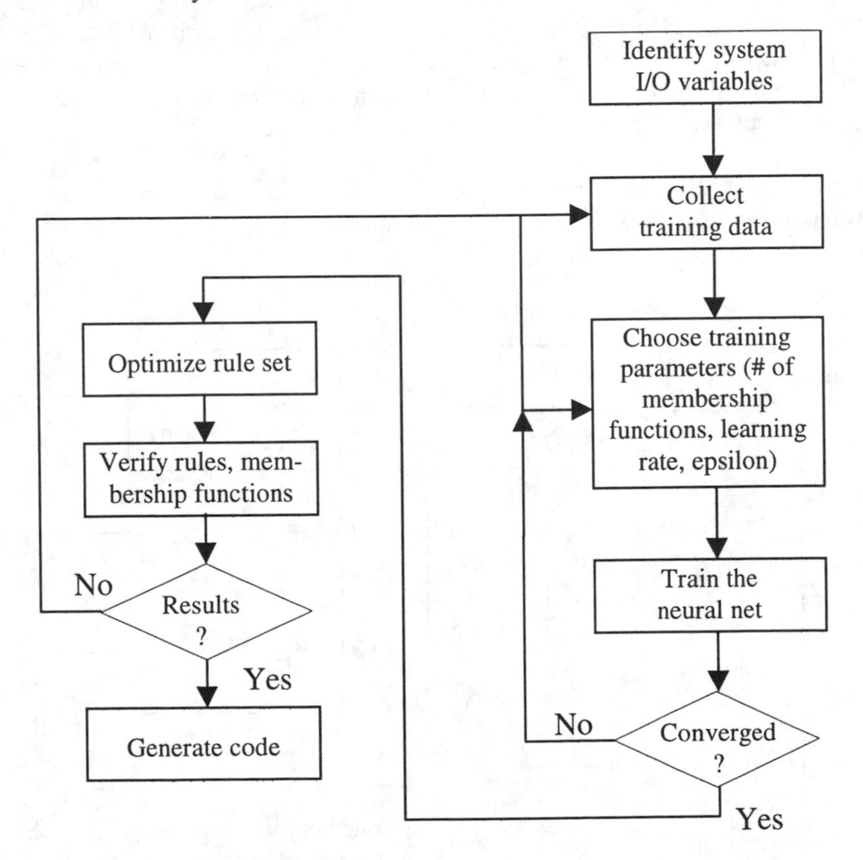

Figure 7. Flow chart for NeuFuz based application development.

The complete flow of developing an application using NeuFuz is shown in Figure 7. This corresponds to how all the components of Neufuz are used in developing an application. Determining the correct set of input-output data for learning is critical, as system performance is dependent on the learning. Input-output data can be obtained from one or more of the following sources: measurements, experience, mathematical models, and simulations.

Measurements provide better data, but this step may be cumbersome and care must be taken in the measurement itself as well as relating measured data. In some cases, direct measurement of data may be very difficult. Figure 8a shows an automated scheme (which avoids direct measurement) to learn the plant (by the model neural net) as well as how to control it with neural fuzzy technology. To learn a plant's model, arbitrary inputs (within the range of interest) can be applied to both plant and model neural nets. The outputs are compared and weights of the model neural net are adjusted using the error e1. The plant's learning is complete once error e1 is minimized. Neural Fuzzy learning then begins using error e2, which is back propagated through the model neural net. Neural Fuzzy learning is complete when error e2 is minimized. Neural Fuzzy then generates fuzzy rules and membership functions (step 2 of Figure 2). These rules then can be verified, optimized, and integrated with other system code. Figure 8b shows how the neural fuzzy solution controls the plant.

Below, we discuss a few specific applications by using the general method of developing a neural fuzzy application as described above.

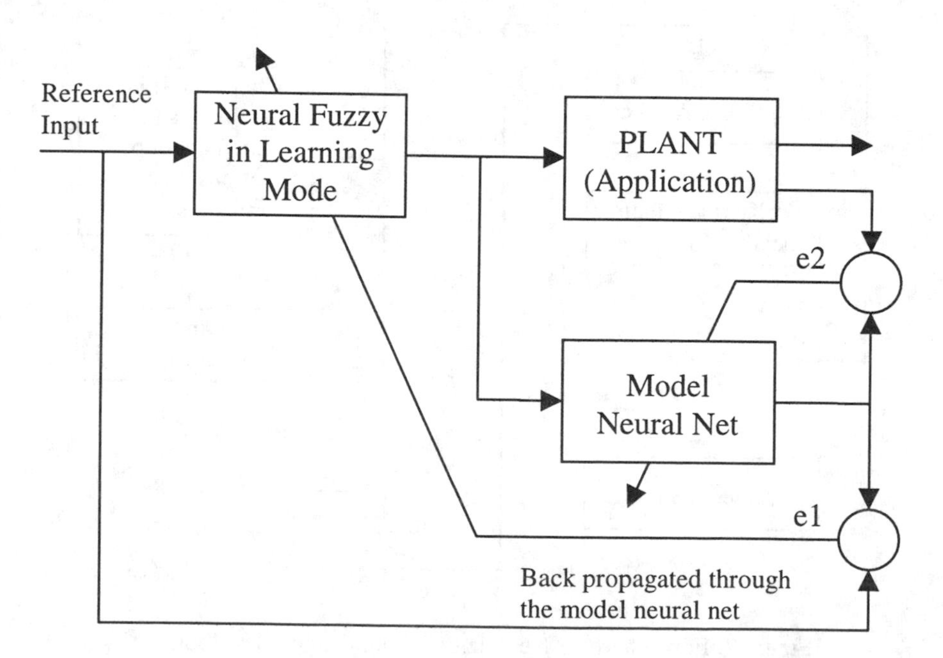

Figure 8a. Learning how to model the plant as well as how to control it.

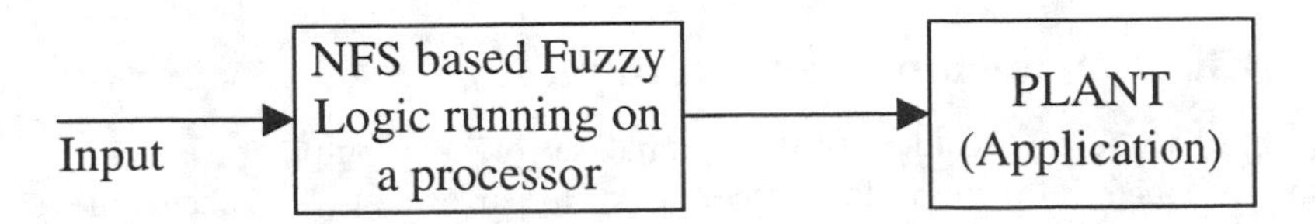

Figure 8b. Controlling the application using NFS generated fuzzy logic design which runs on a processor.

6.1 Motor Control

Motor control for accurate positioning and speed is a very important function for many applications. The algorithm for motor control is the key to the success of the product. Conventional control schemes (e.g., PID – proportional, integral, and derivative – controllers) use a linearized model of a nonlinear system. This results in degraded performance which may be unsatisfactory for highly nonlinear systems. The mathematical manipulations used in this approach are often time consuming and error prone. In this section, the design of a NeuFuz based motor controller is described. The high level of design automation provided by NeuFuz significantly reduces design time and offers increased reliability and accuracy. The performance of the NeuFuz controller is compared with a corresponding conventional PID controller.

The control structure for the motor is shown in Figure 9. A DC motor is used which can operate at a maximum speed of 2500 rpm and generate up to 1/8 HP. The input voltage to the motor ranges from 0V DC to 130V DC. The objective of the motor controller is to regulate the input voltage of the motor to minimize rise time, reduce overshoot, and maintain a desired speed even when the load is varied. The NeuFuz based application development flow shown in Figure 7 is used to develop the controller, C, for the motor, M. The motor is connected to the generator, G, which is connected to an electrical load. The controller, C, is implemented by National Semiconductor's 8-bit COP8 microcontroller.

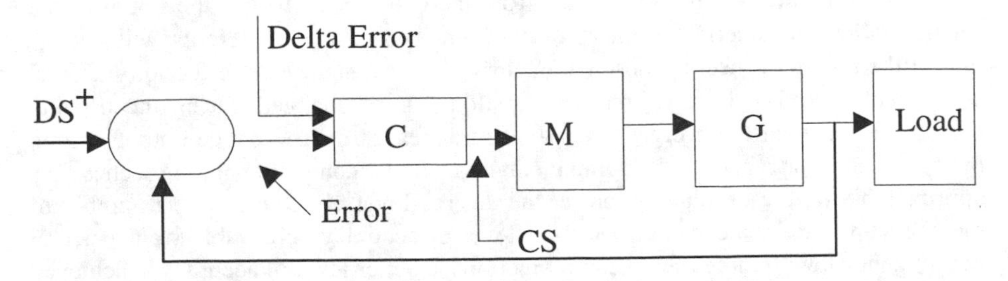

Figure 9. Control structure for the motor control application.

6.1.1 Choosing the Inputs and Outputs

The objective here is to identify the optimal set of input/output variables that will contain adequate information for the controller to perform at a satisfactory level. An improper set of input and output variables may result in undesirable performance and/or higher cost. In this case, the inputs chosen were error and the change in error, and the output chosen was the change in the controller output. Following are the definitions of the inputs and output:

Input 1: error = desired speed - current speed
Input 2: delta error = error - previous error
Output: delta out = required change in the motor input voltage
DS = Desired speed, CS = Current speed

The two input variables mentioned above contain all the information needed to adequately control the motor. At any given instant, the error and the change in error informs the controller of the status of the motor. For example, a positive error implies that the motor is running at a lower speed than desired, thus indicating that the motor input voltage needs to be increased. The change in error input tells the controller at what rate and in what direction the motor speed is being corrected. This information is critical in determining the amount of additional effort (delta out) required to bring the motor to the desired speed optimally (without overshoots).

6.1.2 Data Collection and Training

Once the optimal set of inputs and outputs has been identified, the next step is to collect sample input and output data to train the neural net. As mentioned earlier, various methods exist to collect the training data. For this application, the training data was collected by measurements.

NeuFuz requires several training parameters to be set before training. These parameters are error (convergence) criterion, learning rate, and number of membership functions. Convergence criterion, epsilon, is the maximum allowable deviation of the neural network outputs from the output specified in the training data set. Learning rate determines the rate at which the neural net weights will change during the training process. These parameters can have significant effect on the final system solution. The learning rate and epsilon can be changed during the training process. The network converges when the neural net learns to produce outputs within the specified error range for all training patterns. Unlike conventional approaches, this approach allows the designer to preset the desired level of accuracy. The number of membership functions chosen affects the level of accuracy achievable by the neural net. In general, with more membership functions, a better level of accuracy is achieved at the expense of larger code and slower response time.

In this design, six membership functions were used for each input. The neural network generated nonlinear (exponential) membership functions which were approximated using shouldered trapezoidal membership functions using NeuFuz's function-editing feature [19]. This approximation allows a cost-effective implementation of the fuzzy solution on a low-cost microcontroller.

6.1.3 Rule Evaluation and Optimization

The generated solution was evaluated using the rule evaluation feature of NeuFuz. Accuracy of the solution was acceptable for the entire range of operation. The rule optimization feature of NeuFuz was used to reduce the number of rules by deleting rules whose contributions were insignificant. In this particular case, the optimizer did not delete any rules without significantly degrading the accuracy level. Hence all the generated rules (36 in total) were used.

6.1.4 Results and Comparison with the PID Approach

Several tests were performed to evaluate the performance of the NeuFuz based fuzzy controller and the PID controller. Results are summarized in Figure 10. These tests show that the NeuFuz controller has reduced overshoots and settling time at start up, while maintaining approximately the same rise time. Both controllers produce zero steady state error. However, when load is changed (not shown in the figure), NeuFuz produces considerably less error than the PID approach. The time domain equation for the motor-generator-load is

$$y(t) = 1 - 1.25e^{-1.1t} + 0.25e^{-6.9t}$$

The s-domain transfer function for the PID solution is based on this equation.

6.2 Toaster Control

The key problem in a toaster control is to maintain the desired darkness level for variations in moisture content, types of bread, size of bread and initial temperature. Most conventional toasters use a pre-programmed timer to determine the length of time its coils will be heated. This approach cannot modify the heating time depending on the initial temperature of the toaster. The result is toast with varying degrees of darkness even if the *"darkness"* setting is unchanged.

Figure 11 shows the control structure of a toaster controller [18]. In this figure "C" is the controller and "T" is the toaster. For any darkness setting, the controller will generate an output that will be used by the toaster to control the heat applied to the slice of bread. The heat applied to the bread can be calculated from the instantaneous temperature of the toaster. This is then fed back to the controller. Based on the darkness setting and how much heat has been applied, the controller decides how much more heat is required and generates appropriate inputs to the toaster. In this example, "C" can be a PID, neural controller, fuzzy controller, or a neural fuzzy (e.g., NeuFuz) controller. The hardware structure is shown in Figure 12.

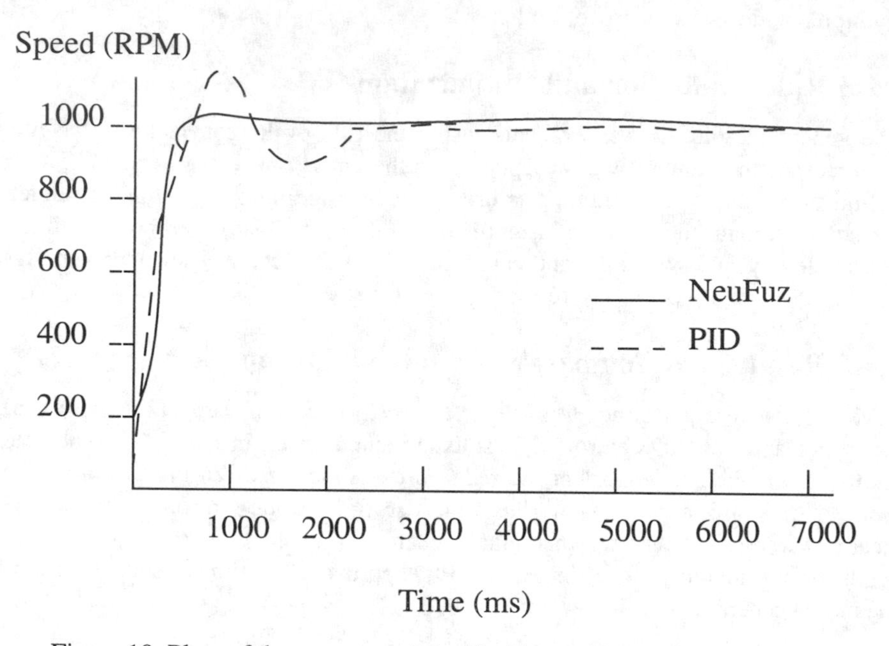

Figure 10. Plots of the response of the NeuFuz and PID controllers for a desired speed of 1000 rpm.

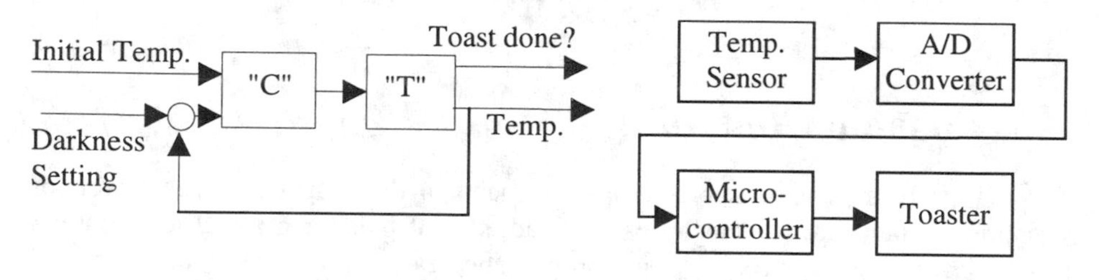

Figure 11. Control structure for a toaster.

Figure 12. Block diagram for the hardware.

There are several variables that determine the darkness of the toast. For example, moisture content of the bread, darkness setting, initial temperature of the toaster, bread size, type of bread, etc. For simplicity, the initial temperature of the toaster and the darkness setting were varied while the other variables were kept constant.

The NeuFuz based solution uses 52 rules and 2, 4, and 7 membership functions for darkness setting, initial temperature and energy-to-be-applied inputs, respectively. The solution provides the desired darkness level. The assembly language code for the fuzzy logic module requires approximately 1K bytes of memory. Some additional memory is required for the code to read temperature, calculate energy, and control toaster output.

6.3 Speech Recognition using RNFS

Speech recognition, perception, and understanding have been active research fields since the 1950s. Over the years many technological innovations have boosted the level of performance. However, for more and more difficult tasks the performance currently achieved by state-of-the-art systems is not yet at the level of a mature technology [2], mainly because of the complexity of tasks, especially considering continuous speech and any simple speech under a noisy environment. The dominant technology, today, is hidden Markow model (HMM) and some combinations of HMM and Neural net [16, 2] which have shown better performance than HMM itself. This is believed to be the right trend for medium and large vocabulary systems. For small vocabulary systems, HMM implementation is not usually cost effective and hence researchers started exploring fuzzy logic and neural fuzzy approaches [5, 13]. In [13], Recurrent Fuzzy Logic (RFL) based on Recurrent Neural Fuzzy System (RNFS) is used to do the word recognition. Below is a brief description of this RNFS based small vocabulary speech recognition system.

6.3.1 Small Vocabulary Word Recognition

Figure 13 shows an isolated word recognition system. The recurrent neural fuzzy technique is used for the recognition step. Speech signals are coded using LPC cepstrum and vector quantization (VQ) is used. A VQ codebook size of 256, 9 pole LPC, 16 KHz sampling rate (with 16 bit speech amplitudes), and 300 samples per frame are used. The TIMIT [3] data base (a speech database developed by Texas Instruments and Massachusetts Institute of Technology, and sponsored by United States Defense Advanced Research Projects Agency) is used to develop the code book as well as to train the RNFS. The recurrent neural net is trained with about 200 speakers from different U.S. regions using 11 words from SA (dialect) sentences. Testing is also done using the TIMIT database (using speakers from both test and train directories). The recognition accuracy is 90%, comparable to HMM based recognition.

6.3.2 Training and Testing

Sampled speech signal from the TIMIT database is applied to the LPC cepstrum block which produces cepstrum coefficients. A VQ code book is generated by grouping (using binary split codebook generation algorithm, [17]) cepstrum coefficients from many speakers into basic speech units (in this case 256, i.e., the size of the code book). Each basic speech unit is represented by a cepstrum vector of dimension 9. The next step of the training phase is to train the RNFS. Sampled speech data from TIMIT (but from the files that have desired word data to be recognized) is used. This time, the LPC cepstrum coefficients are directly applied to the VQ index generation block.

The VQ index generator compares the input coded speech (i.e., cepstrum coefficients) with the contents (i.e., basic speech units) in the VQ codebook generated by the VQ generation block. The VQ index generation block determines the basic speech unit of VQ codebook that has the closest match with the input coded speech and outputs the corresponding address which is known as the VQ codebook index. VQ indices for many speakers for the desired words are then used to train the RNFS.

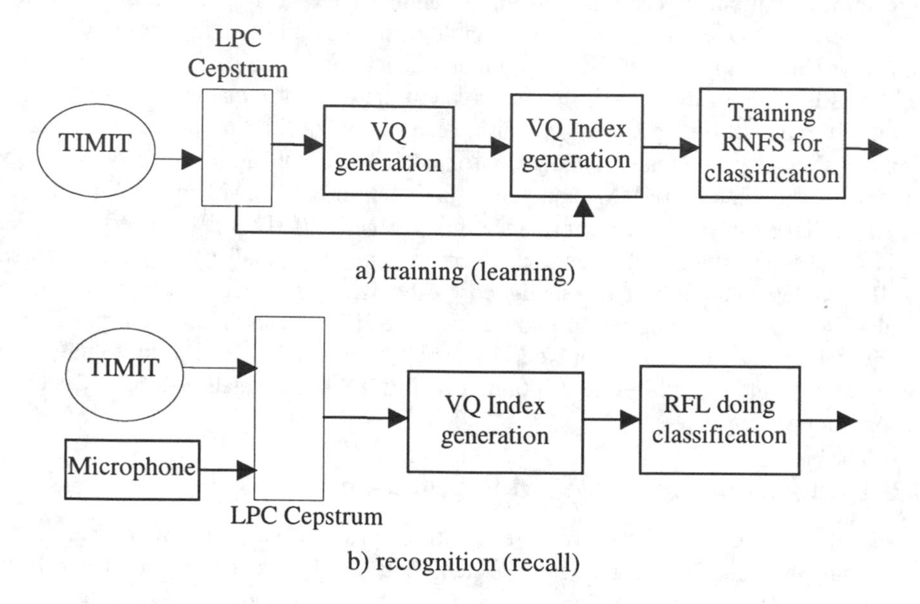

Figure 13. Isolated word recognition using Recurrent Fuzzy Logic.

In the recognition phase, sampled speech data from the TIMIT data base or from the microphone is applied to the LPC cepstrum generator whose output is used by the VQ index generation block. The generated VQ indices for a word are applied to the RFL classifier. The RFL classifier then recognizes the input word. Note that there is one RFL for each word and properly labeled. The RFL that provides maximum output (or lowest error) indicates the recognized word. Note that after the training is completed, RNFS automatically generates recurrent fuzzy logic rules and membership functions which are used by the RFL in the recognition phase.

7. Conclusion

The need to solve highly nonlinear, time variant problems has been growing rapidly as many of today's applications have nonlinear and uncertain behavior which changes with time. Conventional mathematical model based techniques can effectively address linear, time invariant problems. However, their capabilities to address more complex nonlinear time variant problems are limited. Currently, no model based method exists that can effectively address complex, nonlinear, and time variant problems in a general way. These problems, coupled with others (such as problems in decision making, prediction, etc.) have inspired a growing interest in intelligent techniques including Fuzzy Logic, Neural Networks, Genetic Algorithms, Expert Systems, and Probabilistic Reasoning. Intelligent Systems, in general, use various combinations of these techniques to address real world complex problems. In this chapter, we have addressed the intelligent systems based on various combinations of neural nets and fuzzy logic, called Neural Fuzzy Systems (NFS). The rationale to combine fuzzy logic with neural nets is emphasized to alleviate the limitations of each of these technologies while adding their advantages. We have presented elegant algorithms to combine neural nets with fuzzy logic, resulting in both feed forward and recurrent neural fuzzy systems. NFS provide several key advantages/features which were highlighted and discussed. Because of the added features, NFS can address problems in wide application areas including control, battery charging, handwriting recognition, speech recognition, language translation, decision making, and forecasting.

NFS techniques are applied to solve many real world problems and we reported a few, namely, motor control, toaster control, and speech recognition. For the motor control application, NFS has reduced overshoots and settling time at start up, while maintaining approximately the same rise time. Both PID and NFS controllers produced zero steady state error. However, when load is changed, NFS produced considerably less error than the PID approach. For the toaster control problem, NFS essentially solved the key problem of maintaining the desired darkness level for variations in the moisture content, types of bread, size of bread, and initial temperature. For speech recognition, which is a more complex problem, the performance of recurrent NFS was found to be comparable to the conventional approaches. Application of NFS to speech recognition problems is relatively new and we believe that in the future NFS will significantly help improve the performance of speech recognition.

References

[1] V. Aisa et al., "Fuzzy Logic and Neural Networks for a New Generation of Household Appliances," An internal paper at Merloni, Italy, 1995.

[2] H. Bourlard and N. Morgan, *Connectionists Speech Recognition*, Kluwer Academic Press, 1994.

[3] "DARPA TIMIT – Acoustic-Phonetic Continuous Speech Corpus," National Institute of Standards and Technology document NISTIR 4930, 1993.

[4] M. Gupta, "Fuzzy Neuron with Somatic and Synaptic Operations," *Proceedings of SPIE - The Society of Optical Engg.*, Vol 1750, pp. 489-496, 1992.

[5] C. Hale et al., "Voice Command Recognition using Fuzzy Logic," *Proceedings of the WESCON*, pp. 608-613, 1995.

[6] Y. Hayashi et al., "Fuzzy Neural Network with Fuzzy Signals and Weights," *Proceedings of the IJCNN92*, Vol. 2, 1992.

[7] J. Hertz, "Introduction to the Theory of Neural Computation," Addison-Wesley, 1991.

[8] E. Khan et al., "NeuFuz: Neural Network Based Fuzzy Logic Design Algorithms," *Proceedings of the FUZZ-IEEE93*, Vol. 1, p. 647, March 1993.

[9] E. Khan, "Neural Network Based Algorithms For Rule Evaluation and Defuzzification in Fuzzy Logic Design," *Proceedings of the World Congress on Neural Nets (WCNN)*, July 1993.

[10] E. Khan, "NeuFuz: An Intelligent Combination of Neural Nets with Fuzzy Logic," *IJCNN*, pp. 2945-2950, 1993.

[11] E. Khan et al., "Recurrent Fuzzy Logic using Neural Networks," Lecture Notes in Artificial Intelligence 1011 - Advances in Fuzzy Logic, Neural Nets and Genetic Algorithms, Springer, pp. 48-55, 1994.

[12] E.Khan, "Recurrent Fuzzy Logic and Recurrent Neural Fuzzy Systems," *Proceedings of the World Congress on Neural Nets (WCNN)*, July 1995.

[13] E. Khan, "Recurrent Fuzzy Logic in Speech Recognition," *Proceedings of the WESCON*, pp. 602-607, 1995.

[14] B. Kosko, *Neural Nets & Fuzzy Systems*, Prentice Hall, 1992

[15] C. Lin, "Neural Network Based Fuzzy Logic Control and Decision System," *IEEE Tr. on Computers*, Vol. 40, no. 12, 1991.

[16] D. Morgan, *Neural Network and Speech Processing*, Kluwer Academic Press, 1991.

[17] L. Rabiner and B. Juang, *Fundamental of Speech Recognition*, Prentice Hall, 1993.

[18] S. Rahman, "Neural-Fuzzy in Consumer Appliance Application," *Proceedings of the Fuzzy Logic 93 organized by Computer Design*, San Francisco, July 1993.

[19] S. Rahman et al., "NeuFuz Approach to Motor Control," Application note from National Semiconductor Corporation, Santa Clara, California, U.S.A.

[20] Z. Ullah et al., "Fast Intelligent Battery Charging: Neural Fuzzy (NeuFuz) Approach," *Proceedings of the National Semiconductor Technology Conference*, 1995.

[21] L. Zadeh et al., *Fuzzy Theory and Applications*, Collection of Zadeh's good papers, 1986.

Chapter 6:

Vehicle Routing through Simulation of Natural Processes

VEHICLE ROUTING THROUGH SIMULATION OF NATURAL PROCESSES

Jean-Yves Potvin
Centre de Recherche sur les Transports
and
Département d'Informatique et de Recherche Opérationnelle
Université de Montréal
Canada

Sam R. Thangiah
Department of Ccomputer Science
Slippery Rock University
U.S.A.

Vehicle routing problems hold a central place in distribution management. Their economic importance has motivated both academic researchers and private companies to find ways to efficiently perform the transportation of goods and services. A vehicle routing problem requires the allocation of each customer to a particular vehicle and the ordering of these customers on each route to minimize the transportation costs, subject to various constraints such as vehicle capacity and route time duration. In this chapter, we review recent attempts at solving those problems, using methods motivated by natural processes, like genetic algorithms and neural networks.

1. Introduction

Neural networks and genetic algorithms both come from the realization that simulating natural processes, even at a high level of abstraction, can bring valuable insights into complex real-world problems. Due to the success of these paradigms in many diverse application areas [5, 21], researchers have recently tackled the field of combinatorial optimization, where an optimal solution is sought among a finite or a countable infinite number of alternatives. An abundant literature may be found, in particular, on applications of these methods for solving the well-known Traveling Salesman Problem (TSP) [24, 25].

In this chapter, we report about recent attempts at solving a class of extensions to the TSP, known as vehicle routing problems. These problems are pervasive in the real

0-8493-9804-5/99/$0.00+$.50

world as they hold a central place in transportation systems, like school bus routing, and in logistics systems where goods or services are distributed from central facilities (plants, warehouses) to customers. These problems require the construction of vehicle routes, that is, the allocation of each customer to a particular vehicle and the ordering of these customers on each route to minimize the total routing costs, subject to a variety of constraints such as vehicle capacity and route time duration.

To introduce this subject, the chapter is organized along the following lines. A classification of prominent vehicle routing problems is first presented in Section 2. Then, applications of neural networks and genetic algorithms for some of these problems are reported in Sections 3 and 4, respectively. Finally, conclusions are drawn in Section 5.

2. Vehicle Routing Problems

An abundant literature may be found on useful abstractions or models of routing problems found in practice. One of the most popular models is the vehicle routing problem (VRP) which is formally defined on a directed graph G=(V,A), where $V=\{v_0,...,v_n\}$ is the vertex set and $A=\{(v_i,v_j): v_i, v_j \in V, i \neq j\}$ is the arc set. Vertex v_0 is a central depot housing a fleet of identical vehicles, while the remaining vertices represent customers to be serviced. A nonnegative distance or cost matrix $D=\{d_{ij}\}$ is defined on A, where d_{ij} is the distance or cost to travel from vertex v_i to vertex v_j. The VRP then consists of determining m vehicle routes of total minimum cost, each starting and ending at the depot, such that every customer is visited exactly once. Figure 1 illustrates a solution to this problem using $m = 3$ vehicles. In this figure, the black square stands for the central depot and the white circles are customers to be serviced.

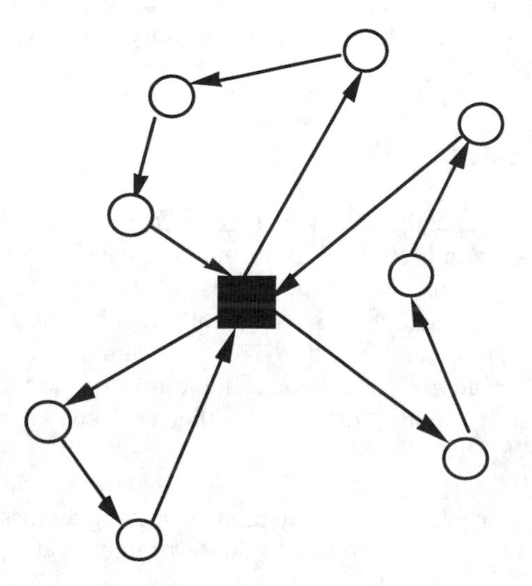

Figure 1. Three vehicle routes servicing eight customers

Different versions of this model are reported in the literature depending on the side constraints to be satisfied.

- Capacitated VRP: a nonnegative demand q_i is associated with each vertex and the total demand on a route may not exceed vehicle capacity Q.
- Distance (time) constrained VRP: the matrix D stands for distances (travel times) between vertices, and the total distance (duration) of any route may not exceed a prespecified bound.
- VRP with precedence constraints: a precedence constraint between two vertices v_i and v_j exists if it is required to visit v_i before v_j. In the VRP with backhauls, for example, mixed routes servicing both pick-up and delivery locations must be constructed. Vehicles are loaded at a central depot and must first service the delivery locations. The vehicles are then allowed to pick up (backhaul) goods en route back to the depot.
- VRP with time windows: each vertex v_i must be visited within a time interval $[a_i, b_i]$, where a_i and b_i are the earliest and latest service times, respectively. Typically, a vehicle is allowed to wait if it arrives at a customer location before its earliest service time.

Though the VRP is simple to state, yet, it is extremely complex to solve and belongs to the class of NP-complete problems [8, 20]. That is, the computation time to obtain the optimum increases exponentially with the number of vertices. Exact algorithmic procedures for this problem do exist, but they are computationally expensive and can be applied only to relatively small instances. Another disadvantage of exact approaches is their unpredictability as they can produce an optimal solution to one instance of the problem but fail to terminate on a smaller instance of the same problem. To obtain solutions that are close to the optimum in a reasonable amount of time, heuristic search strategies are widely used. In the following, we will focus on heuristic methods inspired by nature to address some of the vehicle routing problems mentioned above, as well as other variants found in practice.

3. Neural Networks

Artificial neural networks (ANNs) are inspired by nature in the sense that they are composed of elements or units that perform in a manner that is analogous to neurons in the human brain. These units are richly interconnected through weighted connections: a signal is sent from one unit to another along a connection and is modulated by the associated weight (which stands for the strength or efficiency of the connection). Although superficially related to their biological counterpart, ANNs possess characteristics associated with human cognition. In particular, they can learn from experience and induce general concepts from specific examples; when presented with a set of inputs, they self-adjust to produce the desired response.

Neural networks were originally designed for tasks well adapted to human intelligence and where traditional computation has proven inadequate, such as speech understanding, artificial vision, and handwritten character recognition. Starting with the pioneering work of Hopfield and Tank [17], they have recently been applied to combinatorial optimization problems as well. For instance, applications of the Hopfield model [16], elastic net [7], and self-organizing map [19] for the TSP are well documented, although it is fair to say that the results are not yet competitive with those reported in the operations research community [18].

Although neural network models can handle spatial relationships among vertices to find good routes, they cannot easily handle side constraints, such as capacity constraints and time windows, that often break the geographic or geometric interpretation of the problem. Accordingly, the literature on the application of neural networks to vehicle routing problems is rather scant. Different variants of the self-organizing map were recently applied to the capacitated VRP [10, 11, 12, 22, 32]. The work of Ghaziri will be used in the following to illustrate how this model can be applied to an academic problem. Then, we discuss the application of a feedforward neural network model with backpropagation learning for a real-world dispatching application found in a courier service [33].

3.1 Self-Organizing Maps

Self-organizing maps are instances of the so-called competitive neural network models [19]. They are composed of a layer of input units fully connected to a layer of output units, the latter units being organized in a particular topology, such as a ring structure. These models "self-organize" (without any external supervision) through an iterative adjustment of their connection weights to find some regularity or structure in the input data. They are typically used to categorize or cluster data. In Figure 2, T_{11} and T_{21} denote the weights on the connections from the two input units I_1 and I_2 to output unit 1, and $T_1 = (T_{11}, T_{21})$ is the weight vector associated with output unit 1.

Assuming an ordered set of *n* input vectors of dimension *p* and a self-organizing map with *p* input units and *q* output units on a ring, the algorithm for adjusting the connection weights can be summarized as follows.

Step 1. (Initialization). Set the connection weights to some initial values.

Step 2. (Competition). Consider the next input vector *I*. If all input vectors are done, start again from the first input vector. Compute the output value o_j of each output unit as the weighted sum of its inputs, namely,

$$o_j = \Sigma_{i=1,\dots,p} T_{ij} I_i \qquad j=1,\dots,q,$$

where T_{ij} is the connection weight between input unit *i* and output unit *j*. The winning output unit j^* is the unit with maximum output value.

Step 3. (Weight adjustment). Modify the connection weights of each output unit, as follows:

$$T_j^{new} = T_j^{old} + f(j,j^*)\ (I - T_j^{old}),$$

where $T_j = (T_{ij})_{i=1,...,p}$ is the weight vector of output unit j, and f is a function of j and j^*.

Step 4. Repeat Step 2 and Step 3 until the weight vectors stabilize. At the end, each input vector is assigned to the output unit that maximizes the weighted sum in Step 2.

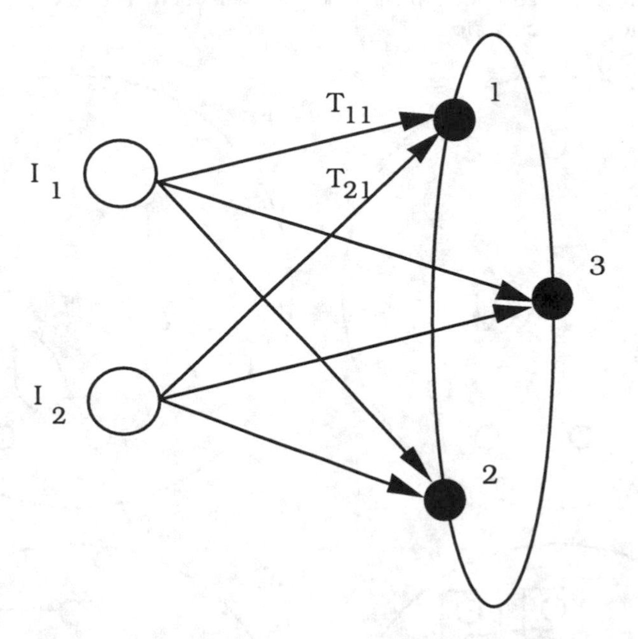

Figure 2. A self-organizing map with two input units and a ring of three output units

This algorithm deserves some additional comments. In Step 3, f is typically a decreasing function of the lateral distance between units j and j^* on the ring (i.e., if there are k units on the ring between the two units, the lateral distance is k+1) and its range is the interval [0,1]. Thus, the weight vector of the winning unit j^* and the weight vectors of units that are close to j^* on the ring move towards the input vector I, but with decreasing intensity as the lateral distance to the winning unit increases. Typically, function f is modified as the learning algorithm unfolds to gradually reduce the magnitude of the weight adjustment. At the start, all units that are close to the winning unit on the ring "follow" that unit in order to move into a neighboring area. At the end, only the weight vector of the winning unit significantly moves toward the current input vector, to fix the assignment.

Self-organizing maps produce topological mappings from high-dimensional input spaces to low-dimensional output spaces. In the case of a ring structure, the p-dimensional input vectors are associated with the position of the winning unit on the ring. The mapping is such that two input vectors that are close in the input space are assigned to units that are close on the ring.

3.1.1 Vehicle Routing Applications

In a VRP context, the input vectors are the two-dimensional coordinate vectors of the vertices and a different ring is associated with each route. The weight learning algorithm is also slightly modified. For academic vehicle routing problems based on the Euclidean metric, the weight vector of the winning output unit is the closest to the coordinate vector of the current vertex in Euclidean distance (rather than the one with the largest correlation with the input vector; see maximization of the inner product in Step 2). Also, the algorithm typically stops when there is a weight vector (unit) sufficiently close to each vertex.

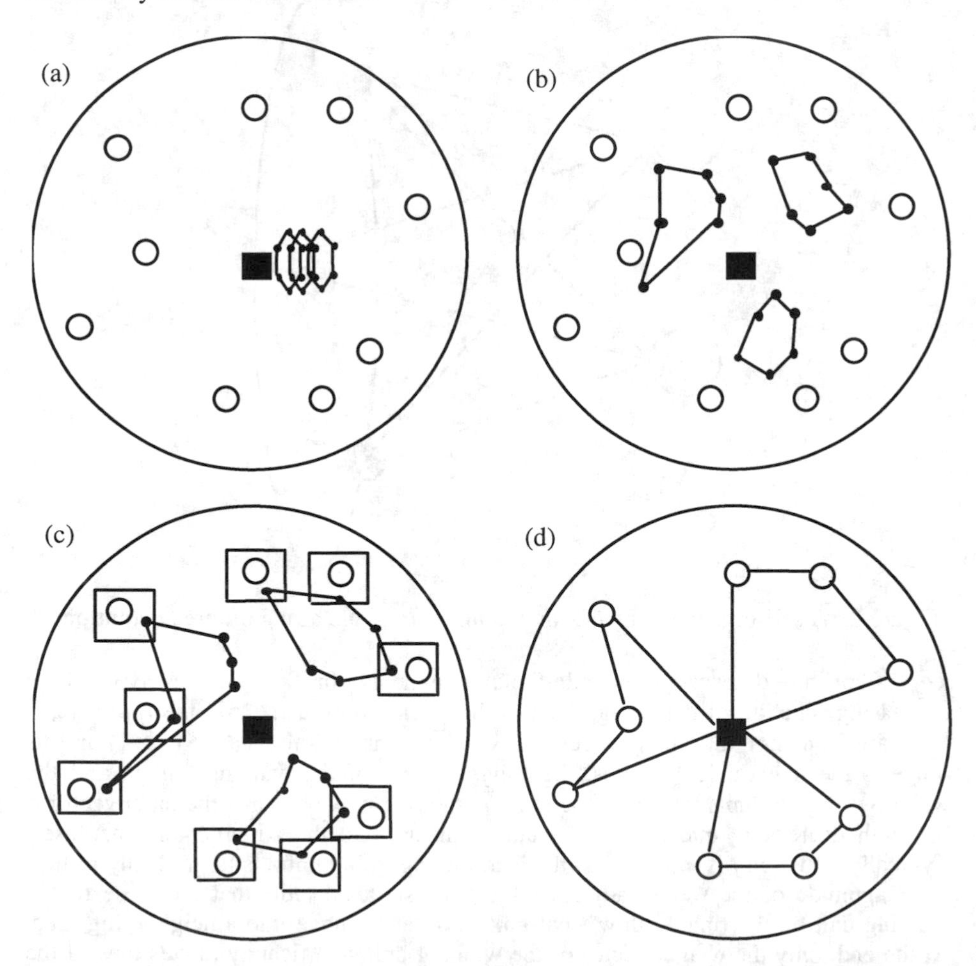

Figure 3. Evolution of three rings in (a), (b), (c), and the final solution (d)

Given that multiple rings are used, the winning unit is determined among all units over all rings. Then, units located on the ring of the winning unit are moved toward the current vertex. At the end, each vertex is assigned to the closest unit. Through this assignment, the ordering of the units on each ring defines an ordering of the vertices on each route. Furthermore, since neighboring units on a given ring tend to migrate toward neighboring vertices in the Euclidean plane, short routes are expected to emerge. Figure 3 depicts the evolution of the rings in a two-dimensional Euclidean plane during the application of the weight adjustment algorithm, starting from the initial configuration illustrated in Figure 3(a). The vertices are large white nodes and the units are small black nodes; the location of each unit in the Euclidean plane is determined by its current weight vector.

At the start, no solution of the problem can be inferred. However, as the weight vectors migrate towards the vertices, the solution progressively emerges. The final solution is determined during the final assignment of vertices to units on the rings (c.f., Step 4).

3.1.2 The Hierarchical Deformable Net

The Hierarchical Deformable Net (HDM) [10, 11] illustrates the application of a self-organizing map to the VRP. The model involves a competition at two different levels: a deterministic competition among units on each ring, and a stochastic competition among the rings. The number of rings is fixed, and this model is applied with an increasing number of rings (routes) until a feasible solution is found. The HDM model can be summarized as follows.

Step 1. *Initialization*. Randomly order the vertices. Then, create one ring per route. Each ring contains 2×ceiling(n/m) units, where n is the number of vertices and m is the number of routes. Initialize the parameters h and G to 1 and 4×ceiling(n/m), respectively (see below).

Step 2. Repeat Step 3 to Step 8 until there is a weight vector sufficiently close to each vertex.

Step 3. Select the next vertex v and set the current input vector I to its coordinate vector (if all vertices are done, restart at the first vertex).

Step 4. *Competition within each ring*. Determine the winning unit u_{j^*k} on each ring r_k. This unit is the one whose weight vector is closest to the current input vector I.

Step 5. *Competition among the rings*. Assign the following winning probability to each ring r_k:

$$p(r_k) = \frac{f(r_k,h)}{\Sigma_{k=1,\ldots,m} f(r_k,h)}$$

$$f(r_k,h) = \frac{e^{-d(T_{j^*k},I)/h}}{1 + e^{-\Delta Q/h}}$$

where h is a parameter of the algorithm, m is the number of rings, $d(T_{j^*k}, I)$ is the Euclidean distance between the coordinate vector I and the weight vector T_{j^*k} of winning unit u_{j^*k} on ring r_k, ΔQ is the difference between the capacity of the vehicle and the total demand on ring r_k (including vertex v).

Step 6. Select the winning ring r_{k^*}, using a selection probability for each ring r_k proportional to $p(r_k)$. Vertex v is assigned to unit $u_{j^*k^*}$.

Step 7. Move the weight vector T_{jk^*} of each unit j on the winning ring r_{k^*} toward the coordinate vector I of vertex v, as follows:

$$T_{jk^*}^{new} = T_{jk^*}^{old} + f(j,j^*,k^*)\,(I - T_{jk^*}^{old}),$$

$$f(j,j^*,k^*) = (1/\sqrt{2})\, e^{-dL(j,j^*,k^*)^2/G^2},$$

where $dL(j,j^*,k^*)$ is the lateral distance between unit j and the winning unit j^* on ring r_{k^*}.

Step 8. Slightly decrease the value of parameters h and G, by setting $h^{new} = 0.99\, h^{old}$ and $G^{new} = 0.9\, G^{old}$.

In Step 5, parameter h is used to modify the probability distribution over the rings. When the value of h is high, the selection probability of each ring is about the same. When the value of h is small, the probability distribution favors the ring closest to the current vertex among all rings that satisfy the capacity constraint (with the addition of the current vertex). In Step 7, parameter G controls the magnitude of the moves. When its value is reduced, the magnitude of the moves is also reduced. Hence, the values of both parameters h and G are gradually lowered to favor convergence toward good feasible solutions.

In a later study [12], Ghaziri extended the HDN to address the time-constrained VRP. In this case, the selection probabilities are adjusted through the introduction of an additional component that takes into account the time limit constraint. As opposed to the capacity constraint, however, the time limit constraint relates to the total duration of a route which, in turn, depends on the sequence of vertices on the route. At each iteration, a TSP procedure is thus applied to the currently assigned vertices on each ring (including the current vertex) to identify a plausible ordering. This ordering is then used to compute the selection probabilities.

Ghaziri has applied the HDN to a few VRPs taken from the test set in [2]. Although the solutions produced by HDN are of good quality, they are not competitive with the best solutions produced with alternative problem-solving methods, in particular, tabu search [9, 35].

3.2 Feedforward Models

The work presented here [33] is very different from the developments reported in Section 3.1 with respect both to the selected application, namely, a courier service, and the type of neural network model used. First, the problem is dynamic in the sense that the customers (or vertices) to be serviced are not known in advance, but occur in a continuous fashion. Hence, a solution cannot be constructed beforehand; new customers are incorporated in the planned routes as these routes are executed. Second, the neural network model is based on supervised learning, that is, desired outputs or responses are provided to the network to guide its weight adjustment procedure.

In the following, we first characterize the problem to be investigated. After a brief review of feedforward models with backpropagation, we explain how this paradigm is exploited to address the courier service application.

3.2.1 Dynamic vehicle routing and dispatching

Dynamic vehicle routing and dispatching refer to a wide range of problems where information on the problem is revealed to the decision maker concurrently with the determination of the solution [29]. These problems have recently emerged as an active area of research due to recent advances in communication and information technologies that now allow real-time information to be quickly obtained and processed. The courier service application considered here is an instance of this class of problems and can be described as follows.

A courier service receives customer calls for the pick-up and delivery of express mail in a local area. Each customer specifies a pick-up and a delivery location, as well as a due date for the delivery. Since most customers demand fast service, the routing and scheduling is done in real time. In such a demand responsive context, the dispatching situation is looked at when a new customer calls into the dispatch office. At that time, some of the earlier customers have been serviced and are no longer considered. Other customer requests have been assigned to vehicles, and are either waiting to be picked up or "en route" to their delivery location. The problem is to determine the assignment of each new request to a particular vehicle, as well as the new planned route for that vehicle (i.e., the ordering of previously assigned but yet unserviced requests, plus the new request). In solving this problem, the dispatcher must find a compromise between two conflicting objectives: minimizing the operations cost, like the total distance traveled, and maximizing customer satisfaction, in particular, satisfaction of the desired due dates.

Currently, this problem is solved by human dispatchers who represent a key element in courier service companies. Specific skills are required to adequately perform the dispatching task; thus, it takes a long time to train dispatchers. Furthermore, their professional career lasts only a few years due to the high level of stress associated with reacting properly and quickly to a dynamically changing environment. Given the difficulty to formalize the knowledge of dispatchers using classical decision rules, and the absence of a clearly-defined objective function to be optimized, a neural network dispatching system was developed to learn the dispatcher's decision procedure through previous decision examples. The basic neural network model is presented in the following; then, its application to the courier service problem is discussed.

3.2.2 Feedforward Neural Network Model with Backpropagation

Feedforward neural networks with backpropagation [31] constitute one of the most popular neural network models. Their supervised learning procedure is based on a simple idea: if the network gives a wrong answer, the weights on the connections are adjusted to reduce this error and increase the likelihood of a correct response in the future.

The model is typically composed of three layers of simple units, namely, the input layer, the hidden layer, and the output layer. Each layer is fully connected to the previous and next layer through weighted connections that propagate the signal in a forward direction from the input to the output (see Figure 4). No processing takes place at the input layer. That is, the input vector provided to the input layer is directly propagated to the hidden layer through the weighted connections. Assuming p input units and q hidden units, each hidden unit j computes its activation level a_j, according to the following formula:

$$a_j = 1/(1 + e^{-S_j}) \quad \text{where} \quad S_j = \Sigma_{i=1,\ldots,p} T_{ij} I_i + \theta_j, \; j=1,\ldots,q.$$

In this formula, I_i is component i of input vector I (or, equivalently, the activation level of input unit i), T_{ij} is the connection weight between input unit i and hidden unit j, and θ_j is the input bias of unit j. Note that the activation level of each hidden unit is in the interval [0,1] due to the sigmoidal transformation. These activation levels are then propagated to the output units through the weighted connections between the hidden and output layers. The same processing takes place at each output unit and the final activation levels of these units (or output vector) represent the response of the network to the input vector I.

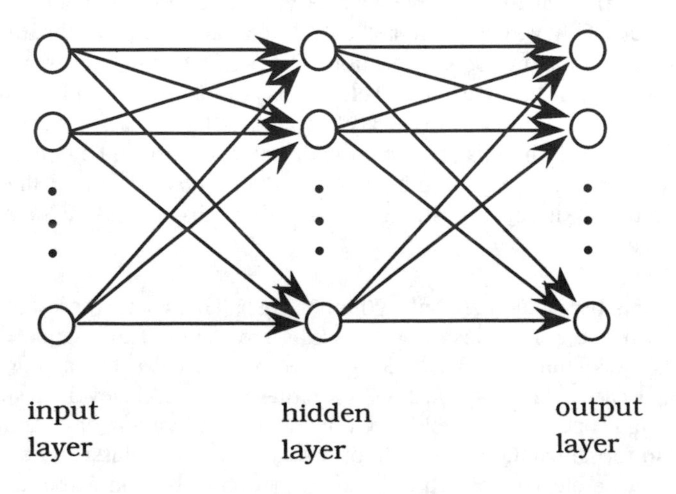

Figure 4. A three-layer feedforward model

The power of this model comes from its ability to adjust the connection weights to perform a particular task. To do this, the backpropagation learning algorithm uses a training set of examples that characterizes the task: in this set, each input vector is associated with a desired response or output vector. These examples are processed one by one by the network and the weights are adjusted accordingly. The mean-squared error between the outputs calculated by the network and the desired outputs (over all examples in the training set) is typically used to monitor the network's performance. The training stops when this error measure stabilizes. We will not go into the details of the backpropagation algorithm; the interested reader is referred to [31] for further details.

3.2.3 An Application for a Courier Service

In the case of the courier service application, the neural network model is used to dispatch new requests to appropriate vehicles. To this end, the network processes each candidate vehicle and produces a numerical evaluation in output. Clearly, vehicles with high evaluations are the best candidates for servicing a new request.

The input vector associated with a given vehicle is composed of attribute values that characterize this vehicle with respect to the new service request: detour in time to service the request, pick-up time, delivery time, additional lateness to customers already assigned to the vehicle route (but not yet serviced) due to the inclusion of the new request, etc. It is worth noting that for computing such an input vector, an insertion place for the new request is first chosen by minimizing an insertion cost over all possible insertion places in the vehicle route. Hence, the neural network does not handle the routing component of the problem. Its capabilities are limited to the dispatching task, namely, what vehicle should service the new request given the default insertion locations.

The output vector contains a single output unit; its activation level corresponds to the evaluation of the vehicle whose description has been provided in input. Using a set of examples composed of dispatching situations where the vehicle selected by the dispatcher is known, the neural network model is trained to assign high evaluations to these vehicles. Ideally, for each dispatching situation, the vehicle selected by the dispatcher should be ranked first by the neural network (i.e., it should get the highest evaluation among all available vehicles).

Using 140 requests collected from a courier service company operating a fleet of 12 vehicles in the Montreal area, the authors demonstrated the potential of the neural network model. After training the network with 90 requests, the model was tested on the 50 remaining requests. In 47 cases out of 50, the vehicle selected by the dispatcher was ranked first, second, third, or fourth by the neural network (out of twelve possible ranks). When the vehicle chosen by the expert did not get the highest evaluation, it was often quite close. Furthermore, the vehicle with the highest evaluation was typically a good alternative choice.

In this work, the network was trained in batch mode from previously collected data. However, it would be possible to "link" the neural network model with the dispatcher during one or more operations per day in order to adjust the connection weights in real time after each new decision. At the end of the training period, the network would then be able to mimic the dispatcher's behavior. Such a network could be used, for example, to facilitate the dispatcher's task by displaying the neural network evaluation of the best vehicles or to assist less experienced dispatchers when confronted with difficult situations. In each case, it is assumed that the final decision is left to the human expert.

4. Genetic Algorithms

The Genetic Algorithm (GA) is an adaptive heuristic search method based on population genetics. The basic concepts were developed by Holland [15], while the practicality of using the GA to solve complex problems was demonstrated in [6, 14]. Under this paradigm, a population of chromosomes evolves over a number of generations through the application of genetic operators, like crossover and mutation, that mimic those found in nature. The evolution process allows the best chromosomes to survive and mate from one generation to the next.

More formally, the GA is an iterative procedure that maintains a population of *P* candidate members over many simulated generations. The population members are artificial chromosomes made of fixed length strings with a binary value (allele) at each position (locus). The *genotype* is the string entity and the *phenotype* is the decoded string entity (search point, solution). In parameter optimization problems, for example, the string could encode numerical parameter values in base-2 notation. The real parameter values would then correspond to the phenotype. A fitness value is also associated with each chromosome; this value is typically related to the quality of the associated search point or solution.

Denoting the population of chromosomes at a given generation as M(Generation), the flow of the GA can be summarized as follows:

Genetic Algorithm (GA):

```
        GA1:    Generation = 0;
        GA2:    Initialize M(Generation);
                Evaluate M(Generation);
        GA3:    While (GA has not converged or terminated)
                        Generation = Generation + 1;
                        Select M(Generation) from M(Generation - 1);
                        Crossover M(Generation);
                        Mutate M(Generation);
                                Evaluate M(Generation);
                End (while)
        GA4:    Terminate the GA.
```

First, an initial population of chromosomes is created at Generation 0, either randomly or through heuristic means, and the fitness value of each chromosome is evaluated. A new generation is then created, by selecting the best fit chromosomes from the previous generation. A typical randomized selection procedure for the GA is the proportional or roulette wheel selection, where the selection probability of each chromosome is proportional to its fitness. To create new chromosomes (new solutions in the search space), crossover and mutation operators are applied to a certain proportion of chromosomes in the new population. Crossover creates two new offspring by merging different parts of two parent chromosomes. The one-point crossover is illustrated in Figure 5; a random cut point is selected on the parent chromosomes and their end parts are exchanged. Other variants exist, like two-point crossover where the substring located between two randomly selected cut points is exchanged, or uniform crossover where the parent that contributes its bit value to the first offspring is randomly selected at each position (while the other parent contributes its bit value to the second offspring). Mutation is a secondary operator that changes a bit to its complementary value with a small probability at each position.

parent 1	:	**1**	**1**	**1** |	**1**	**1**	**1**
parent 2	:	0	0	0 |	0	0	0
offspring 1	:	**1**	**1**	**1**	0	0	0
offspring 2	:	0	0	0	**1**	**1**	**1**

Figure 5. The one-point crossover.

Together, selection and crossover effectively search the problem space by combining bit patterns found on good chromosomes to produce offspring with potentially higher fitness values (building block hypothesis). Mutation acts as a background operator to prevent the premature loss of bit values at particular loci. The genotypes that survive, over time, will be those that have proven to be the most fit. The termination criteria of a GA are convergence within a given tolerance or completion of a predefined number of generations. Note that the sequence of populations generated by the GA can be viewed as parallel search trajectories in the space of admissible strings.

Although theoretical results that characterize the behavior of the GA have been obtained for bit-string chromosomes [15], not all problems lend themselves easily to this representation. This is the case, in particular, for sequencing problems where an integer representation is often more appropriate. In the case of the VRP, an integer could stand for a particular customer, while a string of integers would represent a vehicle route. For example, the string 3 1 2 would mean that, starting from the central depot, customer 3 is visited first, customer 1 is visited second, and customer 2 is visited last.

However, a straightforward application of the GA on this integer representation would not be effective, because traditional crossover and mutation operators are not designed to preserve orderings. For example, assuming $m = 1$ vehicle and $n = 5$ customers numbered from 1 to 5, the one-point crossover always produces infeasible offspring from the integer strings 1 2 3 4 5 and 5 4 3 2 1; at each cut point, customers are duplicated while others are missing from the resulting offspring routes (see Figure 6).

parent 1	:	**1**	**2**	**3** |	**4**	**5**
parent 2	:	5	4	3 |	2	1
offspring 1	:	**1**	**2**	**3**	2	1
offspring 2	:	5	4	3	**4**	**5**

Figure 6. One-point crossover does not produce valid orderings

A number of studies have thus been conducted to address the VRP with GAs based on innovative representation schemes and special crossover operators that preserve sequential information. The approaches presented in the next section have been successful in solving VRPs with complex side constraints [43]. They are illustrative of different coding schemes, either binary- or integer-based, as well as different ways of handling side constraints through penalties, decoders, and repair operators.

4.1 Genetic clustering

The methods presented in this section follow the cluster-first, route second problem-solving approach, where clusters of customers that naturally fit in the same vehicle route (due, for example, to spatial proximity) are first identified. Then, the customers are sequenced within each cluster. Here, the GA is used in the clustering phase, while the routing or sequencing is done using classical operations research approaches.

4.1.1 Genetic Sectoring (GenSect)

GenSect [36] is modeled after the sweep heuristic of Gillett and Miller [13] for the capacitated VRP. Starting from some seed point and using the central depot as a pivot, the ray from the depot to the seed is swept clockwise or counter-clockwise. Customers are added to a sector as they are swept until the capacity constraint forbids the inclusion of the next customer. This customer then becomes the seed point for the next sector. This is repeated until all customers are assigned to a particular sector (see Figure 7). Routes are then constructed within each sector. One of the drawbacks of this algorithm is its sensitivity to the selection of the first seed point. Although the algorithm tries to compensate for this by exchanging customers between adjacent sectors at the end, it does not look at all customers during the sweep. GenSect tries to compensate for this by using a GA to adaptively look at all customers in a global manner.

We assume a set of n customers with planar coordinates (x_i, y_i) and a polar coordinate angle p_i. The GenSect method assumes that the initial number of vehicles m required to solve the problem is known. In the case of the capacitated VRP, this number can be obtained using, for example, the Clarke and Wright [3] or the Gillett and Miller algorithm. The Gensect method divides the customers into m clusters by identifying a set of seed angles, $s_0, ..., s_m$ and by drawing a ray from the depot in the direction of each seed angle. The initial seed angle s_0 is assumed to be 0°. The first sector will thus lie between seed angles s_0 and s_1, the second sector will lie between seed angles s_1 and s_2, and so on. GenSect assigns customer *i* to sector *k* if $s_{k-1} < p_i \leq s_k$, $k = 1, ..., m$.

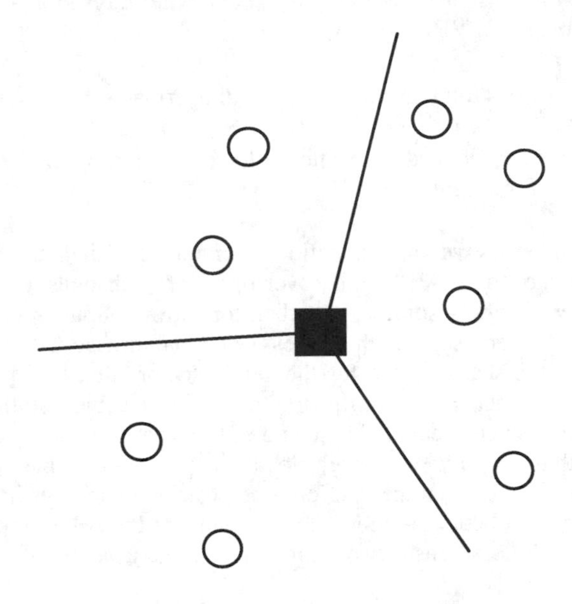

Figure 7. Three sectors identified by the sweep algorithm

Each seed angle s_k is computed from the previous one by adding a fixed angle *F* and an offset e_k. That is,

$$s_k = s_{k-1} + F + e_k, \quad k = 1, ..., m.$$

The fixed angle corresponds to the minimum angular value for a sector; it assures that each sector gets represented. The fixed angle is computed by taking the customer with maximum polar angle and dividing this value by 2*m*. The offset is the extra region from the fixed angle that allows the sector to encompass a larger or a smaller area. If the addition of the fixed angle F and the offset e_k to s_{k-1} exceeds 360°, then s_k is set to 360°, thereby allowing the method to consider less than *m* vehicles to service the customers. Once the sectors are identified, the sequencing of the customers within each sector to produce the final routes is done with a cheapest insertion heuristic [30].

The GA is used to search for the offsets that result in the best routes. A set of *m* offsets is encoded in base-2 notation on each chromosome. A chromosome that uses four bits to encode each offset with *m* = 3 is shown below.

$$\underset{e_1}{\underline{0110}} \quad \underset{e_2}{\underline{0011}} \quad \underset{e_3}{\underline{1110}}$$

The decimal conversion of each offset is then mapped between 0° and 360°. The fitness of the chromosome is the total routing costs obtained with the cheapest insertion heuristic when applied to the sectors encoded on the chromosome. Since there is no guarantee that feasible routes can be produced, the routing costs correspond to the total distance traveled plus a penalty term for violated constraints. In the case of the capacitated VRP, we have

$$\alpha \times \text{total distance} + \beta \times \text{total overload}, \ \alpha + \beta = 1.$$

Hence, chromosomes associated with infeasible solutions are penalized and are less likely to be selected for mating.

Classical one-point crossover and mutation operators are then used to evolve the population of chromosomes. As the crossover operator exchanges a portion of the bit string between two chromosomes, partial information about sector divisions is exchanged. The GenSect system thus uses selection, crossover, and mutation to adaptively explore the search space for the set of sectors that will produce the best possible routes. Note that there is no guarantee that a feasible solution will ever be produced during the genetic search. At the end of the search, the best solution returned by the GA is thus improved through local reoptimization methods that move customers from one route to another or from one position to another in the same route. Through this process, a better solution may be found (if the solution produced by the GA is already feasible) or a feasible solution may emerge from an infeasible one.

Gensect was applied to different types of vehicle routing problems with time windows [36, 39, 40, 41]; it has led to the development of GenClust, which is described in the following.

4.1.2 Genetic Clustering with Geometric Shapes (GenClust)

Quite often, it is desirable to have routes with particular shapes, like petal or circular shapes when each vehicle starts at the depot, visits a set of customers, and returns to the depot. In [4], it is shown that geometric shapes can be used to interactively design routes for the VRP. In fact, geometric shapes are indicated if those shapes are good approximations of the route shapes that result from the minimization of the routing costs.

GenClust [42] follows the general principles of GenSect, but uses geometric shapes to cluster customers. A particular geometric shape can be described by a set of attributes. For example, two attributes may be used for a circle, namely, the origin (x,y) and the radius r. Similar to GenSect, a chromosome encodes the attribute values associated with a given shape in base-2 notation. In GenSect, each chromosome encodes m different circles, one for each cluster, and the GA is then used to search for the set of circles that lead to the best solution. Different heuristic rules are used to associate a customer with a particular cluster, when it is not contained in exactly one circle (i.e., none or many). For example, when a customer is not contained in any circle, it is associated with the closest one.

GenClust was applied to a multi-depot extension of the VRP, where customers can be serviced from more than one depot. It has found the best known solution to 12 of the 23 benchmark problems found in the literature [44].

4.1.3 Real-World Applications

Both GenSect and GenClust are generic methods that can be applied with success to real-world vehicle routing problems with specific side constraints. For example, GenSect was used in the context of school bus routing, where students are transported from predefined locations to a school using a fleet of buses with different capacities. The routes obtained in this application were superior to those in use in two school districts [37].

Another complex routing problem is the multi-commodity transshipment problem (MCTP). In the MCTP, each node in a transportation network is a source or sink for a particular commodity and there is an upper bound on the maximum amount of commodities that can be transported on the arcs. The cost of transporting a commodity is proportional to the distance traveled by the commodity from its source node to its sink node. The objective is to maximize the amount of commodities transported and, for this amount, to minimize the fleet size and transportation costs. A natural application is found in airline fleet operation where each major city is a node and passengers traveling between two cities are commodities. A system called MICAH [38], which couples the GenSect method with a MCTP solver, was used to solve a real world MCTP; a solution was found where 13224 units of 54 different commodities are transported between 54 different nodes using a fleet of 18 vehicles. In this application, the MICAH system increased the total transportation of commodities by 38% in comparison with the solution obtained with an MCTP solver combined with manual routing.

4.2 Decoders

A decoder is a hybrid system where a GA is coupled with a simple insertion heuristic. In a vehicle routing context, an insertion heuristic adds customers one by one in the routes, each time using the feasible location that minimizes some cost measure, like the detour. For example, the detour would be $d_{iu} + d_{uj} - d_{ij}$ in Figure 8.

The insertion order (not the route) is specified by a string of integers, where each integer stands for a particular customer. For example, the string 3 1 2 5 4 means that customer 3 should be the first to be inserted in the solution, followed by customers 1, 2, 5, and 4. At any point, a new route can be created to service a given customer if there is no feasible insertion place in the current routes (due to the side constraints). In the case of the string 3 1 2 5 4, a route will first be created to service customer 3. Then, customers 1 and 2 will be inserted in the newly created route. At this point, we assume that there is no feasible insertion place for customer 5. Thus, a new route is created to service this customer. Finally, customer 4 will be inserted in one of the two routes, depending on its best feasible insertion place. Clearly, different insertion orders are likely to lead to different routes. The role of the GA is thus to find an ordering of customers for the insertion heuristic that will lead to the best solution. This problem-solving approach is attractive, because the GA does not need to explicitly consider side constraints; these are handled by the insertion heuristic. It is thus possible to apply this method to problems with different types of side constraints through simple modifications to the insertion heuristic.

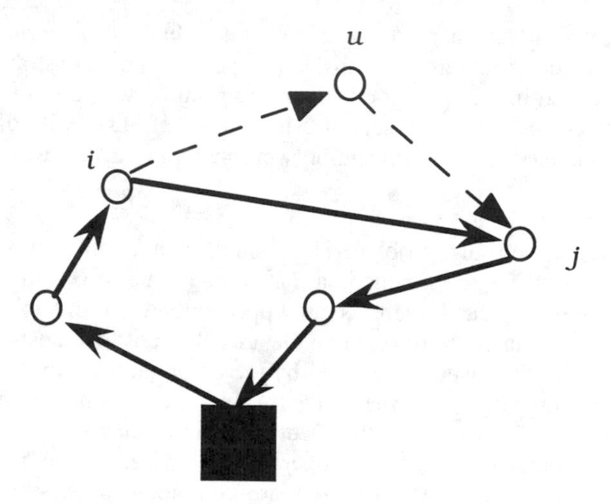

Figure 8. Inserting customer u between customers i and j.

In a decoder system, the GA searches the space of possible insertion orderings using specialized crossover and mutation operators that produce valid offspring orderings from two parent orderings. Many operators of this type have been designed, in particular, for the Traveling Salesman Problem (see Chapter 10 of [23] and [25]). These operators can be applied to insertion orders or to the routes themselves when these routes are represented as strings of integers.

In [1], a decoder system for the VRP with time windows is described. This system uses two order-based crossover operators, MX1 and MX2, to produce new insertion orders. Since it is generally desirable to insert customer i before customer j in the solution if the time window at i occurs before the time window at j, the two operators are strongly biased by this precedence relationship.

In Figure 9, we show how the MX1 operator produces a valid ordering from two parent orderings. This example assumes five customers numbered from 1 to 5 and the time window precedence relationship $1 < 2 < 3 < 4 < 5$ (i.e., the time window at customer 1 is the earliest and the time window at customer 5 is the latest). Starting at position 1, the customers on the two parent chromosomes are compared and the customer with the earliest time window is added to the offspring. Then, the procedure is repeated for the next position, until all positions are considered. In the example provided in Figure 9, customers 1 and 4 are first compared. Since the time window at customer 1 occurs before the time window at customer 4, customer 1 is selected and takes the first position on the offspring (a). Now, in order to produce a valid ordering, customers 1 and 4 are swapped on parent 2. Customers 3 and 5 are then compared at position 2, and customer 3 is selected to occupy the second position on the offspring (b). Customers 5 and 3 are swapped on parent 1, etc. The resulting offspring is shown in (e). Note that MX1 tends to push customers with early time windows to the front of the resulting string. These customers are thus the first to be inserted in the solution by the insertion heuristic.

(a)

parent 1 :	**1**	**5**	**2**	**3**	**4**
parent 2 :	4	3	1	2	5
offspring :	**1**	-	-	-	-

(b)

parent 1 :	**1**	**5**	**2**	**3**	**4**
parent 2 :	1	3	4	2	5
offspring :	**1**	3	-	-	-

(c)

parent 1 :	**1**	**3**	**2**	**5**	**4**
parent 2 :	1	3	4	2	5
offspring :	**1**	3	**2**	-	-

(d)

parent 1 :	**1**	**3**	**2**	**5**	**4**
parent 2 :	1	3	2	4	5
offspring :	**1**	3	**2**	4	-

(e)

parent 1 :	**1**	**3**	**2**	**4**	**5**
parent 2 :	1	3	2	4	5
offspring :	**1**	3	**2**	4	**5**

Figure 9. MX1 crossover

Specialized order-based mutation operators have also been designed [25]. The Remove-and-Reinsert (RAR) operator, for example, randomly selects a customer on the chromosome and moves it to another position. The customers located between the two positions are shifted accordingly. In Figure 10, customer 3 is moved from position 2 to position 4. Swapping operators, where two customers exchange their position, are also widely used.

chromosome:	1	3	2	4	5
chromosome:	1	2	4	3	5

Figure 10. RAR mutation

The generality of the decoder approach has led to its application to complex vehicle routing problems with many side constraints (which prevent the application of more specific problem-solving methodologies). For example, a successful implementation for a vehicle routing problem with backhauls and time windows is described in [28]; solutions within 1% of the optimum, on average, have been produced for problems with up to 100 customers.

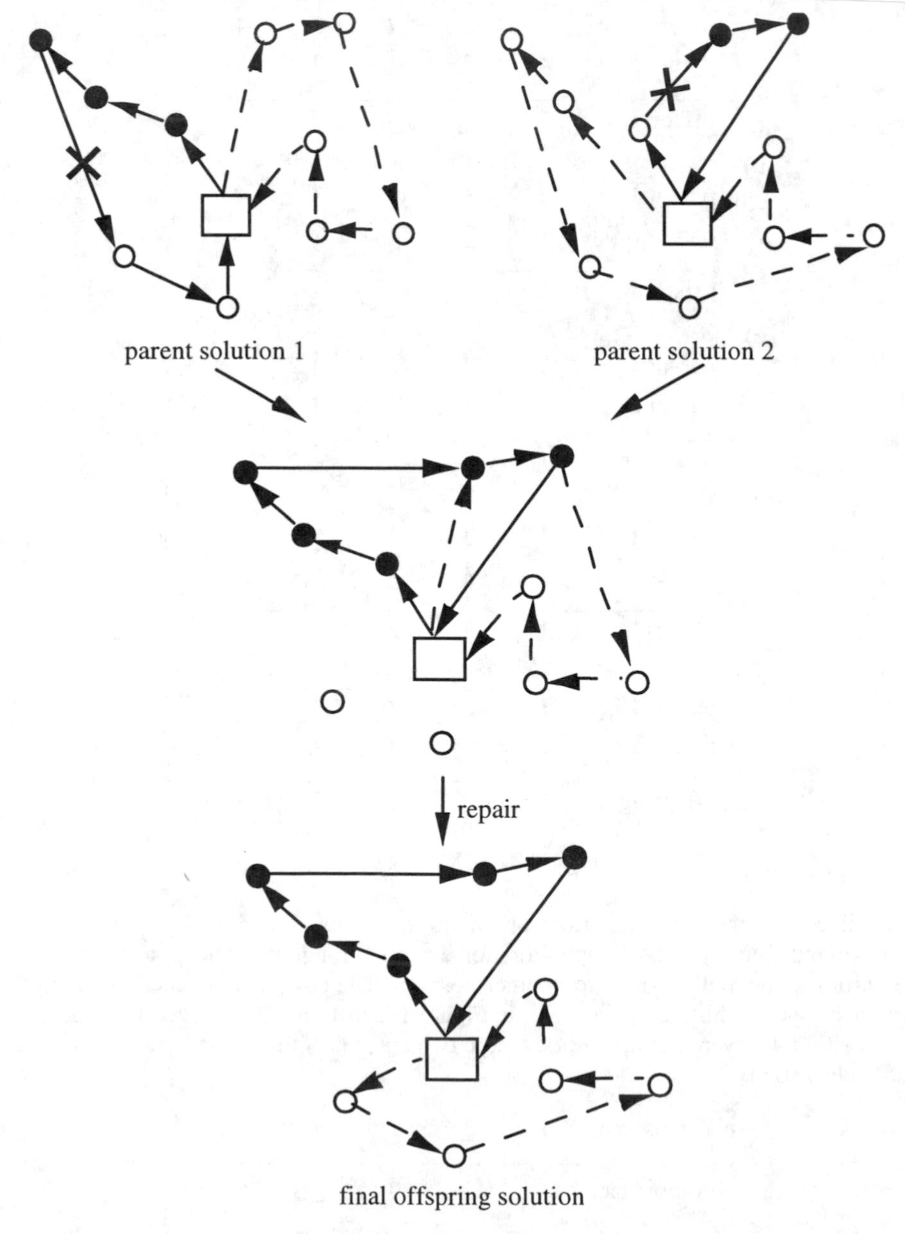

Figure 11. The SBX operator

4.3 A Nonstandard GA

A different approach is taken in [26], as the authors apply the general principles of genetic search on the routes themselves and do not address the encoding issue. Hence, a population of solutions (not chromosomes) evolve from one generation to the next through the application of operators that are similar in spirit to crossover and mutation. For example, the Sequence-Based Crossover (SBX) merges the first part of a route in a given solution to the end part of another route in another solution. This operator is illustrated in Figure 11. First, a link is randomly selected and removed from each parent solution. Then, the customers that are serviced before the break point on the route of the first parent solution are linked to the customers that are serviced after the break point on the route of the second parent solution (c.f. black nodes). This new route then replaces the old one in the first parent solution. A second offspring can be created by inverting the role of the parents. In the process, customers are likely to be duplicated or eliminated from the resulting offspring solution. Hence, repair operators are used to transform an infeasible solution into a feasible one; a duplicate is eliminated by removing the customer from one of the routes, while customers that were left apart by the SBX operator are reintroduced in the solution through a simple insertion mechanism.

In the GENEROUS system, this algorithm is coupled with powerful local reoptimization methods that improve the offspring solutions by exchanging customers between routes or by moving customers from one position to another on the same route. With such local reoptimization methods, the GENEROUS system was able to produce many best known solutions on a set of benchmark problems with time windows [34], outperforming other competing approaches based on simulated annealing and tabu search.

5. Conclusion

This chapter has reported some interesting developments in the field of vehicle routing using neural networks and genetic algorithms. Genetic algorithms, in particular, have been successful for different types of academic VRPs with complex side constraints as well as for real-world applications. This is probably related to the fact that GAs work with a population of solutions, rather than a single solution, thus achieving a wider exploration of the solution space.

Neural network models have not obtained the same success on some classical vehicle routing problems. Since the human mind is not particularly well suited for solving complex combinatorial optimization problems, the fundamental analogy upon which these models rely do not really hold. However, the vehicle dispatching problem reported in Section 3 is an interesting application for this paradigm, since the problem is ill-defined and requires a fair amount of human expertise. Such carefully identified problems should provide a "niche" for the application of neural networks.

Hybrid approaches combining genetic algorithms and neural networks could also provide interesting research avenues in the future. An example is found in [27], where a neural network is used to find good seed points for a set of vehicle routes; the GA then tunes the parameters of the following insertion phase.

Acknowledgments

The first author was funded by the Canadian Natural Sciences and Engineering Research Council (NSERC) and by the Quebec Fonds pour la Formation de Chercheurs et l'Aide à la Recherche (FCAR). The second author was partially supported by the Slippery Rock University Faculty Development Grant. This support is gratefully acknowledged.

References

[1] Blanton, J.L. and Wainwright, R.L. (1993), "Multiple Vehicle Routing with Time and Capacity Constraints using Genetic Algorithms," in *Proceedings of the 5th Int. Conf. on Genetic Algorithms*, Urbana-Champaign, IL, pp. 452-459.

[2] Christofides, N., Mingozzi, A., and Toth, P. (1979), "The Vehicle Routing Problem," in *Combinatorial Optimization*, N. Christofides, A. Mingozzi, P. Toth, and C. Sandi (Eds.), Wiley: Chichester, pp. 315-338.

[3] Clarke, G. and Wright, J. (1964), "Scheduling of Vehicles from a Central Depot to a Number of Delivery Points," *Operations Research 12*, pp. 568-581.

[4] Cullen, F.H. (1984), "Set Partitioning Based Heuristics for Interactive Routing," Ph.D. Dissertation, Georgia Institute of Technology, Georgia.

[5] Davis, L. (1991), Handbook of Genetic Algorithms, Van Nostrand Reinhold: New York.

[6] DeJong, K.A. (1975), "An Analysis of the Behavior of a Class of Genetic Adaptive Systems," Ph.D. Dissertation, University of Michigan.

[7] Durbin, R. and Willshaw, D. (1987), "An Analogue Approach to the Traveling Salesman Problem using an Elastic Net Method," *Nature 326*, pp. 689-691.

[8] Garey, M.R. and Johnson, D.S. (1979), Computers and Intractability: A Guide to the Theory of NP-Completeness, Freeman: San Francisco.

[9] Gendreau, M., Hertz, A., and Laporte, G. (1994), "A Tabu Search Heuristic for the Vehicle Routing Problem," *Management Science 40*, pp. 1276-1290.

[10] Ghaziri, H. (1991), "Solving Routing Problems by a Self-Organizing Map," in *Artificial Neural Networks*, T. Kohonen, K. Makisara, O. Simula and J. Kangas (Eds.), North-Holland: Amsterdam, pp. 829-834.

[11] Ghaziri, H. (1993), "Algorithmes Connexionnistes pour l'Optimisation Combinatoire," Thèse de doctorat, Ecole Polytechnique Fédérale de Lausanne, Lausanne, Switzerland (in French).

[12] Ghaziri, H. (1996), "Supervision in the Self-Organizing Feature Map: Application to the Vehicle Routing Problem," in *Meta-Heuristics: Theory & Applications*, I.H. Osman and J.P. Kelly (Eds.), Kluwer:Boston, pp. 651-660.

[13] Gillett, B. and Miller, L. (1974),), "A Heuristic Algorithm for the Vehicle Dispatch Problem," *Operations Research 22*, pp. 340-379.

[14] Goldberg, D.E. (1989), Genetic Algorithms in Search, Optimization, and Machine Learning. Addison-Wesley: Reading, MA.

[15] Holland, J.H. (1975), Adaptation in Natural and Artificial Systems, University of Michigan Press: Ann Arbor.

[16] Hopfield, J.J. (1982), "Neural Networks and Physical Systems with Emergent Collective Computational Abilities," in *Proceedings of the National Academy of Science 79*, pp. 2554-2558.

[17] Hopfield, J.J. and Tank, D.W. (1985), "Neural Computation of Decisions in Optimization Problems," *Biological Cybernetics 52*, pp. 141-152.

[18] Johnson, D.S. and McGeoh, L.A. (1997), "The Traveling Salesman Problem: A Case Study," in *Local Search in Combinatorial Optimization*, E. Aarts and J.K. Lenstra (Eds.), Wiley: Chichester, pp. 215-310.

[19] Kohonen, T. (1988), Self-Organization and Associative Memory, Springer: Berlin.

[20] Lenstra, J. and Rinnooy Kan, A. (1981), "Complexity of Vehicle Routing and Scheduling Problems," *Networks 11*, pp. 221-227.

[21] Maren, A., Harston, C., and Pap, R. (1990), Handbook of Neural Computing Applications, Academic Press: San Diego.

[22] Matsuyama, Y. (1991), "Self-Organization via Competition, Cooperation and Categorization Applied to Extended Vehicle Routing Problems," in *Proc. of the Int. Joint Conf. on Neural Networks*, Seattle, WA, pp. I-385-390.

[23] Michalewicz, Z. (1996), Genetic Algorithms + Data Structures = Evolution Programs, Third Edition, Springer: Berlin.

[24] Potvin, J.Y. (1993), "The Traveling Salesman Problem: A Neural Network Perspective," *ORSA Journal on Computing 5,* pp. 328-348.

[25] Potvin, J.Y. (1996), "Genetic Algorithms for the Traveling Salesman Problem," *Annals of Operations Research 63*, pp. 339-370.

[26] Potvin, J.Y. and Bengio, S. (1996), "The Vehicle Routing Problem with Time Windows - Part II: Genetic Search," *INFORMS Journal on Computing 8*, pp. 165-172.

[27] Potvin, J.Y., Dubé, D., and Robillard, C. (1996), "A Hybrid Approach to Vehicle Routing using Neural Networks and Genetic Algorithms," *Applied Intelligence 6*, pp. 241-252.

[28] Potvin, J.Y., Duhamel, C., and Guertin, F. (1996), "A Genetic Algorithm for Vehicle Routing with Backhauling," *Applied Intelligence 6*, pp. 345-355.

[29] Psaraftis, H.N. (1995), "Dynamic Vehicle Routing," *Annals of Operations Research 61*, pp. 143-164.

[30] Rosenkrantz, D., Sterns, R., and Lewis, P. (1977), "An Analysis of Several Heuristics for the Traveling Salesman Problem," *SIAM Journal of Computing 6*, pp. 563-581.

[31] Rumelhart, D.E., Hinton, G.E., and Williams, R.J. (1986), "Learning Internal Representations by Error Propagation," in *Parallel Distributed Processing - Explorations in the Microstructure of Cognition*, Volume 1, D.E. Rumelhart and J.L. McClelland (Eds.), The MIT Press: Cambridge, pp. 318-364.

[32] Schumann, M. and Retzko, R. (1995), "Self-Organizing Maps for Vehicle Routing Problems - Minimizing an Explicit Cost Function," in *Proceedings of the Int. Conf. on Artificial Neural Networks*, F. Fogelman-Soulie (Ed.), EC2&Cie, Paris, France, pp. II-401-406.

[33] Shen, Y., Potvin, J.Y., Rousseau, J.M., and Roy, S. (1995), "A Computer Assistant for Vehicle Dispatching with Learning Capabilities," *Annals of Operations Research 61*, pp. 189-211.

[34] Solomon, M.M. (1987), "Algorithms for the Vehicle Routing and Scheduling Problems with Time Window Constraints," *Operations Research 35,* pp. 254-265.

[35] Taillard, E. (1993), "Parallel Iterative Search Methods for Vehicle Routing Problems," *Networks 23*, pp. 661-673.

[36] Thangiah, S.R., Nygard, K., and Juell, P. (1991), "GIDEON: A Genetic Algorithm System for Vehicle Routing with Time Windows," in *Proceedings of the 7th IEEE Conference on Artificial Intelligence Applications*, Miami, FL, pp. 322-328.

[37] Thangiah, S.R. and Nygard, K. (1992), "School Bus Routing using Genetic Algorithms," in *Proceedings of the SPIE Conference on Applications of Artificial Intelligence X: Knowledge Based Systems*, Orlando, FL, pp. 387-397.

[38] Thangiah, S.R. and Nygard, K. (1992), "MICAH: A Genetic Algorithm System for Multi-Commodity Transshipment Problems," in *Proceedings of the 8th IEEE Conference on Artificial Intelligence Applications*, Monterey, CA, pp. 240-246.

[39] Thangiah, S.R. (1993), "Vehicle Routing with Time Windows using Genetic Algorithms," Technical Report SRU-CpSc-TR-93-23, Computer Science Department, Slippery Rock University, PA.

[40] Thangiah, S.R., Osman, I.H., Vinayagamoorthy, R., and Sun, T. (1993), "Algorithms for Vehicle Routing with Time Deadlines," *American Journal of Mathematical and Management Sciences 13*, pp. 323-355.

[41] Thangiah, S.R., Vinayagamoorthy, R., and Gubbi, A. (1993), "Vehicle Routing with Time Deadlines using Genetic and Local Algorithms," in *Proceedings of the 5th Int. Conf. on Genetic Algorithms*, Urbana-Champaign, IL, pp. 506-513.

[42] Thangiah, S.R. (1995), "An Adaptive Clustering Method using a Geometric Shape for Vehicle Routing Problems with Time Windows," in *Proceedings of the 6th Int. Conf. on Genetic Algorithms*, Pittsburgh, PA, pp. 536-543.

[43] Thangiah, S.R. (1996), "Vehicle Routing with Time Windows using Genetic Algorithsm," in *Practical Handbook of Genetic Algorithms: New Frontiers*, Volume II, Lance Chambers (Ed.), CRC Press, FL.

[44] Thangiah, S.R., Salhi, S., and Rahman, F. (1997), "Genetic Clustering: An Adaptive Heuristic for the Multi-Depot Vehicle Routing Problem," Technical Report SRU-CpSc-TR-97-33, Computer Science Department, Slippery Rock University, PA.

Chapter 7:

Fuzzy Logic and Neural Networks in Fault Detection

FUZZY LOGIC AND NEURAL NETWORKS IN FAULT DETECTION

B. Köppen-Seliger and P. M. Frank
Gerhard-Mercator-Universität - GH Duisburg
FB 9, Meß- und Regelungstechnik
Bismarckstr. 81, BB
D-47048 Duisburg, Germany

This chapter introduces advanced supervision concepts and presents examples showing how fuzzy logic and neural networks can be applied in model-based fault diagnosis.

Emphasis is placed on a combined quantitative/knowledge-based concept incorporating fuzzy logic for residual evaluation and the application of certain types of neural networks for residual generation and evaluation. Realistic simulation studies at a wastewater plant and an actuator benchmark prove the applicability of the proposed schemes.

1. Introduction

Due to the increasing complexity of modern control systems and the growing demand for quality, cost efficiency, availability, reliability, and safety, the call for fault tolerance in automatic control systems is gaining more and more importance. Fault tolerance can be achieved by either passive or active strategies. The **passive** approach makes use of robust control techniques to ensure that the closed-loop system becomes insensitive with respect to faults. In contrast, the **active** approach provides fault accommodation, i.e., the reconfiguration of the control system when a fault has occurred. While robust control can tolerate small faults to a certain degree, the reconfiguration concept is absolutely inevitable when serious faults occur that would lead to a failure of the whole system.

0-8493-9804-5/99/$0.00+$.50

In order to accomplish fault accommodation a number of tasks have to be performed. One of the most important and difficult of these tasks is the early **diagnosis** of the faults. Besides this, fault diagnosis is needed as part of the supervision of complex control systems that incorporate artificial intelligence.

Fault diagnosis has thus become an important issue in modern automatic control theory and, during the last two and a half decades, an immense deal of research was done in this field, resulting in a great variety of different methods with increasing acceptance in practice. The core of the fault diagnosis methodology is the so-called **model-based** approach (see also Section 2). For literature on model-based techniques, the reader is referred to comprehensive survey papers as, for example, [5], [6], [7], [12], [35] and [43] or the book by [34].

In the case of fault diagnosis in **complex** systems, one is faced with the problem that no, or insufficiently accurate, mathematical models are available. The use of **knowledge-model-based** or **data-model-based** techniques, either in the framework of diagnosis expert systems or in combination with a human expert, is then the only feasible way to proceed.

This contribution provides a combined analytical- and knowledge-based approach applying fuzzy logic and a data-model-based approach based on neural networks as system models and pattern classifiers. Application results from realistic simulation studies at a wastewater plant and an industrial actuator benchmark problem are given to illustrate the different methods.

2. Fault Diagnosis

A permanent goal in operating technical processes is to ensure safety and reliability due to the general aim of increasing economic efficiency. This gives rise to the current demand for modern supervision concepts basically based on fault diagnosis schemes. In this context the term fault incorporates any kind of malfunctioning up to a complete failure of a system component, actuator, or sensor. The aim of fault diagnosis is to detect the faults of interest and their causes early enough so that failure of the overall system can be avoided.

The faults can be commonly described as additional inputs whose time of occurrence and size is unknown. In addition there is always modeling uncertainty due to unmodeled disturbances, noise, and model mismatch. This may not be

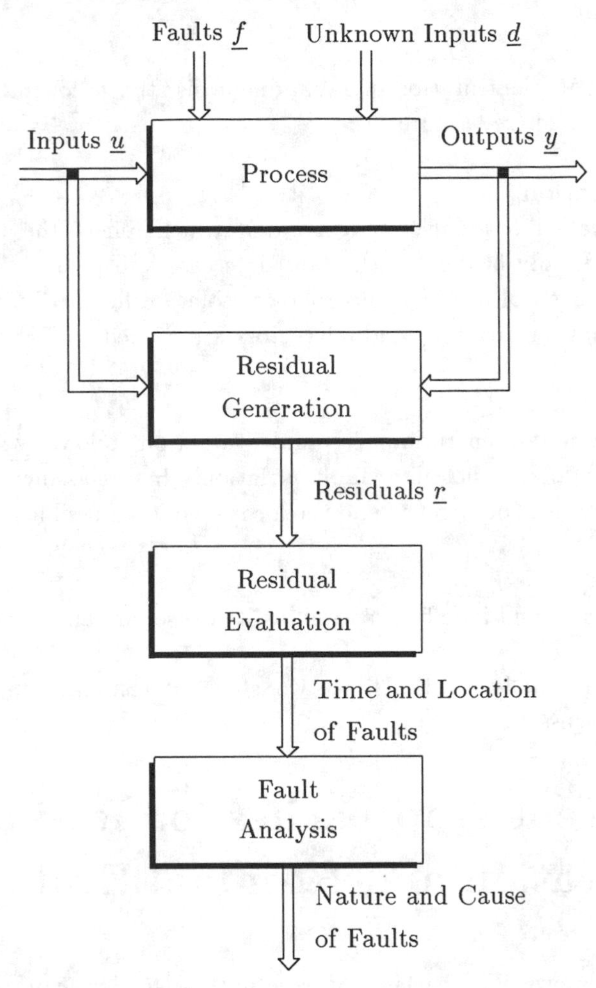

Figure 1: General scheme for fault diagnosis.

critical for the process behaviour but may obscure the fault detection by raising false alarms.

2.1 Concept of Fault Diagnosis

The basic tasks of fault diagnosis are to **detect** and **isolate** faults and to provide **information** about their size and source. This has to be done on-line in the face of the existing unknown inputs and without, or with only very few, false alarms. As a result the overall concept of fault diagnosis consists of the three subtasks; fault **detection**, fault **isolation**, and fault **analysis**.

For the practical implementation of fault diagnosis, the following three steps are usually performed (see Figure 1) :

Residual generation
Signals, the so-called residuals, are generated which reflect the faults. The residuals should ideally be zero in the fault-free case and deviate from zero in case of an occurrence of a fault. In order to isolate different faults, properly structured residuals or directed residual vectors are needed.

Residual evaluation
Subsequent to residual generation, residual evaluation follows, with the goal of fault detection and, if possible, fault isolation. In a classification process a decision on the time of occurrence and the location of the possible fault is made.

Fault analysis
In this step the fault and its effect, as well as its cause, are analyzed.

In this chapter methods for the first two steps of residual generation and evaluation are discussed.

2.2 Different Approaches for Residual Generation and Residual Evaluation

The commonly known approaches for residual generation can basically be divided into two categories of signal-based and model-based concepts with a further subdivision as shown in Figure 2. The main research emphasis of the last two decades has been placed on the development of model-based approaches starting from analytical models and leading to the recently employed data-based models, such as neural networks.

Residual evaluation techniques can be principally divided into threshold decisions, statistical methods, and classification approaches (Figure 3). The methods of fuzzy and neural classification will be introduced and applied in the following sections.

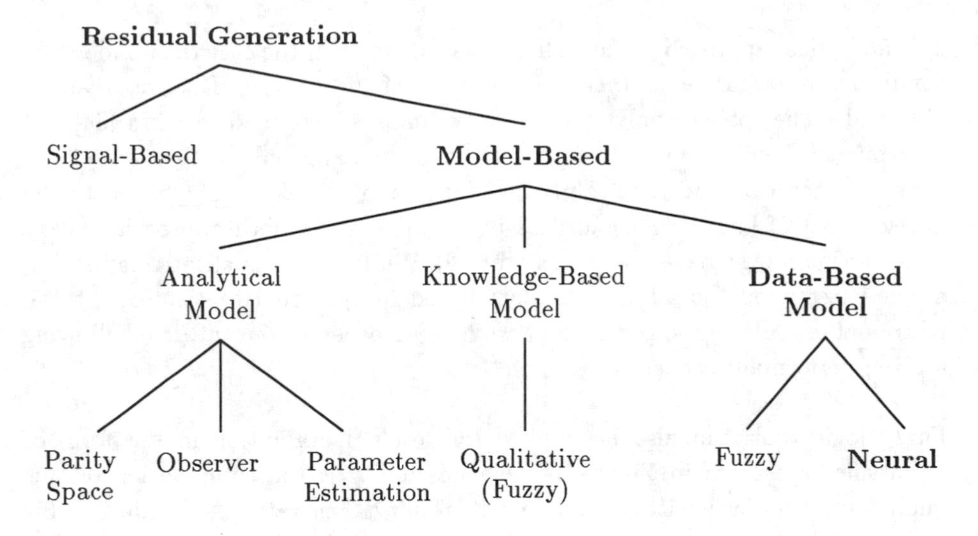

Figure 2: Classification of different residual generation concepts

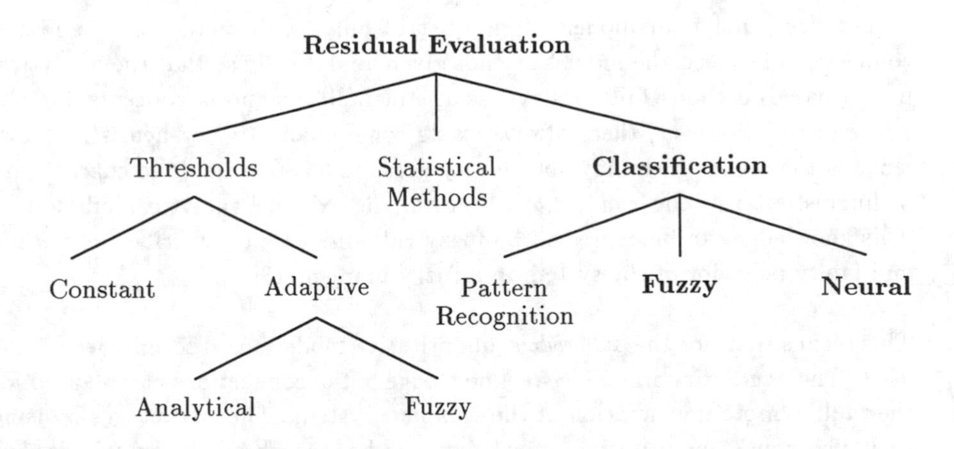

Figure 3: Classification of different residual evaluation concepts

3. Fuzzy Logic in Fault Detection

The analytical approach to fault diagnosis suffers from the fact that under real conditions no accurate mathematical models of the system of interest can be obtained. The robust analytical design techniques described, e.g., in [34] can overcome this deficiency only to a certain degree and only with great effort. This consideration and the evolution of fuzzy and neural techniques led to the development of knowledge- and data-based models. In both approaches fuzzy logic can be integrated as seen in Figure 2. While in the qualitative approach a rule-based model is set up, the data-based fuzzy model consists of a fuzzy relational module whose parameters are trained by input-output data following a given performance criterion.

Fuzzy logic tools can also be applied for residual evaluation in the form of a classifier as shown in Figure 3. One possibility is the combination of this qualitative approach with a quantitative residual generating algorithm. This idea is motivated and developed in the following section.

3.1 A Fuzzy Filter for Residual Evaluation

In practice, analytical models often exist of only *parts* of the plant and the connections between the models are not given analytically so that the analytical model-based-methods fail to serve as useful fault diagnosis concepts for the *whole* plant. However, there always exists some qualitative or heuristic knowledge of the plant which may not be very detailed but is suitable to characterize in linguistic terms the *connections* between the existing analytical submodels. This knowledge can be expressed by fuzzy rules in order to describe the normal and faulty behavior of the system in a fuzzy manner [18].

This means that, for the *submodels*, quantitative model-based techniques can be used. The *qualitative* and *heuristic* knowledge of the connections can be used for the fault symptom generation of the *complete* system. The advantages of using such a combined quantitative/knowledge-based approach can be summarized as follows:

- It is not necessary to build an analytical model of the complete process. It is sufficient to have analytical models of the subparts.

- The connections between the submodels can be described by *qualitative* or *heuristic* knowledge. This is often easier because some *qualitative* or *heuristic* description of the plant or the interconnections between the submodels is normally known.
- The mathematical effort, compared to using a model of the complete plant, is significantly reduced.
- The causes and effects of the faults can be transferred more easily into the fault diagnosis concept.

3.1.1 Structure of the Fuzzy Filter

The fuzzy residual evaluation is a process that transforms *quantitative* knowledge (residuals) into *qualitative* knowledge (fault indications). Residuals generated by analytical submodels, as described above and depicted in Figure 4 [17], represent the inputs of the Fuzzy Filter which consists of the three basic components:

- Fuzzification
- Inference
- Presentation of the fault indication

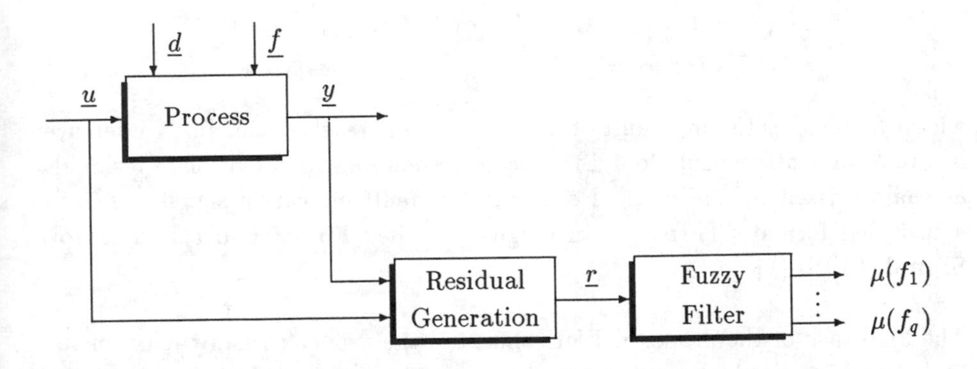

Figure 4: General structure of the Fuzzy-Filter-based diagnostic concept.

As a first step, a knowledge base has to be built which includes the definition of the faults of interest, the measurable residuals (symptoms), the relations between the residuals and the faults in terms of IF/THEN rules, and the representation of the residuals in terms of fuzzy sets, for example, "normal" and "not normal."

The process of **fuzzification** includes the proper choice of the membership functions for the fuzzy sets and is defined as the assignment of a suitable number of fuzzy sets to each residual component r_i with [14].

$$\underline{r} = \{r_1, \ldots, r_i, \ldots, r_n\} \tag{1}$$

but not for the fault symptoms f_i. This procedure can mathematically be described for the residuals as.

$$\begin{aligned} r_i &\mapsto r_{i1} \circ r_{i2} \circ \ldots \circ r_{ir} \\ r &\to [0,1] \end{aligned} \tag{2}$$

where r_{ij} describes the j_{th} fuzzy set of the i_{th} residual and $\circ$ describes the fuzzy composition operator.

This part is very important because the coupling or decoupling of the faults, respectively, will be significantly influenced by this procedure.

The task of the FDI system is now to determine, from the given rule base, indication signals for the faults with the aid of an inference mechanism [18]. The **inference** can be appropriately carried out by using so called *Fuzzy Conditional Statements*,

$$\begin{aligned} &IF\,(effect = r_{11})\,AND\;IF\,(effect = r_{12})\ldots \\ &THEN\,(cause = f_m) \end{aligned}$$

where f_m denotes the m_{th} fault of the system. The result of this fuzzy inference is a fault indication signal found from a corresponding combination of residuals as characterized by the rules. Note that this fault indication signal is still in a fuzzyfied format. Therefore, this signal is called **Fuzzy Fault Indication Signal** (FFIS) [15].

The final task of the proposed FDI concept is the proper **presentation** of the fault situation to the operator who has to make the final decision about the appropriate fault handling [18]. Typical for the fault detection problem is that the output consists of a number of fault indication signals, one for each fault, where these signals can take only the values one or zero (yes or no). For a fuzzy representation this means that it is not necessary to have a number of fuzzy sets to represent the output, as in control. Rather, each fuzzy fault indication signal FFIS is, by its nature, a singleton, the amplitude of which characterizes the degree of membership to only one, preassigned fuzzy set "$fault_m$." This degree is characterized by the FFIS, i. e., the signal obtained as a result of the inference. Specific for this approach is that it refrains from defuzzification and represents the fault indication signal for each fault to the operator in the fuzzy format, i.e., in terms of the FFIS, which represents the desired degree of membership to the set "$fault_m$."

There are some advantages to this procedure as far as the computational expenditure is concerned. There is no need for a representation of the output by a number of fuzzy sets. All available information about the appearance of a fault can be incorporated into the definition of the fuzzy sets of the inputs, in our case, the residuals. And, finally, one can dispense with the defuzzification of the signals obtained after the inference has been performed. To be more specific, instead of using the standard format of the statement

$$\ldots THEN\; fault = big \tag{3}$$

where *big* is defined as one of a number of fuzzy sets that characterizes the output, the following format of the statement is used:

$$\ldots THEN\; fault = fault_1 \tag{4}$$

where "$fault_1$" is the only existing fuzzy set of fault f_1. This applies in a similar way to all faults under consideration. Note, that the fuzzy set "$fault_i$" has a degree of membership which is identical to the aggregated output of the evaluated residuals.

As a result, one of the key issues of the fuzzy inference approach is that the representation of the result of the residual evaluation concept is different from the conventional concepts in that it directly provides the human operator with the FFIS, leaving to him or her the final decision of whether or not a fault has occurred.

This combination of a human expert with a fuzzy FDI toolbox allows us to avoid false alarms, because the fault situation can be assessed on the basis of a fuzzy characterization of the fault situation together with the human expertise and experience.

The key issue of this kind of residual evaluation approach is the design of the Fuzzy Filter. To simplify this design problem, an algorithm is presented in the following section, which provides a systematic support by efficiently reducing the degrees of freedom in the design process.

3.1.2 Supporting Algorithm for the Design of the Fuzzy Filter

Problem Formulation

There are two possible uses of the supporting algorithm for the design of the Fuzzy Filter in the residual evaluation process [16]. The first possibility starts with an *empty* rule base. That means that the designer has to generate a *complete* rule base using this algorithm. This ensures that the generated rule base is consistent and complete with respect to the fault detection scheme. Therefore, all rules have to be consistent and unique in order to represent each fault under consideration. The suggested algorithm automatically checks whether or not these conditions are fulfilled.

The second possibility starts with a given, possibly inconsistent, rule base. The task is now to check which part of the given rule base is inconsistent and/or incomplete. This part of the rule base should be modified as described in the next section. It should be mentioned that these two possibilities use the same algorithm, just the initial conditions of these two possibilities are different. To use this algorithm, the so-called *Fuzzy Switching Functions* have to be defined in order to simplify the procedure.

Fuzzy Switching Functions

Fuzzy switching functions are introduced and modified by [13], [25], [31]. In contrast to the Boolean logic, where the assignment of a value is just the assignment to the values zero or one ($x \in \{0,1\}$), a fuzzy formula assigns a value to a number of the set ranging from zero to one ($x \in [0,1]$). These fuzzy formulas have to be defined using a fuzzy algebra as an extension to the boolean algebra [13]. In order to use this fuzzy algebra, some definitions are given.

Definition 1 (Fuzzy Algebra) A system Z is called a fuzzy algebra if the following conditions are met ($x, y, z \in Z$):

1. $x + x = x$ and $x * x = x$
2. $x + y = y + x$ and $x * y = y * x$
3. $(x + y) + z = x + (y + z)$ and $(x * y) * z = x * (y * z)$
4. $x + (x * y) = x$ and $x * (x + y) = x$
5. $x + (y * z) = (x + y) * (x + z)$ and $x * (y + z) = x * y + x * z$
6. For every $x \in Z$ the complement must be unique $\overline{x} \in Z$ with $\overline{\overline{x}} = x$
7. $\forall x \in Z \, \exists e_+ \in Z$, such that $x + e_+ = e_+ + x = x$
8. $\forall x \in Z \, \exists e_* \in Z$, such that $x * e_* = e_* * x = x$

9. $\overline{x+y} = \overline{x} * \overline{y}$ and $\overline{x*y} = \overline{x} + \overline{y}$

Definition 2 (Fuzzy Variable) A fuzzy variable x is defined as the degree of membership of a membership function $\mu(x)$.

This provides the advantage that a fuzzy variable can be treated like a "normal" logic variable as defined for the boolean algebra.

The next step is the generation of a fuzzy formula. This provides the possibility of combining different values together to form one fuzzy expression. This allows the transformation of a fuzzy rule into a fuzzy formula with identical meaning.

Definition 3 (Value of a Fuzzy Formula) The value of a fuzzy formula is uniquely defined using the following rules:
1. $\mu(A) = 0$, if $A = 0$
2. $\mu(A) = 1$, if $A = 1$
3. $\mu(A) = \mu(x)$, if $A = x$
4. $\mu(A) = 1 - \mu(B)$, if $A = \overline{B}$
5. $\mu(A) = min(\mu(B), \mu(C))$, if $A = B * C$
6. $\mu(A) = max(\mu(B), \mu(C))$, if $A = B + C$

Definition 4 (Fuzzy Switching Function) A fuzzy switching function is a mapping of $[0,1] \rightarrow [0,1]$ described by a fuzzy formula.

Definition 5 (Clause) A clause is a disjunctive combination of at least two variables. (Combination with +)

Definition 6 (Phrase) A phrase is a conjunctive combination of at least two variables. (Combination with *)

Definition 7 (Disjunctive Normal Form DNF) A fuzzy formula is in DNF if $S = P_1 + P_2 + \ldots + P_m$ and every P_i is a phrase.

Definition 8 (Conjunctive Normal Form CNF) A fuzzy formula is in CNF if $S = C_1 + CP_2 + \ldots + C_m$ and every C_i is a clause.

Using these definitions above it is now possible to describe every fuzzy formula in DNF or CNF.

To define whether a rule and therefore a fuzzy formula (fuzzy switching function) is valid, the terms *fuzzy–valid* and *fuzzy–inconsistent* have to be defined.

Definition 9 (Fuzzy–Consistent) A fuzzy formula $F(x)$ is fuzzy–consistent if the output of the formula is $\geq 0.5\ \forall x$.

Definition 10 (Fuzzy–Inconsistent) A fuzzy formula $F(x)$ is fuzzy–inconsistent if it is not consistent.

These definitions ensure that every rule of the knowledge base can uniquely be transformed into fuzzy switching functions. With these fuzzy switching functions, the algorithm can be described.

Design of the Algorithm

It is assumed that certain *initial* rules have been generated. That means that for each value of the residual a number of fuzzy sets are defined. From this definition of the fuzzy sets, the number of rules are determined. The key question is now [16]: is it possible to distinguish between all defined faults using the given rules ? To answer this question it is necessary to prove whether or not a distinction between the faults can be made. If all premises of two fault descriptions have the same fuzzy values, a distinction is not possible. Two faults are distinguishable if they have at least one different definition in the premise of the rule. To illustrate this fact, the following example is given:

If Res_1 is positive and Res_2 is positive then f_1
If Res_1 is positive and Res_2 is normal then f_2

For these two rules, a unique distinction between f_1 and f_2 is possible. In addition, it can be mentioned that f_1 and f_2 **cannot** occur at the same time because Res_2 can just be positive *or* normal.

To illustrate two distinguishable faults, which can occur at the same time, the following example is given:

If Res_1 is positive and Res_2 is positive then f_1
If Res_1 is positive and Res_3 is negative then f_2

These rules allow the unique decision whether f_1 or f_2 has occurred, but both faults can occur at the same time, because f_1 and f_2 use two different residuals for the second premise, in this case Res_2 and Res_3 while the first premise uses Res_1 identically for both faults. As a consequence, the rules have to be modified so that the fuzzy set *positive* in *residual 1* is divided into two *different* fuzzy sets, f_1 and f_2. Now both faults are uniquely distinguishable and cannot occur at the same time.

From these examples it can be seen that the rule base has to be checked for such inconsistencies. If there are certain rules which lead to inconsistent fault behavior, these fuzzy sets have to be subdivided into (at least) two fuzzy sets. The task of the algorithm is now to check *all* faults against all others. For the first fault, this can be described as follows:

f_1 occurs, but none of the other faults

This has to be defined for all faults and leads to the following description:

f_k occurs, but none of the other faults

To handle this with the algorithm, each rule has to be transformed into a fuzzy switching function. If the result of the fuzzy switching function is ≥ 0.5 then the rule is fuzzy–consistent. If the result is < 0.5 the rule is fuzzy–inconsistent. That means that for fuzzy-inconsistent rules the compatibility degree is < 0.5 for all x. This implies that there exists at least one phrase P_i in the fuzzy switching function with the following structure:

$$x_i * \overline{x_i} \tag{5}$$

The task can now be specified as follows:

Find phrases with the form shown in Equation (5)

These phrases must be modified to make the fuzzy switching functions fuzzy-consistent. The complete design algorithm can be summarized as follows [16]:

Step 1: Define the number of faults which are of interest.

Step 2: For each residual component, two fuzzy sets have to be defined as an initial definition. These two fuzzy sets are *normal* and *not normal.*

Step 3: The rules are transformed into fuzzy switching functions. As a simple example for this transformation consider the following rule:

If Res_1 is normal and Res_2 is negative *or* If Res_1 is positive and Res_2 is negative then f_1

The corresponding fuzzy switching function is given as

$$f_1 = x_{11} * x_{21} + x_{12} * x_{21} \tag{6}$$

The definition is based on the assumption that the first index indicates the residual and the second index indicates the fuzzy set of this residual.

Step 4: Based on the fuzzy sets defined in Step 3 and the faults defined in Step 1, the resulting number of rules has to be generated.

Step 5: Prove whether or not it is possible to distinguish between *all* faults. That means that the fuzzy switching functions have to be checked for phrases described by Equation (5). This procedure must be performed for all faults. This leads to the following scheme:

$$f_k * \overline{f_{k+1} + f_{k+2} + \cdots + f_p} \qquad \forall k \in [1, p-1]$$

In this formula to check for $fault_k$, just the terms for k to p are considered, because the previous steps checked that fault k is fuzzy consistent with respect to $fault_1, \cdots, fault_{k-1}$.

Step 6: If the distinction is possible, all faults can be detected *and* isolated and the procedure is terminated.

Step 7: If a perfect distinction with this choice of the distribution of the fuzzy sets is not possible, one or more fuzzy sets have to be modified. This means, for example, that instead of the fuzzy set *"not normal"* two fuzzy sets *"slightly deviating"* and *"strongly deviating"* may be discriminated. The fuzzy set that has to be changed is a result of the reduced switching function. It is the fuzzy sets that lead to a fuzzy–inconsistency described by Equation (5).

Step 8: Now carry out the algorithm again and repeat the procedure until a unique distinction is possible.

This algorithm ensures that all faults are detectable *and* distinguishable. If a perfect distinction of the residuals is not possible, the algorithm indicates these inconsistencies. This helps the operating personnel to evaluate signals giving consideration to the inconsistency. To prove the algorithm it was applied to a part of a wastewater plant.

3.2 Application of the Fuzzy Filter to a Wastewater Plant

3.2.1 Description of the Process

The anaerobic digestion process is a self–regulating biological process that converts organic matter into gas. The metabolic pathway of the process is shown in Figure 5 [15]. The output of the process is carbon–dioxide (CO_2) and methane (CH_4). To run such a biological process under optimal conditions, which means to reduce the waste concentration in the effluence and yield a maximum methane rate, some requirements have to be met. To include all effects, the model of the process has to be highly nonlinear and of high order. Because of these facts some researchers have tried to simplify the model [1], [28] by reducing the number of acids in the process. The complete model contains three kinds of acids:

- Acetic acid
- Propionic acid
- Butyric acid

Because acetic acid is the most significant part some models contain just this kind of acid, see, e.g., [36]. The next step is the integration of the propionic

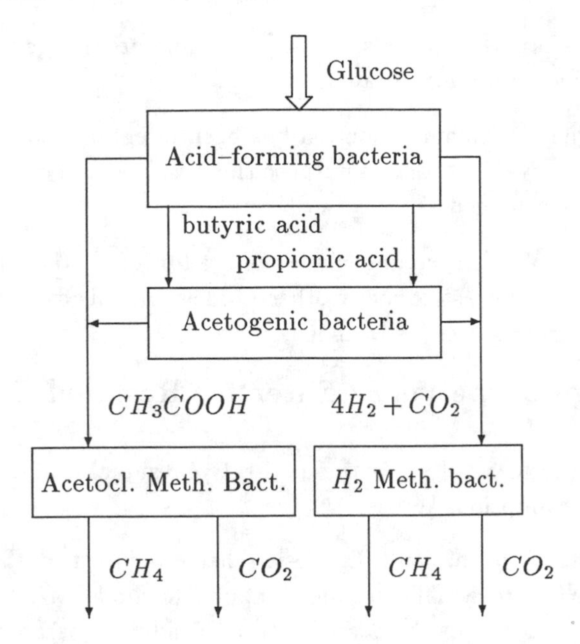

Figure 5: Metabolic pathway of the process

acid [24], [27]. If, on the other hand, the model contains all the acids, the used kinetics and the transformation of the gas from the liquid to the gas–phase are quite simple [30]. In addition most of the models consider only the acids as the source of methane. It has been shown in [8] that approximately 30% of the methane derives from the synthesis of hydrogen (H_2) and carbon dioxide (CO_2). To include this synthesis, the hydrogen and carbon–dioxide balance equations have to be adequately modeled. Finally, most of the models consider an acid as an input, rather than considering a component which is one or two steps above in the metabolic pathway, see Figure 5.

To include all the effects and leave out the above mentioned disadvantages, the model presented in this paper includes the following effects [18]:

- All acids (acetic, propionic, and butyric) have been considered.
- Four different bacteria have been considered.
- The complete production and consumption of H_2 has been considered in the complete metabolic pathway.
- For the acetic and propionic acid forming bacteria, Haldane kinetics have been used.
- The influence of the acetic acid into the propionic acid has been considered.

- The transformation process from the liquid to the gas–phase via gas–bubbles has been incorporated.
- An on–line H^+-balance equation has been integrated to estimate the pH-value of the system. For this purpose the dissolved carbon balance equation has been considered.
- The input of the system includes one additional step of the metabolic pathway, in contrast to many other models. In this case the formation of the acids from glucose is included.

3.2.2 Design of the Fuzzy Filter for Residual Evaluation

The design of the fuzzy filter for the qualitative residual evaluation is based on the following assumptions [15]:

- The structure of the fault diagnosis scheme is based on the topology described in Figure 4. This includes a nonlinear model of the process as well as a linear model for the observer-based residual generator.
- For the design of the fuzzy filter, only a qualitative description of the faults is needed.
- Both the quantitative residual vector and the qualitative description of the fault behavior are used as inputs to the fuzzy filter in order to detect and isolate the faults.

To generate a fault diagnosis scheme using the method described above, a description of the faults of interest has to be generated in order to build a list for the residual evaluation; therefore, different types of faults have been investigated [15]:

- Organic overload

  The reactor is fed with a continuous stream of organic matter. The magnitude of the glucose input is within a certain interval. An organic overload is defined as an *additional* glucose input, which is added to the normal glucose input at a certain, but not *a priori* known, time and with a certain magnitude.

- Hydraulic overload

  To maintain the process, the out–flow of the reactor will be filtered to collect the biomass and feed it back to the process. This ensures a higher biomass concentration inside the reactor resulting in a higher efficiency of the process. The filtered biomass will be transported via a pump to the input of the process. The percentage rate of the input feed of the biomass is a very sensitive parameter, which means that relatively small changes in the *dilution rate [D]* have a great influence on the stability of the system.

- Toxic overload

  A toxic overload is understood to be a toxic input to the input stream which affects the bacteria of the process. Some toxins have an influence on all bacteria and some influence only one kind of bacteria. A toxic overloading affects the stability of the system if it lasts for a long time; therefore, the *detection* of toxins is of high interest.

- Acidification

  A very important value is the measurement of the pH–value, which is also a very important indication of the process because the bacteria cannot survive in an acid environment ($pH < 7$).

- Sensor faults

  In total, 7 states of the complete system will be measured. The measurements, together with the estimated outputs of the observer, are the inputs for the residual evaluation.

To simplify the design of the rule base, fault detection and isolation have been separated into four parts with each part providing the corresponding fault symptoms. The four groups are:

- Hydraulic and organic overload
- Toxic overload
- Acidification
- Sensor faults

For each group, only a subset of the measurements has been used as an input. This way of cascading the total number of faults into different subgroups significantly reduces the amount of rules.

For each of the groups, the following choices have to be made in order to get the fault symptoms:

- Fuzzification

  - Number of fuzzy sets of each residual component
  - Shape of fuzzy sets of each residual component

- Inference

  Number of rules together with the definition of the fuzzy sets of the fuzzification part.

- Presentation of the fault indicators

  Here the number of the faults is equivalent to the number of outputs of the FDI–system.

3.2.3 Simulation Results

To demonstrate the functionality of the proposed method, some simulation results will be shown. In order to show just the essential parts of the simulation, only the fault indication signal that deviates from zero will be presented. In the ideal case this should be the fault symptom that corresponds to the fault under consideration.

Organic Overload

A constant input glucose feed of 20 $mmol/l$ was applied to the system. At $t = 20\,d$ an organic overload was applied to the system, in the form of an additional input glucose feed of 3 $mmol/l$, which is equivalent to an increase of the input feed of 15%. It can be seen from Figure 6 that this type of fault

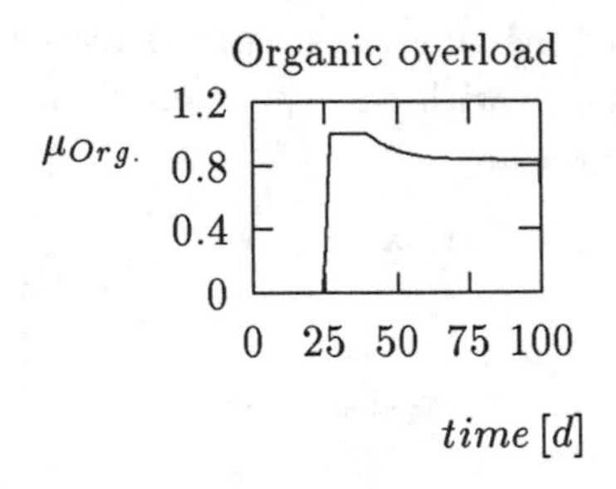

Figure 6: Organic overload with a constant glucose input of $30 mmol/l$.

can be detected and uniquely isolated. The corresponding fault symptom for this fault is the only symptom that deviates from zero. Even if the magnitude of the input feed and the magnitudes of the fault are changed, this type of fault can be detected for all tested simulations and can be uniquely isolated for almost all cases. For those cases where unique isolation does not occur, only one other signal deviates from zero, this being the fault symptom for the hydraulic overload. It can be pointed out that the organic and the hydraulic overload are quite similar and it is therefore difficult to distinguish between these two types of faults. In addition, it has been shown that the distinction between an organic and a hydraulic overload, using purely quantitative model-based techniques, is impossible.

Hydraulic Overload

As a second type of overload, a hydraulic overload has been applied to the system. The glucose input feed is, in this case, 30 $mmol/l$. The magnitude of the fault is 10%. As mentioned before, this type of overload is of special interest, because it has a significant influence on the stability of the system. It can be seen from Figure 7 that this type of overload can be both detected

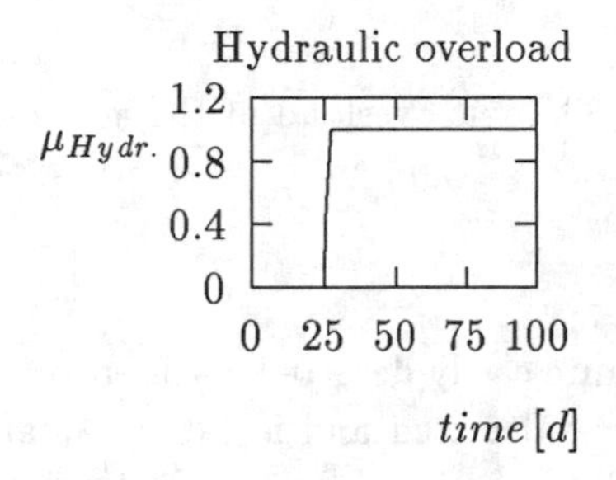

Figure 7: Hydraulic overload with a constant glucose input of 30 $mmol/l$.

and uniquely isolated. Varying the magnitude of the fault and the time of occurrence as well as the input glucose feed also leads to good results. The only fault symptom that is affected, besides the symptom for the hydraulic overload, is the fault symptom for the organic overload. This is again due to the close relationship of the organic and hydraulic overloads.

Toxic Overload

A toxic overload is rather difficult to detect because this type of overload can be treated as an incipient fault, which means that the influence on the overall system is rather small. A long period of time is required for the influence on the system to be significant enough to allow this type of overload to be detected. As an example, the acid bacteria are affected. The magnitude of the fault is 70% of the normal value. The fault is applied to the system at $t = 20\,d$. It can be seen from Figure 8 that the fault is detected at $t = 45\,d$. This relatively long detection time indicates that this type of overload is more incipient than other types. On the other hand, the influence and, therefore, the instability of the system, is also not very strongly affected.

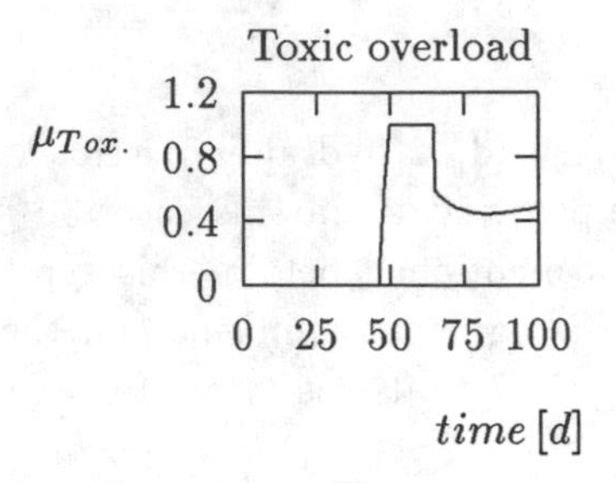

Figure 8: Toxic overload of the acetic bacteria.

Sensor Fault

A sensor fault can be quite easily detected. All sensor faults, at any time during the process states, can be detected and isolated. As an example, a sensor fault on the methane (CH_4) sensor has been applied to the system. The result is

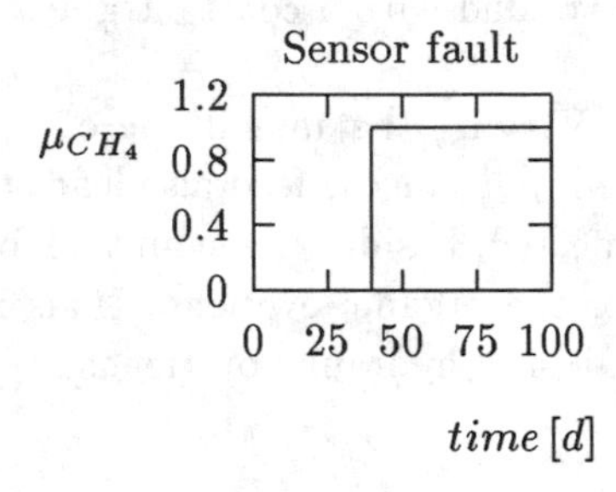

Figure 9: Sensor fault of the methane sensor.

given in Figure 9. The fault is detected immediately after it has been applied to the system.

4. Neural Networks in Fault Detection

Since the late eighties artificial neural networks have been widely reported for model-based fault detection and isolation in slowly varying complex systems where analytical models are not (or not fully) available [9, 10, 40, 41]. Since neural networks have proven their capability in the field of pattern recognition (e.g., image processing, speech recognition), it is an obvious step to apply them to fault detection and isolation applications as well.

Basically, the artificial neural network consists of neurons, simple processing elements that are activated as soon as their inputs exceed certain thresholds. The neurons are arranged in layers which are connected such that the signals at the input are propagated through the network to the output. The choice of the transfer function of each neuron (e.g., sigmoidal function) contributes to the nonlinear overall behavior of the network. During a training phase, a set of parameters of the network is learned which leads to the "best" approximation of the desired behavior. If a neural net is used for fault detection, the training is performed with measurements from fault-free and, if possible, faulty situations.

First applications of neural networks to fault diagnosis can be found in the chemical and process industries with their quasi-stationary processes [11, 39, 42]. Here pattern-recognition-like problems are to be solved when evaluating process signals. This idea has been extended to the task of residual evaluation which can be interpreted as a classification of pre-processed signals (see Figure 3).

Furthermore, since neural networks have successfully been applied to process modeling [4], they became of relevance for residual generation tasks as well (see Figure 2).

4.1 Neural Networks for Residual Generation

For residual generation purposes, the neural network replaces the generally analytical model describing the process under normal operation. This approach is of special interest where no exact or complete analytical or knowledge-based model can be produced, but a large input-output-measurement data base is available.

Before applying the neural network for residual generation, the network has to be trained for this task [21]. For this purpose, an input signal data base and a corresponding output signal data base must exist. These data can either be collected at the process itself, if possible, or with the help of a simulation model that is as realistic as possible. The latter possibility is of special interest for collecting data relating to the different faulty situations in order to test the residual generator, since such data is not generally available from the real process. The training is then commonly performed by applying a supervised learning algorithm. After completing training, the neural network can now be applied for on-line residual generation (Figure 10).

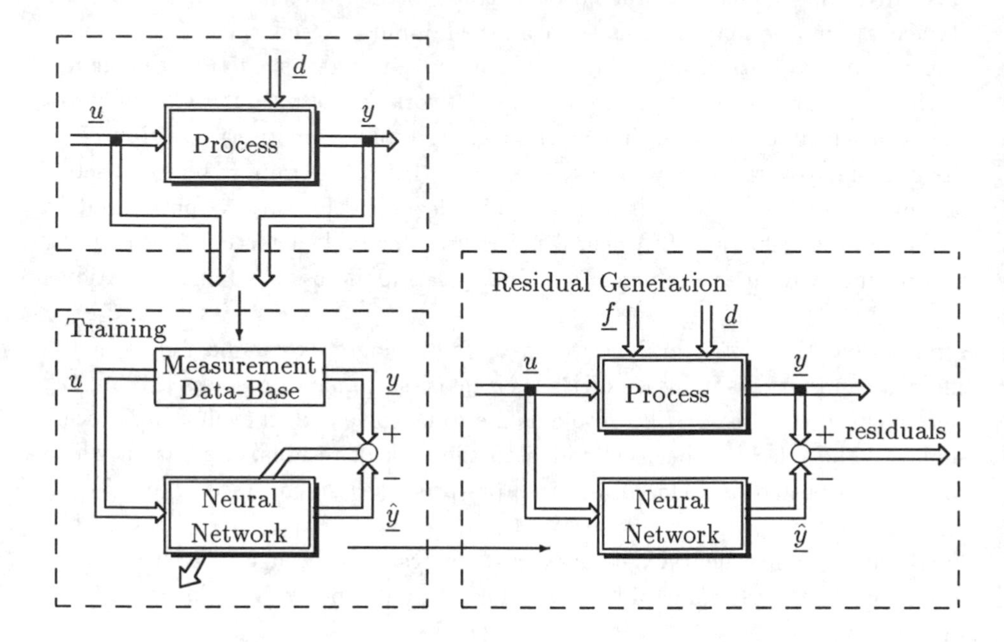

Figure 10: General scheme for off-line neural net training and on-line residual generation.

Two different types of neural networks suitable for residual generation are introduced in the following sections.

4.1.1 Radial-Basis-Function(RBF) Neural Networks

The Radial-Basis-Function Network (RBF-Net) distinguishes itself from other neural networks, e.g., the backpropagation network, by its location [21], [23]. This feature is due to the local behavior of its hidden layer transfer functions. As a consequence the accuracy of nonlinear function approximation is very high within the trained data range, but generalization is rather poor. While other neural nets have a given fixed number of neurons, the RBF-net adds new neurons depending on the complexity of the underlying problem.

The RBF-net consists of two layers: a hidden layer and an output layer [26]. The input vector of dimension N is passed to all L hidden neurons. The output

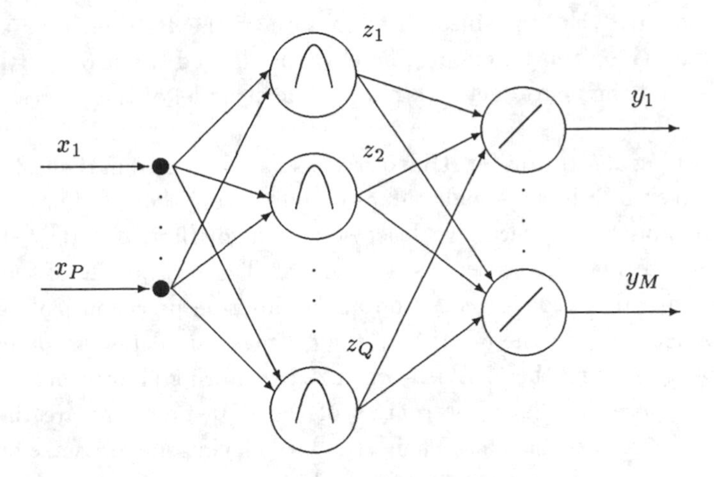

Figure 11: Radial-Basis-Function Neural Network

of each hidden node is calculated as follows:

$$z_l = \exp\left\{-\sum_{n=1}^{N}\frac{(x_n - c_{ln})^2}{\sigma_{ln}^2}\right\} \qquad l = 1, \ldots, L \tag{7}$$

where x_n denotes the n-th element of the input vector $\vec{X}$, c_{ln} the n-th element of the center vector $\vec{C}$ of the l-th hidden neuron, and σ_{ln} the n-th element of the width vector $\vec{\sigma}$ of the l-th hidden neuron. Equation (7) represents the transfer function of the hidden neurons, a bell-shaped graph basically described by its center and width. The output can assume only values between zero and one, depending on the input. The name "Radial Basis Function" stems from the radial symmetry with respect to the center.

The M output neurons exhibit linear transfer behavior:

$$y_m = \sum_{l=1}^{L} w_{ml}\, z_l \qquad m = 1, \ldots, M \tag{8}$$

where w_{ml} denotes the weight between the m-th output neuron and the l-th hidden neuron. Consequently, the output layer performs a weighted linear combination of the hidden layer outputs. Overall, the RBF-net performs a nonlinear transformation from the $\mathcal{R}^N$ to the $\mathcal{R}^M$.

The training of the RBF-net is performed in a supervised manner, i.e., for each training input pattern, a corresponding output pattern (the function to be approximated), is presented. During training, the weights, centers, and widths are adapted such that the linear combination of the RBF-neurons eventually approximates the given function. Thereby, the desired input-output behaviour is achieved and the neural network may act as a model of the process.

Several learning algorithms for RBF-networks have emerged in the last few years. They are all similar in the weight adaption employing the well-known gradient-descent method or the recursive least square algorithm, but they differ very much in how to find the center and the width. The positioning of the center is especially critical for convergence and approximation precision [38]. Often the number of neurons and the position of their centers are determined empirically, e.g., by evenly spacing them. A self-organized clustering algorithm for the input data was proposed in [29]. The centers of the RBF-functions are then placed at the centers of the clusters. Their standard deviations are set equal to the distance to the next nearest center. Since this method does not try to minimize the modeling error, this might not be appropriate for nonlinear system modeling. An adaptive learning algorithm for RBF-networks which optimizes the output errors with respect to the weights, the centers, and the widths was developed in [26]. A gradient-descent method is used to adapt all three vectors,

$$x(k+1) = x(k) - \eta_x(k)\frac{\partial \epsilon(k)}{\partial x} \qquad x = w_{ml}, c_{ln}, \sigma_{ln} \tag{9}$$

with the learning rates η_x and the instant output error index $\epsilon(k)$. By employing one-dimensional optimization for the learning rates, the convergence can be speeded up but the additional amount of computation slows down the training process.

The location property of the RBF-function is utilized by defining active and inactive neurons. If the output of a RBF-node exceeds a certain threshold, i.e., the distance between the input and the neuron's center stays below a certain value, the neuron is called active. Otherwise, it will hardly contribute to the function approximation and will therefore be called inactive. Only the neurons that are active are included in the adaption process. A minimal and maximal number of active neurons is required. Additionally, the area where RBF-neurons can be placed should be limited, depending on the range of input values to be expected. Otherwise, an excessive number of neurons are placed that do not greatly contribute to the approximation of the presented data [23]. Normally, in order to minimize the global approximation error and not only the current output error, randomly distributed input sequences are used. This can be avoided by implementing a multistep learning algorithm as proposed in [26]. A sliding window of size $\mu \in \mathcal{N}$ is used, where all μ input-output data vectors contribute to the error calculation. Thus, the data coming from a process can be used as it becomes available, which saves off-line work and enhances convergence properties.

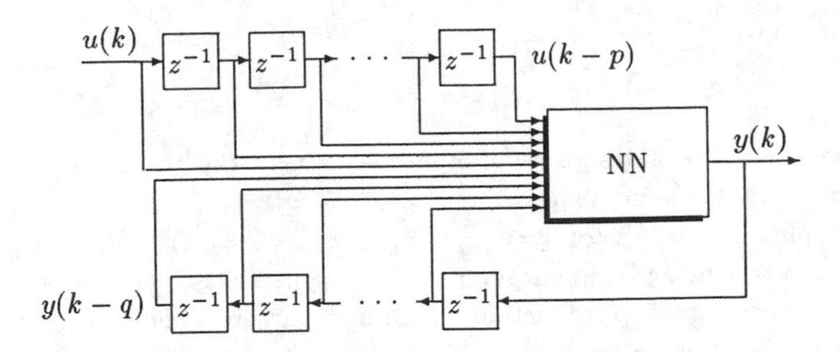

Figure 12: RBF-network structure for identifying the input-output form [19] © 1995 IEEE.

For static neural networks, such as RBF-networks, the input-output form for the nonlinear system description is considered here in order to perform the nonlinear system modeling [32], [26].

$$y(k) = g(y(k-1), ..., y(k-q), u(k), ..., u(k-p)) \tag{10}$$

A different approach to nonlinear system modeling, using dynamic neural networks such as recurrent networks, will be presented in the following section.

4.1.2 Recurrent Neural Networks (RNN)

In this chapter, recurrent neural networks, the second type of neural networks proposed for nonlinear system identification, will be employed for residual generation. This type of neural network offers the possibility of modeling dynamic nonlinear systems as given by the nonlinear state space description (Figure 13) [22], [23].

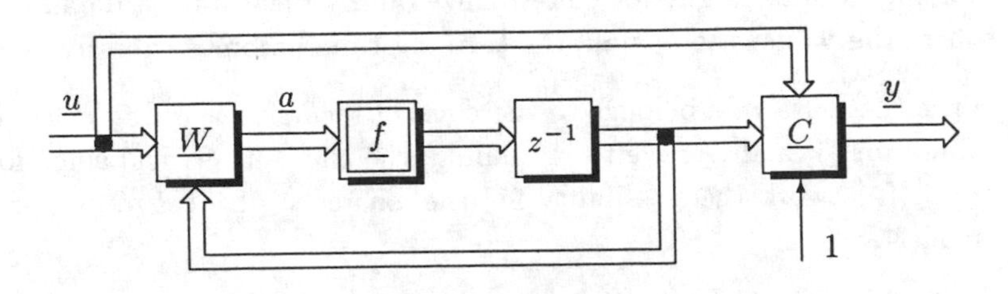

Figure 13: Recurrent neural network in vector form.

$$\underline{x}(t+1) = \underline{f}(\underline{a}(t)) \tag{11}$$
$$\underline{a}(t) = \underline{W} * \underline{\tilde{u}}(t) \tag{12}$$
$$\underline{y}(t) = \underline{C} * \underline{\tilde{x}}(t) \tag{13}$$

with the vector of the states $\underline{x}(t)$, the input vector $\underline{u}(t)$, the output vector $\underline{y}(t)$, the activation $\underline{a}(t)$, the combined input vector $\underline{\tilde{u}} = (x_1, \ldots, x_n, u_1, \ldots, u_m)^T$, and the combined state vector $\underline{\tilde{x}} = (x_1, \ldots, x_n, u_1, \ldots, u_m, 1)^T$. $\underline{W}$ and $\underline{C}$ are known as the weight and the output matrices, respectively. For the transfer function $\underline{f}(\ldots)$, the tanh(...) function has been used here. The outputs of the n neurons represent the n states.

The objective of a learning process is to adapt the elements of the weight matrix $\underline{W}$ and the output matrix $\underline{C}$ such that the desired input-output behaviour of the given dynamic system will be assumed. This training can be performed by applying the delta-rule as shown below for the weight matrix [22].

$$\underline{W}(p+1) = \underline{W}(p) + \Delta\underline{W}(p) \quad \text{with} \quad \Delta\underline{W}(p) = -\eta * \nabla_{\underline{W}} I \tag{14}$$

$$\text{which means for a single component} \quad \Delta w_{i,j}(p) = -\eta * \frac{\partial I}{\partial w_{i,j}} \tag{15}$$

The application of this learning rule involves the following major problems [22], [23]:

1. In recurrent network structures, the computational effort to calculate the gradient $\frac{\partial I}{\partial w_{i,j}}$ is much higher than for networks without feedback. Employing a combined algorithm consisting of *Backpropagation Through Time* and *Real Time Recurrent Learning* [44] the computational effort can be minimized to the order of n^3.

2. The learning algorithm is very sensitive with respect to a proper choice of the learning rate η. In the case where the selected value is too large, existing minima might not be recognized during learning, while in the case where the value is too small, the training time becomes excessively high.

3. Since the delta rule belongs to the class of recursive learning rules, initial values for $\underline{W}$ and $\underline{C}$ have to be found. The choice is critical since for this type of network the possibility for the existence of local minima is very high [2].

In the following, solutions to the above problems are presented which have been applied to the actuator benchmark problem as described in Section 4.3 of this chapter.

Computation of the Learning Rate

Using a search algorithm, the zero of the function $d(\underline{W}, \underline{C}) = I - \xi$ should be calculated [22]. ξ denotes the lower bound of the cost function to be minimized. This bound has to be placed between the actual error I and the minimal achievable error I_{min}. In order to obtain an algorithm for the learning rate in analogy to Newton's method, the difference $d(\ldots)$ is expanded in a Taylor series at the point $\underline{W}_0, \underline{C}_0$ and terminated after the first term.

$$\begin{aligned} d(\underline{W}, \underline{C}) &\approx d(\underline{W}_0, \underline{C}_0) \\ &+ \sum_{i,j} \left.\frac{\partial d}{\partial w_{i,j}}\right|_{\underline{W}_0, \underline{C}_0} * \Delta w_{i,j} + \sum_{k,l} \left.\frac{\partial d}{\partial c_{k,l}}\right|_{\underline{W}_0, \underline{C}_0} * \Delta c_{k,l} \stackrel{!}{=} 0 \end{aligned} \quad (16)$$

One possible solution to this requirement is

$$\Delta w_{i,j} = -\frac{\partial I}{\partial w_{i,j}} * \left.\frac{I(\underline{W}, \underline{C}) - \xi}{\sum_{i,j} \left(\frac{\partial I}{\partial w_{i,j}}\right)^2 + \sum_{k,l} \left(\frac{\partial I}{\partial c_{k,l}}\right)^2}\right|_{\underline{W}_0, \underline{C}_0} \quad (17)$$

$$\Delta c_{k,l} = -\frac{\partial I}{\partial c_{k,l}} * \left.\frac{I(\underline{W}, \underline{C}) - \xi}{\sum_{i,j} \left(\frac{\partial I}{\partial w_{i,j}}\right)^2 + \sum_{k,l} \left(\frac{\partial I}{\partial c_{k,l}}\right)^2}\right|_{\underline{W}_0, \underline{C}_0} \quad (18)$$

This solution for $\Delta w_{i,j}$ and $\Delta c_{k,l}$ provides a structure in which the second term can be interpreted as a computable learning rate [23].

$$\eta(I, \xi, \nabla_{\underline{W}} I, \nabla_{\underline{C}} I) = \frac{I(\underline{W}, \underline{C}) - \xi}{\sum_{i,j} \left(\frac{\partial I}{\partial w_{i,j}}\right)^2 + \sum_{k,l} \left(\frac{\partial I}{\partial c_{k,l}}\right)^2} \quad (19)$$

The bound ξ can be determined as follows [23]. By systematically decreasing the bound, the cost function is brought to a minimum and kept there in the following training steps; thereby, the cost function remains at the minimum.

Initialization by Linearization

In order to compute initial values for the weight and output matrices, the recurrent neural network description is linearized [22]. The nonlinear part of the state space description (11), (12)

$$\underline{x}(t+1) = \tanh(\underline{W} * \underline{\tilde{u}}(t)) \quad (20)$$

has to assume the linear form

$$\underline{x}(t+1) = \underline{W}^* * \underline{\tilde{u}}(t) \quad (21)$$

at the stationary point $\underline{\tilde{u}}_0$. This leads to

$$w^*_{i,j} = \left.\frac{\partial x_i(t+1)}{\partial \tilde{u}_j(t)}\right|_{\underline{\tilde{u}}_o} = (1 - \tanh^2(\sum_{j=1}^{n+m} w_{i,j} * \tilde{u}_{0_j})) * w_{i,j} \tag{22}$$

for the elements of the weight matrix $\underline{W}^*$. Considering the stationary point $\underline{\tilde{u}}_0 = \underline{0}$, this method results in $\underline{W}^* = \underline{W}$. Therefore, if a linear model exists in state space form of the nonlinear process to be investigated

$$\underline{x}_s(t) = \underline{A}_D * \underline{x}_s(t) + \underline{B}_D * \underline{u}(t) \tag{23}$$

$$\underline{y}_s(t) = \underline{C}_D * \underline{x}_s(t) + \underline{D}_D * \underline{u}(t) \tag{24}$$

giving an exact representation of the system behavior at the point of zero energy $\underline{x}_{s0} = \underline{u}_0 = \underline{0}$, then the neural network with the following choice of weight and output matrices represents a good approximation of the system at this stationary point [22].

$$\underline{W} = (\underline{A}_D, \underline{B}_D) \quad , \quad \underline{C} = (\underline{C}_D, \underline{D}_D, 0) \tag{25}$$

Starting from this initialization, further adaption of these matrices during learning leads to substantially improved modeling results.

4.2 Neural Networks for Residual Evaluation

Since residuals generally cannot be generated fully structured and robust for real processes, more sophisticated residual evaluation techniques have to be applied. Different faults may have similar effects on the residuals but should be attributed to the different causes. On the other hand, varying sizes of the same fault should be recognized as only one fault cause [23]. These issues represent problems which can be solved by advanced classification methods such as fuzzy logic or neural networks.

In order to apply neural networks for residual evaluation, first of all residuals have to exist (Figure 14) [23]. They can either be generated by another neural network as described before or by one of the analytical methods such as observers or parameter estimation.

Before applying the neural network for evaluation of these residuals, the network has to be trained for this task. For this purpose a residual data base and a corresponding fault signature data base have to exist.

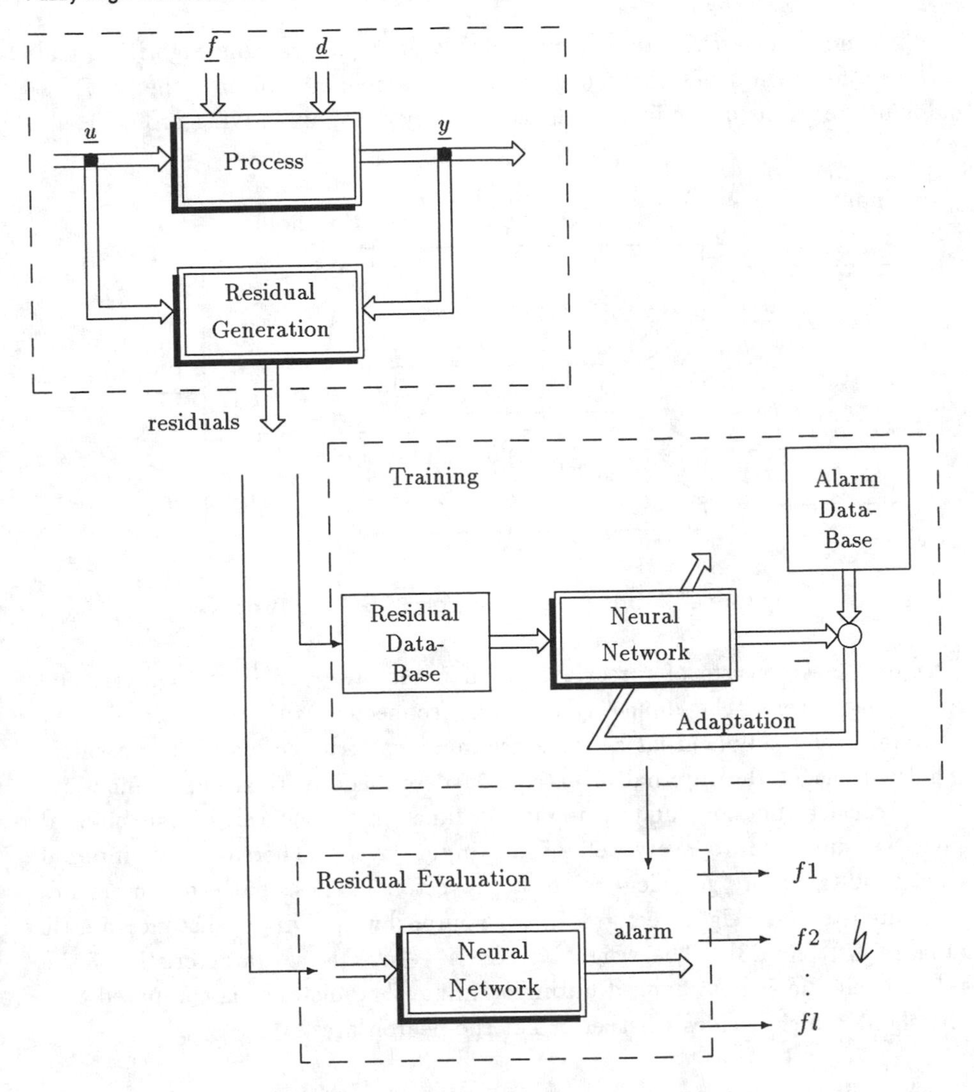

Figure 14: General scheme for off-line neural net training and on-line residual evaluation.

After completing training, the neural network can be applied for on-line residual evaluation, deciding whether or not a fault has occured and isolating which one is the propable cause.

4.2.1 Restricted-Coulomb-Energy(RCE) Neural Networks

The Restricted-Coulomb-Energy Network (RCE-Net) was presented in 1982 by Cooper, Reilly, and Elbaum [37]. The main advantages of this network are the simple and lucid architecture combined with a rapid learning algorithm [19]. While other neural nets have a given fixed number of neurons, the RCE-net adds new neurons depending on the complexity of the underlying problem.

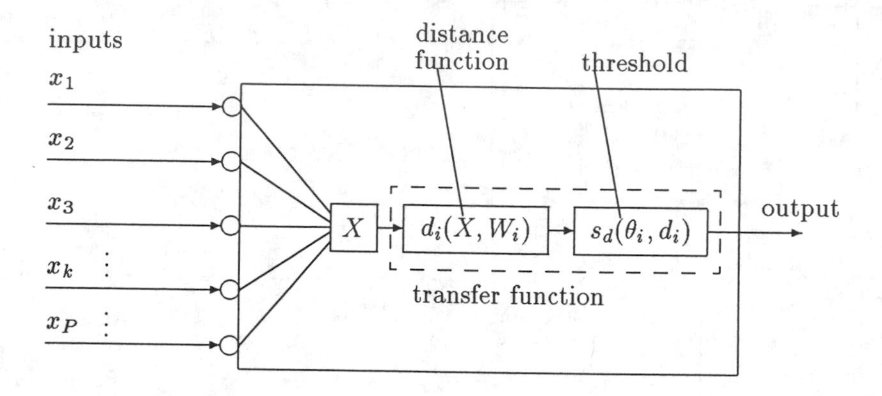

Figure 15: Hidden layer neuron of RCE network.

The RCE-net consists of three layers: an input layer, an internal layer, and an output layer. The input layer is fully connected with the internal layer containing as many neurons as there are input pattern vectors to be classified. The neurons of the internal layer can be described by their input function, their transfer function, and their output function. Each cell of the internal layer is connected to every cell of the input layer. Therefore, all internal cells simultaneously get identical pattern vectors. These pattern vectors are compared to the weight vectors of each neuron by applying a distance metric (Figure 15) [20, 23]. The weight vector $W_i \in \mathcal{R}^n$ is a characteristic of the related cell and is not changed during training. The distance is compared to a threshold, which decides whether or not the neuron fires.

The output layer is sparsely connected to the internal layer; each internal layer neuron projects its output to only one output layer cell. The number of output cells is given by the number of classes to be seperated. The output layer neurons perform a logical OR on its inputs such that, as soon as at least one internal cell is firing, the corresponding output cell will fire.

The training of the RCE-net is performed in a supervised manner, i.e., for each training input pattern, a binary output pattern is given reflecting the categories belonging to this input. Learning can be described by two distinct mechanisms, adaption of thresholds and addition of new cells [23].

4.3 Application to the Industrial Actuator Benchmark Test

The benchmark is based on an electro-mechanical test facility, which has been built at Aalborg University in Denmark [3, 33]. The equipment simulates a speed governor for large marine engines (Figure 16) [20]. The governor determines the amount of fuel loaded into the cylinders by controlling the pump efficiency. The pump efficiency is controlled through the position of a rod that is set by an actuator motor. The position is controlled by a digital controller. The actuator is a brushless synchronous DC motor that connects to a rod through an epicyclic gear train and an arm. To simulate the external load torque a similar arrangement is mounted in parallel. With this load motor a desired load torque can be programmed.

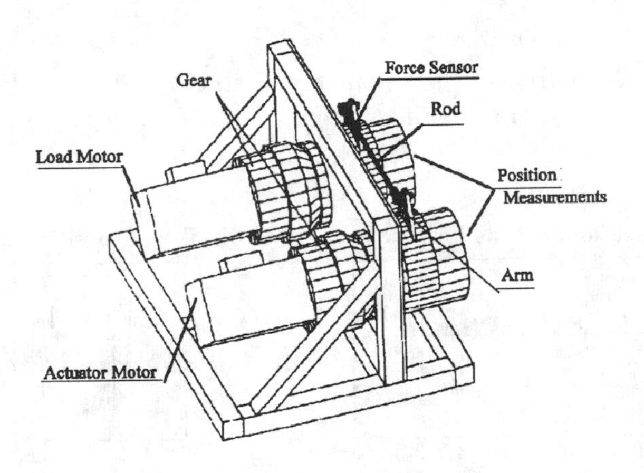

Figure 16: Industrial actuator.

Two types of faults have been considered. The first is a position sensor fault where the wiper of a feedback potentiometer loses contact with the resistance element. The second type is an actuator current fault due to a malfunction of an end-stop switch, as caused by a broken wire or a defect in the switch element due to heavy mechanical vibration. As a result of this fault the power drive can deliver only positive current. Since both faults are intermediate and last for only a very short time, they are difficult to detect by the operating personnel.

4.3.1 Simulation Results for Residual Generation

Radial-Basis-Function Neural Networks

For system modeling and residual generation, two Radial-Basis-Function neural networks, to estimate the process outputs gear output position s_0 and motor shaft velocity n_m, respectively, were designed. Training was performed with noisy data from different reference input situations, different load torque behavior, and different fault appearances with respect to time.

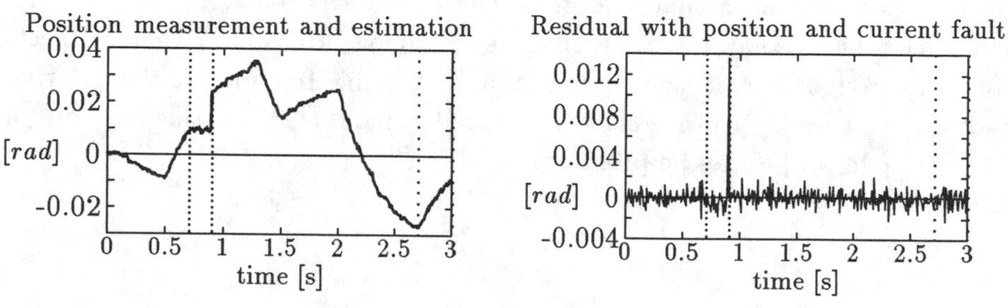

Figure 17: Comparison of position measurement and estimation and residual with position and current fault.

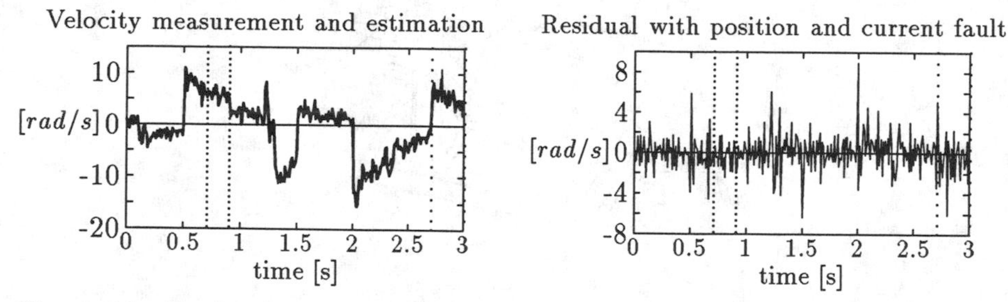

Figure 18: Comparison of velocity measurement and estimation as well as residual with position and current fault.

The results presented in Figures 17 and 18 come from a noisy data set not used during training. A position fault occured from $t = 0.7 - 0.9s$ and the current fault occured at $t = 2.7 - 3.0s$ [23]. Here, and in the following, the thick line corresponds to the measured data and the thin line to the estimate.

In particular, the residuals demonstrate the very good process modeling ability of this type of neural network. Two typical problems in fault diagnosis can also be observed. The first one is that the estimate adapts itself to sensor faults if the model is used in an observer-like structure. This effect can be seen in the case of the position fault. The second one is that a fault that has no effect on the available measurements cannot be detected, which is the case for the current fault in this reference/load situation.

Recurrent Neural Networks

Again two neural networks, this time of recurrent structure, were designed and applied to estimate the process outputs gear output position s_0 and motor shaft velocity n_m, respectively. Each network consisted of only three neurons which essentially reduces the training effort and guarantees on-line applicability, even with an additional algorithm for residual evaluation [22], [23]. Training and testing was performed with noisy data from different reference input situations, different load torque behavior, and different fault appearances with respect to time.

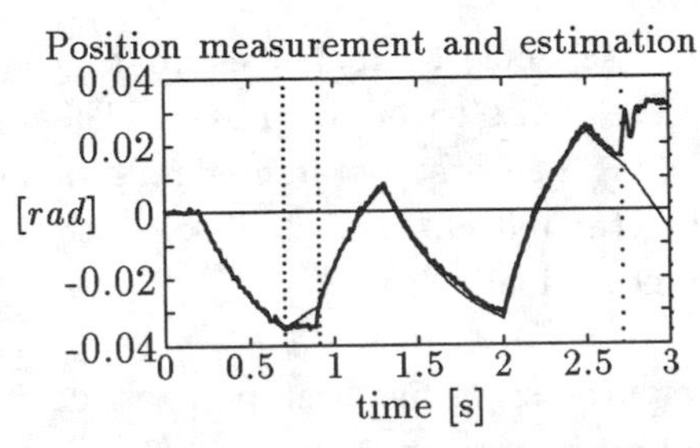

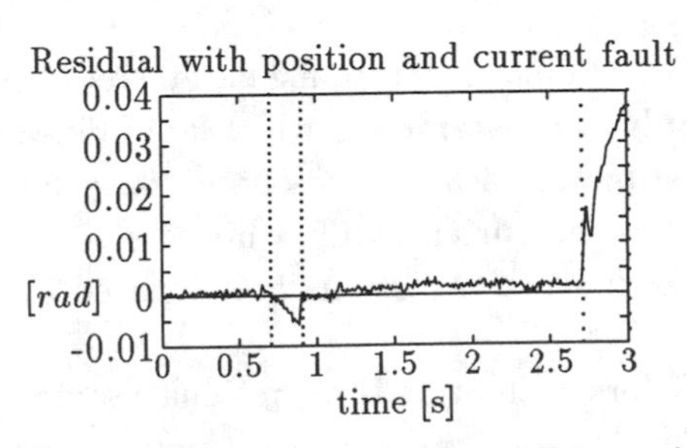

Figure 19: Comparison of position measurement and estimation as well as residual with position and current fault.

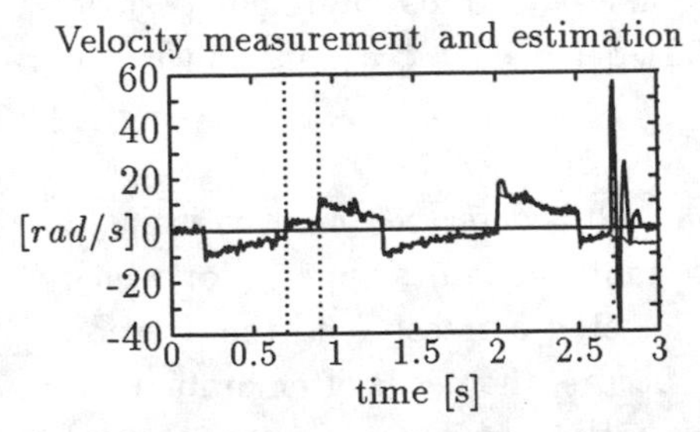

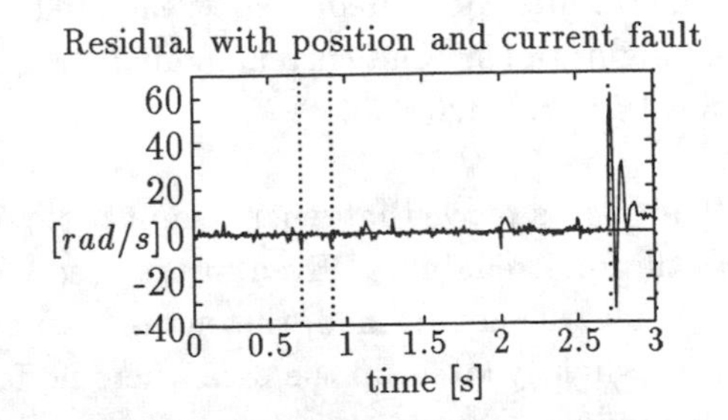

Figure 20: Comparison of velocity measurement and estimation as well as residual with position and current fault.

The results presented show a comparison of the process measurements and the estimated signals and prove the excellent modeling ability of the proposed neural network. This capability is essentially due to the improved learning algorithm and the systematic network initialization as described above. For fault detection purposes, two different faults were implemented in this simulation; a position fault occurring at $t = 0.7 - 0.9s$ and a current fault occurring at $t = 2.7 - 3.0s$ [23]. The results demonstrate that the position fault had only a small effect on the position measurement which leads to a merely observable deviation between the measurement and the estimate. On the other hand the impact of the current fault on both measurements is much higher yielding residuals which would allow a reliable fault detection. In these kinds of cases, where the effect of the faults varies greatly with the current operating conditions, an intelligent residual evaluation using fuzzy logic or, again, neural networks is required for fault detection and isolation [19].

4.3.2 Simulation Results for Residual Evaluation

Restricted-Coulomb-Energy Neural Network

For residual generation a parameter identification scheme was applied estimating only two parameters which are influenced by the faults to be diagnosed. The estimation scheme was based on a linear model. The first parameter reflects only the current fault while the second parameter reflects both faults. Both residuals are additionally influenced by an unknown load torque.

Before evaluating these residuals some signal preprocessing has been performed. The training was carried out using a number of residual time series for different reference input situations, different load torque behavior and different fault appearances with respect to time.

The results presented were generated for different kinds of reference/load configurations. The current fault in Figure 21 ocurred at $t = 2.0s$ and lasted until $t = 2.3s$ [20, 23].

The results prove that when a fault occurs the fault is detected and even isolated with great reliability. Even in the case where multiple faults appear, for which the net was not trained, the current fault is detected correctly and the position fault slightly later. In the case where no fault occurs, the net is often ambiguous due to load changes and some time delay between the time of fault occurence and its appearance in the parameter estimates. Nevertheless, no false alarms were thereby produced, therefore, the ambiguous state can be viewed as belonging to the no-fault case which leads to an excellent diagnostic performance of the proposed scheme.

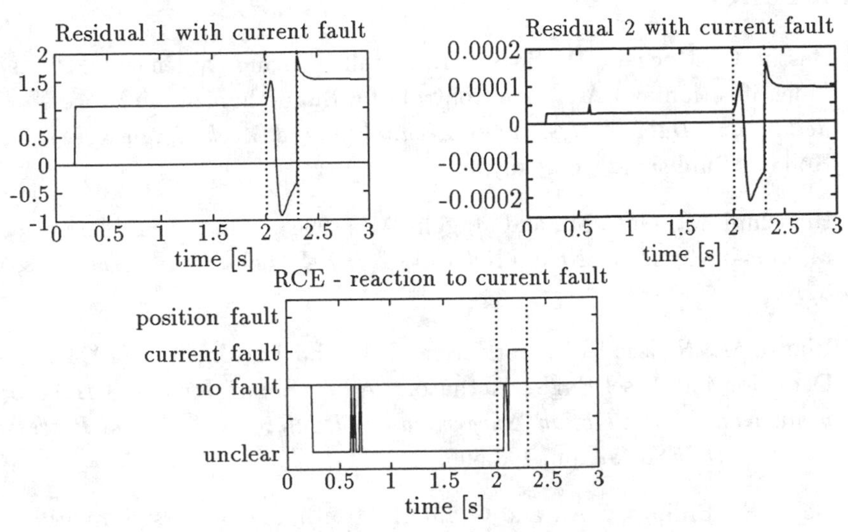

Figure 21: Residuals with current fault and reaction of the RCE-network.

5. Conclusions

Following recent trends in fault diagnosis, the application of fuzzy logic and neural networks to fault detection and isolation is presented. Fuzzy logic is employed in the framework of a combined quantitative/knowledge-based approach. The design algorithm for a residual evaluating fuzzy filter is discussed in general and applied to a wastewater plant.

Neural networks can be used in the context of residual generation as well as residual evaluation. In this contribution two types of neural networks suitable for system modeling and, therefore, for residual generation are described. A third network type is presented for residual evaluation. Their application to an actuator benchmark problem proves the applicability of the proposed schemes.

References

[1] Bastin, G., Dochain, D., Haest, M., Installe, M. and Opdenacker, P. (1983), "Identification and Adaptive Control of a Biomethanization Process," *Modelling and Data Analysis in Biotechnology and Mech. Engineering*, North Holland Publishing Company.

[2] Bianchini, M., Gori, M., and Maggini M. (1994), "On the Problem of Local Minima in Recurrent Neural Networks," *IEEE Trans. Neural Networks*, Vol. 5, No. 2.

[3] Blanke M., Nielsen, S.B., Jørgensen, R.B. and Patton, R.J. (1994), "Fault Detection for Diesel Engine Actuator - A benchmark for FDI," *IFAC Symposium on Fault Detction, Supervision and Safety in Technical Processes - SAFEPROCESS '94*, pp. 498-506.

[4] Chen, S., Billings, S.A., and Grant, P.M. (1992), "Recursive hybrid algorithm for non-linear system identification using radial basis function networks," *International Journal of Control*, Vol. 55, No.51, pp. 1051-1070.

[5] Frank, P.M. (1990), "Fault diagnosis in dynamic systems using analytical and knowledge-based redundancy - A survey," *Automatica*, Vol. 26, pp. 459-474.

[6] Frank, P. M. (1996), "Analytical and qualitative model-based fault diagnosis - A survey and some new results," *European Journal of Control*, Vol.2, No.1.

[7] Gertler, J. (1991), "Analytical redundancy methods in fault detection and isolation," *Proc. of the IFAC/IMACS Symposium Safeprocess*, Baden-Baden, pp. 9-21.

[8] Gujer, W. and Zehnder, A. (1983), "Conversion processes in anaerobic digestion," *Water Science and Technology*.

[9] Himmelblau, D. M., Braker, R. W. and Suewatanakul, W. (1991), "Fault Classification with the aid of artificial neural networks," *IFAC/IMAC-Symposium Safeprocess'91* Baden-Baden, pp 369-373, Sept 10-13.

[10] Himmelblau, D. M. (1992), "Use of artificial neural networks to monitor faults and for troubleshooting in the process industries," *IFAC Symposium On-line fault detection and supervision in the chemical process industry*, Newark, Delaware, U.S.A., April 22-24.

[11] Hoskins, J. C., Kaliyur, K. M., and Himmelblau, D. M. (1991), "Fault diagnosis in complex chemical plants using artifical neural networks," *AIChE J.*, 37, 137-142.

[12] Isermann, R. (1994), "Integration of Fault Detection and Diagnosis Methods," *Safeprocess '94*, pp. 597-612, Helsinki.

[13] Kandel, A. (1986), "Fuzzy mathematical techniques with applications," Addison–Wesley.

[14] Kiupel, N., Köppen-Seliger, B., Schulte Kellinghaus, H. and Frank, P. M. (1995), "Fuzzy Residual Evaluation Concept (FREC)," *IEEE Int. Conf. on Systems, Man, and Cybernetics*, Vancouver, Canada.

[15] Kiupel, N. and Frank, P. M. (1996), "Fuzzy supervision for an anaerobic wastewater plant," *CESA'96 IMACS Multiconference*, Lille, France.

[16] Kiupel, N. and Frank, P. M. (1996), "An algorithm for a filter design for fuzzy supervision," *World Automation Congress, WAC'96*, Montpellier, France.

[17] Kiupel, N. (1997), "Fuzzy-Logik-basierte-Fehlerdiagnose am Beispiel eines anaeroben Abwasserreinigungsprozesses," *Dissertation*, Gerhard-Mercator-Universität- GH Duisburg, Fachgebiet Meß- und Regelungstechnik, VDI Fortschrittberichte, Reihe 8, Nr. 627.

[18] Kiupel, N. and Frank, P. M. (1997), "A Fuzzy FDI Decision Making System for the Support of the Human Operator," *Safeprocess '97*, Hull, U.K.

[19] Köppen-Seliger, B. and Frank, P. M. (1995), "Fault Detection and Isolation in Technical Processes with Neural Networks," *34th Conference on Decision and Control*, New Orleans, pp. 2414 - 2419, December.

[20] Köppen-Seliger, B., Frank, P. M., and Wolff (1995), "Residual Evaluation for Fault Detection and Isolation with RCE Neural Networks," *American Control Conference*, Seattle, pp. 3264 - 3268.

[21] Köppen-Seliger, B. and Frank, P. M. (1996), "Neural Networks in Model-Based Fault Diagnosis," *13th IFAC World Congress*, San Francisco, Vol. N, pp. 67-72.

[22] Köppen-Seliger, B., Schubert, M., and Frank, P. M. (1996), "Recurrent Neural Networks for Fault Detection," *EUFIT'96*, Aachen, pp. 240-244.

[23] Köppen-Seliger, B. (1997), "Fehlerdiagnose mit künstlichen neuronalen Netzen," *Dissertation*, Gerhard-Mercator-Universität- GH Duisburg, Fachgebiet Meß- und Regelungstechnik, VDI Fortschrittberichte, Reihe 8, Nr. 632.

[24] Kus, F. (1993), "Kinetik des anaeroben Abbaus von Essig– und Propionsäure in Bioreaktoren mit immobilisierten Bakterien," Ph.D. thesis, Technische Universität Berlin.

[25] Lee, R.C.T. (1993), "Fuzzy Logic and the Resolution Principle," *Readings in Fuzzy Sets for Intelligent Systems.*

[26] Liang, F. and ElMaraghy, H.A. (1993), "Multistep localized adaptive learning RBF networks for nonlinear system indentification," *European Control Conference ECC'93* Groningen, The Netherlands, June 28–July 1.

[27] Mather, M. (1986), "Mathemathische Modellierung der Methangärung," Ph.D. thesis, Technische Universität München.

[28] Moletta, R., Verrier, D. and Albagnac, G. (1986), "Dynamic Modelling of Anaerobic Digestion," *Wat. Res.*, Vol. 20 No. 4, Pergamon Press.

[29] Moody, J. and Darken, C. (1989), "Fast learning in networks of locally-tuned processing units," *Neural Computation*, Vol. 1, pp.281-294.

[30] Mosey, F. E. (1983), "Mathematical modelling of the anaerobic digestion process: Regulatory mechanisms for the formation of short chain volatile acids from glucose," *Water Science and Technology.*

[31] Mukaidono, M. (1985), "Representation of Fuzzy data with Fuzzy logic expressions," North Holland.

[32] Narendra, K.S. and Parthasarathy, K. (1990), "Identification and Control of Dynamical Systems Using Neural Networks," *IEEE Trans. Neur. Networks*, Vol. 1, No. 1, pp. 4-27.

[33] Nielsen, S.B., Patton, R.J., Blanke, M., and Jørgensen, R.B. (1993), "Industrial actuator benchmark test," Distributed document, Aalborg & York Universities.

[34] Patton, R.J., Frank, P.M. and Clark, R.N. (1989), (Eds.), "Fault diagnosis in dynamic systems, theory and application," Prentice Hall.

[35] Patton, R.J. (1993), "Robustness Issues in Fault Tolerant Control," *TOOLDIAG'93 Int. Conf. on Fault Diagnosis*, Toulouse, France.

[36] Rappl, C. (1994), "Anwendung eines modellgestützten Meß -und Regelungsverfahrens beim anaeroben Essigsäureabbau," Ph.D. thesis, Technische Universität Berlin.

[37] Reilly, D.L. and Cooper, N. (1990), "An Overwiew of Neural Networks: Early Models to Realworld Systems" in *An Introduction to Neural and Electronic Networks*, Academic Press Inc., San Diego.

[38] Saha, A. and Keeler, J.D. (1990), "Algorithms for Better Representation and Faster Learning in Radial Basis Function Networks," *Advances in Neural Information Processing Systems 2*, Morgan Kaufmann Publishers, Inc., pp. 482-489.

[39] Sorsa, T. and Koivo, H.N. (1991), "Application of artifical neural networks in process fault diagnosis," *IFAC/IMAC-Symposium on fault detection supervision and saftey for technical Processes Safeprocess'91*, Baden-Baden, 10-13 Sept. 91.

[40] Sorsa, T., Suontausta, J., and Koivo, H.N. (1993), "Dynamic fault diagnosis using radial basis function networks," *TOOLDIAG'93*, Toulouse, April 5-7.

[41] Tzafestas, S.G. and Dalianis, P.J. (1994), "Fault Diagnosis in Complex Systems using Artificial Neural Networks," *The Third IEEE Conference on Control Applications*, pp. 877-882, Glasgow, August 24-26.

[42] Venkatasubramanian, V. and Chan, K. (1989), "A neural network methodology," *AIChE J.*, 35, 1993-2002.

[43] Willsky, A. S. (1976), "A survey of design methods for failure detection in dynamic systems," *Automatica*, 12, 601-611.

[44] Zell, A. (1994), "Simulation Neuronaler Netze," Addison-Wesley Publ. Comp.

Chapter 8:

Application of the Neural Network and Fuzzy Logic to the Rotating Machine Diagnosis

APPLICATION OF THE NEURAL NETWORK AND FUZZY LOGIC TO THE ROTATING MACHINE DIAGNOSIS

Makoto Tanaka
The Chugoku Electric Power Co., Inc.
Japan

Rotating machines used in power plants and factories require regular maintenance to avoid a failure leading to a halt of activities at the plant. To perform efficient maintenance, a highly accurate diagnosis system is required. To achieve high accuracy diagnosis, an effective feature selection from vibration data and an effective and accurate fault diagnosis algorithm are required. In this chapter, we introduce the application of neural networks and fuzzy logic to rotating machine fault diagnosis.

1 Introduction

Rotating machines are used in turbines, generators, and pumps. In order to avoid catastrophic failure and perform efficient maintenance, many rotating machine diagnosis systems have been developed (Watanabe, et al., 1981; Yamaguchi, et al., 1987; Nakajima, 1987; Yasuda, et al., 1989; Yamauchi, et al., 1992; Hashimoto, 1992).

Rotating machine diagnosis systems are designed to detect an abnormal condition and estimate the cause of a failure. Using a rotating machine diagnosis system, we can expect a change in the maintenance style from traditional TBM (Time Based Monitoring; the maintenance is performed in a constant time cycle) to CBM (Condition Based Monitoring; the maintenance is performed according to the degradation degree of the equipment). However, condition based monitoring is difficult because CBM requires a highly accurate diagnosis which is not available in the current diagnosis systems of rotating machines.

0-8493-9804-5/99/$0.00+$.50

In this chapter, we introduce the rotating machine diagnosis system and the conventional fault diagnosis technique. Then, we describe the application of neural networks and fuzzy logic to the rotating machine diagnosis system.

2 Rotating Machine Diagnosis

Techniques used in rotating machine diagnosis are classified by the type of machine. In this chapter, we describe the diagnosis technique for small rotating machines (for example, motor, fan, pump). Small rotating machines use the rolling element bearing (REB) for holding the rotor and this bearing is degraded with age; therefore, replacement of the REB is required from time to time. The rotating motor load is typically a fan and a pump. If the rotating axis of the load and the motor don't coincide or there is some unbalance in the load, premature failure of the REB is likely.

The types of faults in rotating machines are also discussed. When failure occurs, the vibration signature of the machine changes and the occurrence of a fault can be detected by measuring the mechanical vibration of the machine.

Figure 1 shows the block diagram of a rotating machine diagnosis system, and Figure 2 shows the vibration waveform after occurrence of the fault. The rotating machine diagnosis system selects the feature of the failure from the mechanical vibration data and estimates the degree and cause of the fault.

Usually, the rotating machine diagnosis system has the following main functions:

- Monitoring of the degradation degree of the rotating machine and detection of the fault.
- Identification of the fault.

These two functions correspond to the following diagnosis technique in the field of equipment check.

- Machine Surveillance Technique (MST): the judgment of which machine is normal or faulty.
- Precision Diagnosis Technique (PDT): the judgment of the cause of machine fault.

In order to efficiently monitor and diagnose a large number of rotating machines, a rough monitoring of the fault is performed by MST, followed by a PDT performed on only the equipment diagnosed as faulty. In the following section, we briefly describe the fault diagnosis techniques for MST and PDT.

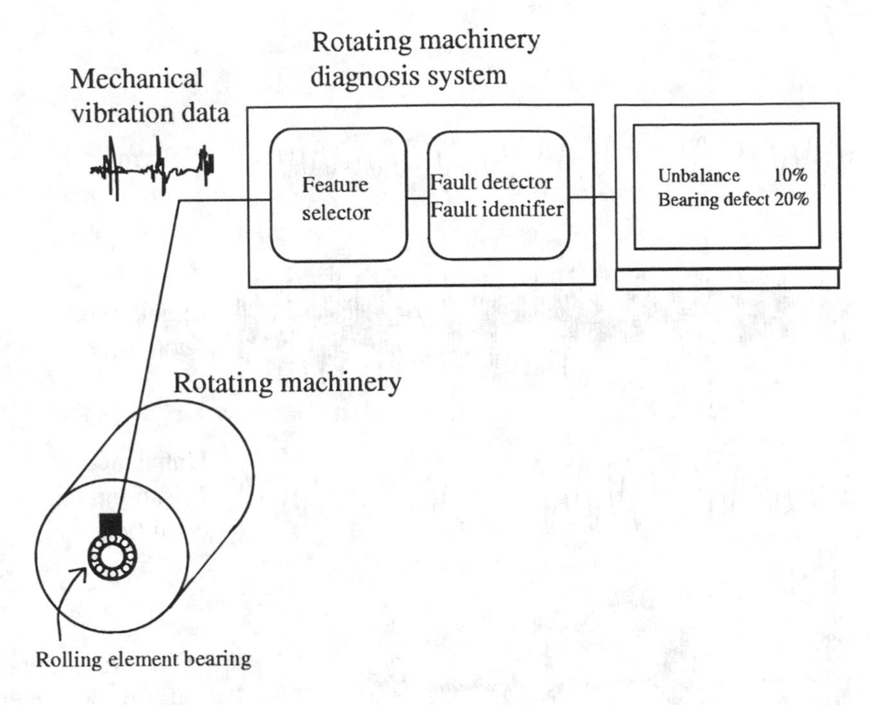

Figure 1: The rotating machine diagnosis system.

2.1 Fault Diagnosis Technique for Rotating Machines

In machine surveillance, the degree of the fault of the rotating machine is approximately represented by the magnitude of the vibration, and the root-mean-square value of the vibration data is generally used. Figure 3 shows a typical change in magnitude of vibration data with time. If the degree of the fault increases, then the magnitude of the vibration will increase. Therefore, we can measure the fault degree of the equipment by cyclic monitoring of the vibration magnitude.

In order to judge the normal or fault conditions, we check the absolute level or the trend of the vibration data. MST uses a simple algorithm for judgment and a high-speed diagnosis.

On the other hand, in the precision diagnosis technique, we use frequency analysis where the vibration data is transformed to the power spectrum by Fourier-Transform, and some features of the fault are selected.

Rotating machines may have many causes of fault. Vibration power concentrations in the frequency domain of a fault were analyzed in detail in past research (ISIJ, 1986). For example, when an unbalance condition occurs, the power of the

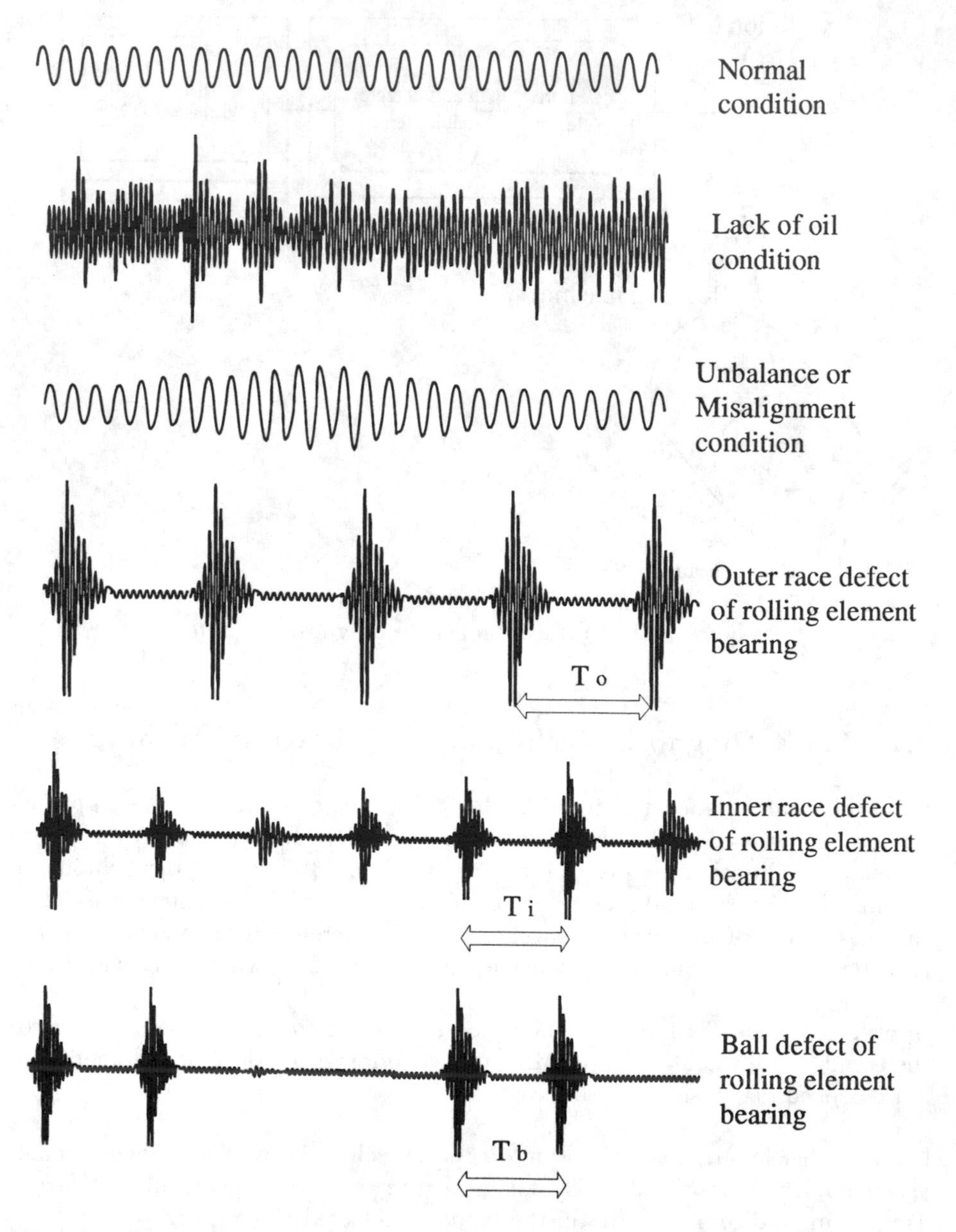

Figure 2: The vibration waveform of the rotating machine in the fault condition.

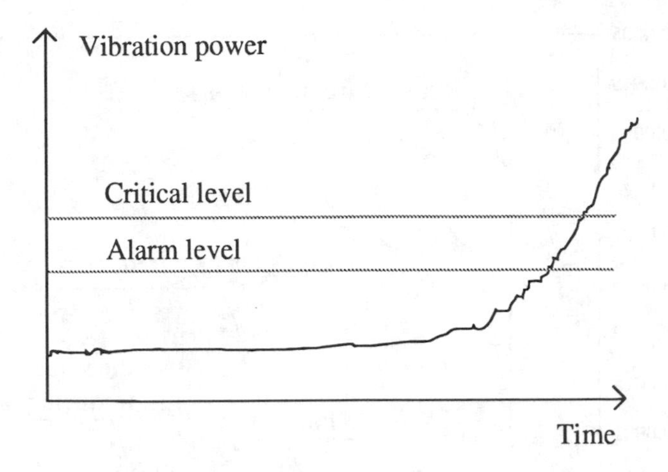

Figure 3: The change of the vibration magnitude in time.

rotating frequency component increases. When a misalignment condition occurs (the phenomenon where the rotating axle of the rotating machine shifts from the mechanical center), power of the second harmonic of the rotating frequency increases. When the defect occurs in the rolling element bearing, if we assume that there is only one spot defect, the frequency component of the impulse vibration which is calculated by Equations (1) ~ (3) increases.

- For the inner race defect of the rolling element bearing

$$f_i = \frac{f_r}{2}\left(1 + \frac{d}{D}\cos\alpha\right)z \tag{1}$$

- For the outer race defect of the rolling element bearing

$$f_o = \frac{f_r}{2}\left(1 - \frac{d}{D}\cos\alpha\right)z \tag{2}$$

- For the ball defect of the rolling element bearing

$$f_b = \frac{f_r}{2}\frac{D}{d}\left(1 + \frac{d}{D}\cos\alpha\right)z \tag{3}$$

where, f_r is the rotating frequency of the axle (inner race) (Hz), D is the diameter of the pitch circle of the rolling element bearing (mm), d is the diameter of the ball (mm), α is the contacting angle(deg), and z is the number of balls.

The rotating machine diagnosis system performs the feature selection of the fault through theoretical and phenomenological analysis. The mechanical vibration

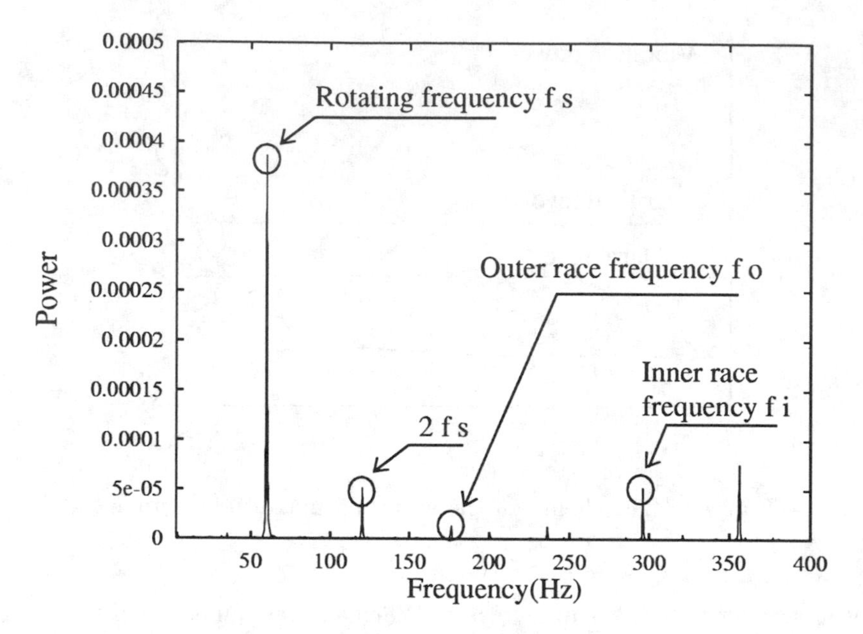

Figure 4: The feature selection from the power spectrum.

data is transformed by Fourier-Transform, and from its power spectrum, we select the components corresponding to the rotating frequency, to the frequency corresponding to Equations (1) $\sim$ (3), and to some multiple of them (refer to Figure 4).

In order to detect the rolling element bearing defect, frequency analysis of the enveloping filter (Figure 5) is used. This method has a low detection accuracy for detecting unbalance and misalignment conditions, but is useful for detecting a bearing defect, which has impulse vibrations represented by Equation (1) $\sim$ Equation (3) (refer to Table 1).

These selected features are diagnosed by the fault diagnosis method using the causal matrix, and the cause of fault is estimated. The causal matrix is shown in Table 2, which represents the correspondence of the cause of fault and the selected feature.

Selected power spectra are used for the estimation of the type of fault using the diagnosis algorithm shown in Figure 6. The diagnosis algorithm performs the multiplication and summation for the selected feature power spectra and the causal matrix parameters. Moreover, it can estimate the type of fault using the total calculated value for the fault. The fault with the highest total calculated value point is evaluated with the highest certainty as the cause of the fault.

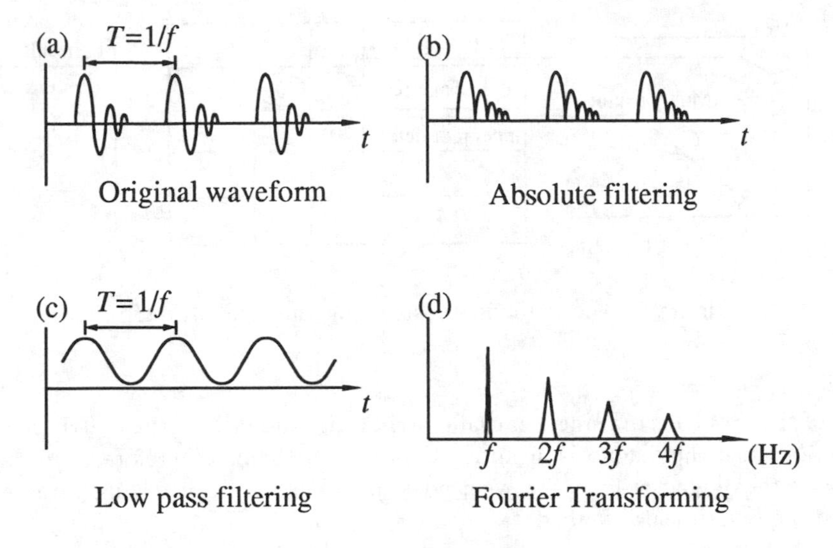

Figure 5: Signal processing method of the envelope filter.

Table 1: Corresponding to the type of fault and the feature selection method

The type of fault	Feature selection method	
	FFT	Envelope filter + FFT
Unbalance	○	×
Misalignment	○	×
Bearing inner race defect	△	○
Bearing outer race defect	△	○
Bearing ball defect	△	○
Lack of oil	○	○

○ : Detectable with high accuracy △ : Detectable with low accuracy × : Not detectable

Table 2: An example of the causal matrix

The type of fault	$0 \sim f_s$	f_s	$2f_s$	$3f_s$	f_i	f_o	$3f_s \sim 8f_s$	$8f_s \sim$
Unbalance	0	80	20	0	0	0	0	0
Misalignment	0	20	80	0	0	0	0	0
Bearing inner race defect	0	0	0	0	100	0	0	0
Bearing outer race defect	0	0	0	0	0	100	0	0
Lack of oil of bearing	0	0	0	0	0	0	50	50

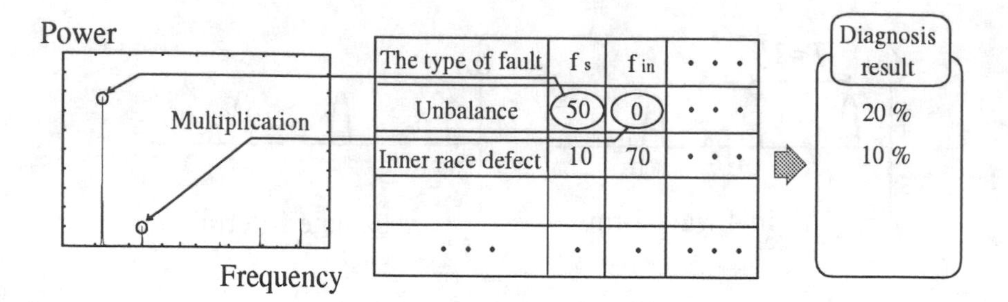

Figure 6: The fault diagnosis using the causal matrix.

One of the problems in using a causal matrix is dependence on the matrix parameters, because the diagnosis accuracy depends on them. Therefore, in order to improve the diagnosis accuracy, we need some method for optimizing parameters based on experimental fault data.

3 Application of Neural Networks and Fuzzy Logic for Rotating Machine Diagnosis

As mentioned above, in rotating machine diagnosis, the fault diagnosis is performed using selected features from the vibration data. Fault detection is generally performed by the simple judgment method using the absolute threshold value, and the fault identification is performed by a linear classifier using the causal matrix.

In the fault diagnosis, diagnosis accuracy depends upon the suitability of the selected feature and the accuracy of the classifier. The suitability of the selected feature is important to detect the incipient fault, whereas, the performance of the classifier is important to estimate the type of the fault with high certainty.

In this section, we present several examples of using neural network and fuzzy logic applied to the feature selection and the fault identification methods in the rotating machine diagnosis. Neural networks can learn the vibration data from several fault conditions, and we can construct a highly convenient diagnosis system with high accuracy using a neural network. Moreover, we can construct a high reliably diagnosis algorithm using fuzzy logic which can treat vagueness. In the following section, we briefly describe these techniques.

3.1 Fault Diagnosis Using a Neural Network

To improve the accuracy of rotating machine diagnosis, many parameters of the causal matrix must be optimized. Each machine has its own causal matrix, because the vibration characteristics differ from one rotating machine to the other, and the parameter adjustment requires skill and manpower.

A neural network can perform a role similar to the causal matrix, since it can learn the vibration data of fault conditions.

In this section, we describe fault diagnosis methods using the feed-forward network, which is widely used for the rotating machine diagnosis.

Figure 7 shows the feed-forward neural network structure which is generally used in rotating machine diagnosis (Iwatsubo, et al., 1992; Aleguindigue, et al., 1993). The power spectrum of the vibration data is input to the input layer, and the signal with the fault condition is given as the supervised signal to the output layer. We use the digital signal (0:non-fault, 1:fault) as output data and the mechanical vibration data obtained from experimental studies of artificial faults as input data. As a learning method, we use the error back-propagation method, where the network is trained until the error has decreased sufficiently.

The diagnosis algorithm using a feed-forward network has the ability to recognize the learned data. However, learning the power spectrum pattern requires a lot of the time, because the number of the input nodes is high (several hundred nodes).

To overcome this problem, we may use neural network to reduce the number of features, as follows.

A recirculation network is used for the feature selection (Aleguindigue, et al., 1993). Figure 8 shows the simplest version of a recirculation network in which two layers operate as signal buffers (input layer and output layer) and two layers are trainable (visible layer and hidden layer). We supply the same power spectrum to input and output layers, and the network learns to operate as the identifier of mapping. After the learning, the compressed representation of the input signal is present in the hidden layer.

This compressed information is given to the classifier network as mentioned above. Original input data has a dimension of several hundreds, and a recirculation network can compress the amount of information to approximately one third of the original amount.

Fault diagnosis using neural networks has a diagnosis ability better than or equal to the conventional method using a theoretical feature selection and a causal matrix. In neural network fault diagnosis, the parameters are automatically

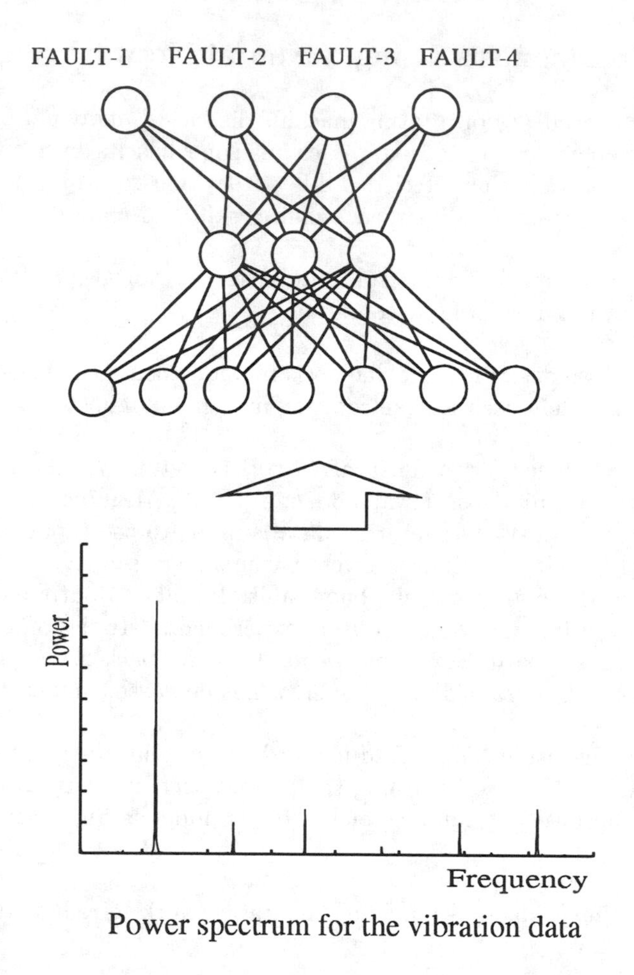

Figure 7: The fault diagnosis using a feed-forward network.

adjusted using the learning ability of the neural network, which reduces the manpower requirements.

In order to improve the diagnosis accuracy, we have to prepare a large amount of learning data; especially, that the fault condition of the actual machine is more complex than the experimental fault condition, and the actual degree of fault is slightly smaller. Therefore, the data collection process of the actual machine is important.

To summarize, diagnosis systems using the conventional diagnosis algorithm with feature selection and the causal matrix requires manual adjustment of the causal matrix parameters. If we use neural networks, this adjustment becomes

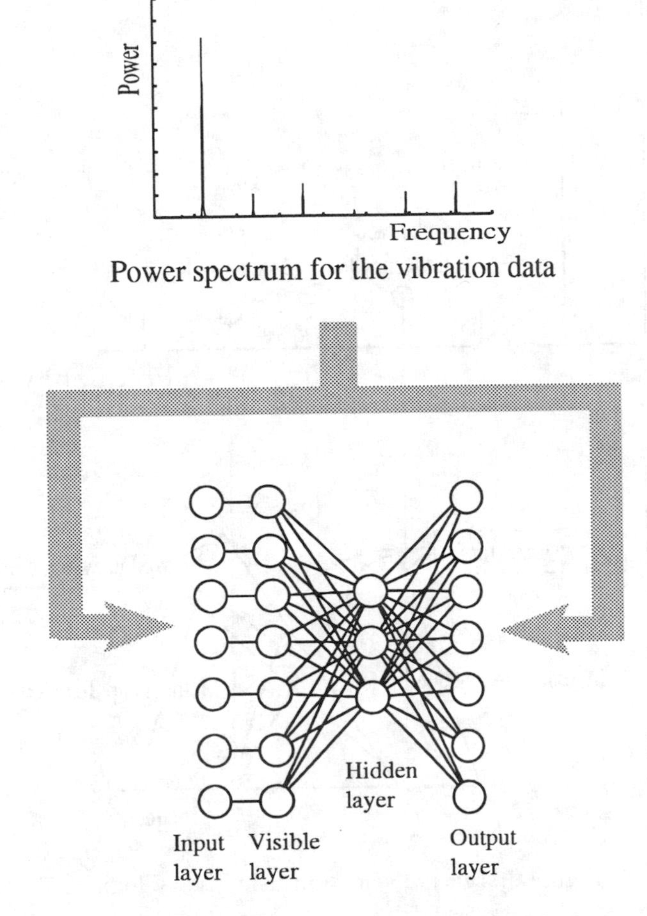

Figure 8: The data compression using the recirculation network.

automatic, leading to a better diagnosis, but neural networks require collection of a large amount of fault condition data for their learning.

3.2 Fault Diagnosis Using Fuzzy Logic

For diagnosis, we need to select features from the power spectrum using the theoretical frequency calculated by Equations (1) $\sim$ (3), rotating frequency, and so on. However, the rotating speed of the machine is changed by the load, and the power spectrum shifts in the frequency domain. Therefore, a simple feature selection logic which searches for only the frequency position has the possibility of missing features.

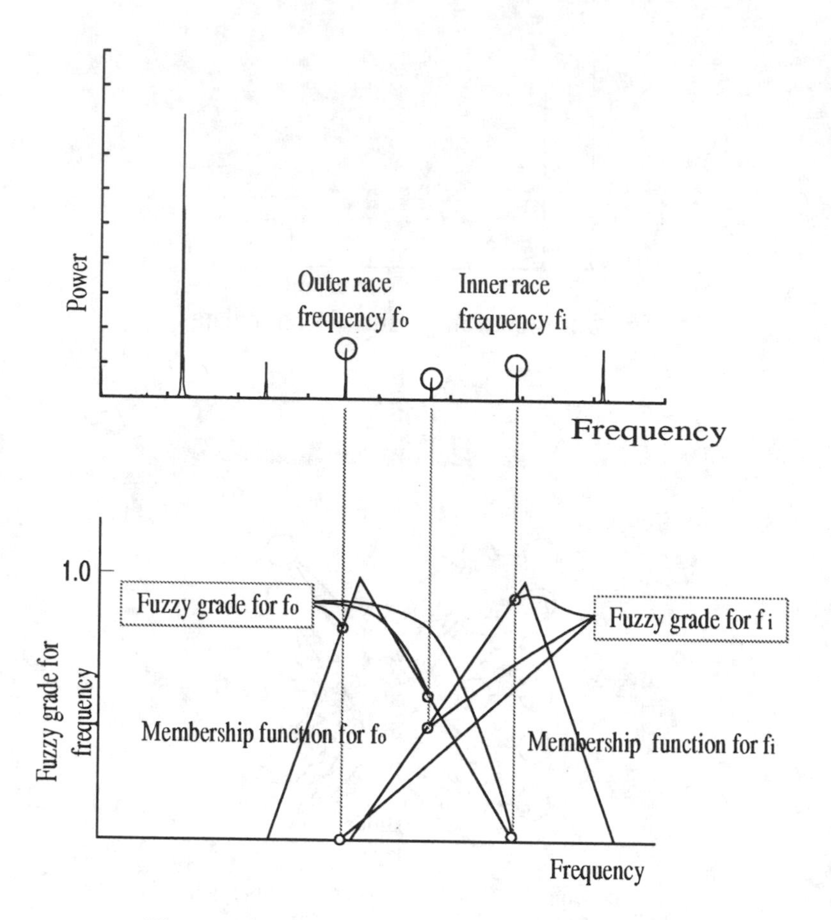

Figure 9: Feature selection using fuzzy logic.

In such a case, fuzzy logic may be used as a feature selection method which can select features from the frequency shifted power spectrum (Hinami, et al., 1991). The fuzzy grade of each frequency position of a peak vibration power in the power spectrum is calculated using the membership function as shown in Figure 9.

Similarly, the fuzzy grade of the fault degree of each peak power is calculated using the membership function in Figure 10. For each peak power, we multiply its fuzzy grade for frequency by its fuzzy grade of fault degree. Then, for each feature, we calculate the feature fuzzy grade as the summation of those outcomes (refer to Figure 11). In this method, the final calculated feature directly corresponds to the fault. Therefore, the final feature diagnoses with certainty the vibration data for one fault.

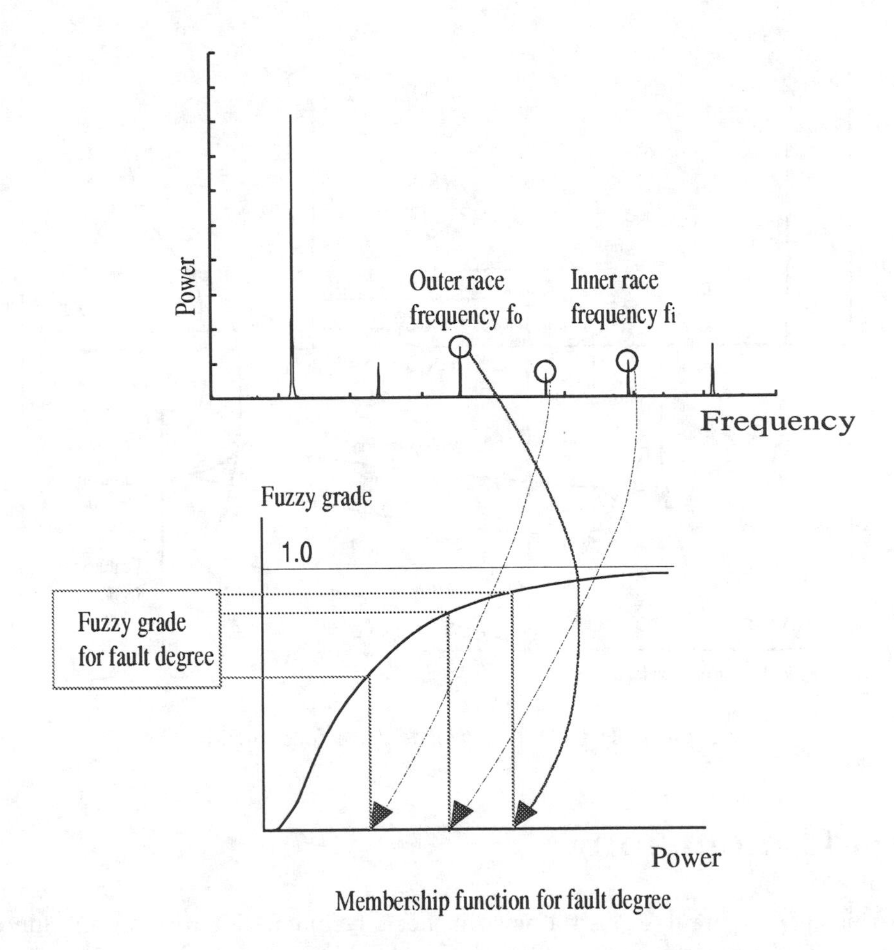

Figure 10: Calculation of the fault degree using fuzzy logic.

Fault diagnosis using fuzzy logic can accommodate frequency shifting of the power spectrum depending on the operating condition of the machine. However, the diagnosis ability depends on the shape of the membership function. Therefore, deciding the shape of the membership function is important, which is a problem similar to the conventional diagnosis method using the causal matrix. Therefore, in order to improve the diagnosis accuracy, the adjusting of fuzzy parameters using fault condition data of the actual machine is required.

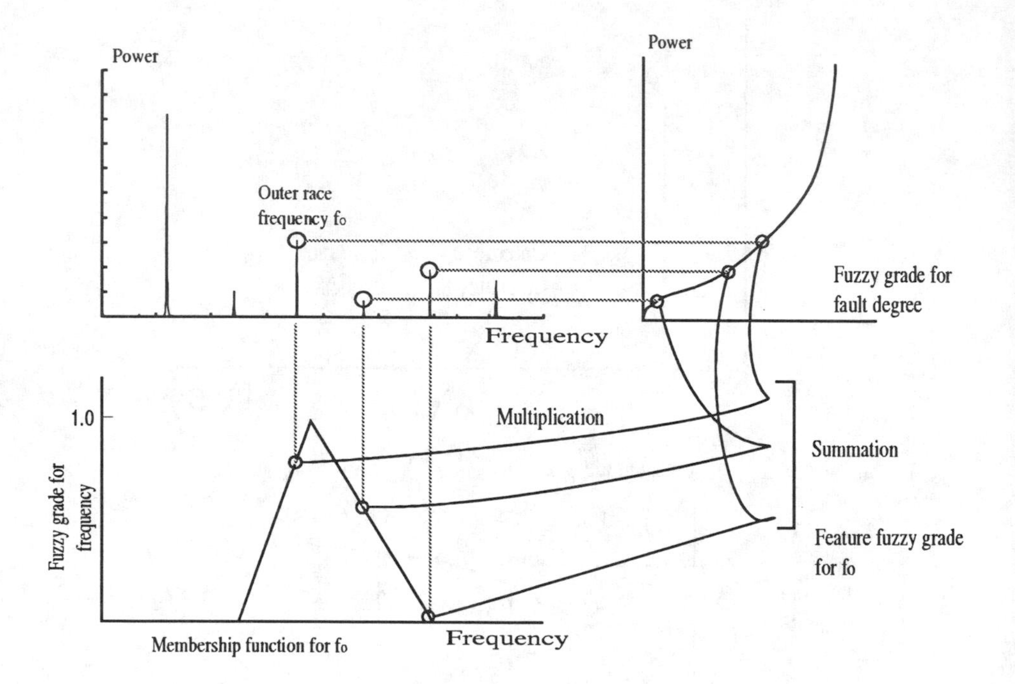

Figure 11: Calculation of feature fuzzy grade.

4 Conclusion

In this chapter, we described a new diagnosis technique for rotating machines using neural networks and the fuzzy logic. The diagnosis method using neural networks can simplify the adjustment of diagnosis parameters by learning from real vibration data. However, the improvement of the diagnosis accuracy requires obtaining a huge amount of fault condition data for training the networks. We believe that improvements in data collection and its processing will play a key role in the successful implementation of the diagnosis system.

References

[1] S.Watanabe and T.Sueki, (1981), "Vibration Diagnostic System for Rotating Machinery," *Toshiba Review*, Vol. 36, No. 5, pp. 462-467 (in Japanese).

[2] H.Yamaguchi, M.Higuchi, C.Yasuda, T.Sasaki, H.Masada, and R.Itou, (1987), "Diagnostic System Based on Vibration Measurement for Rotating Machines," *Mitsubushi Juko Giho*, Vol. 24, No. 5, pp. 445-450 (in Japanese).

[3] H.Nakajima, M.Tokuhira, S.Hisa, T.Noda, and T.Mozai, (1987), "Rotor

Vibration Diagnosis System for Large Steam Turbine," *The Thermal and Nuclear Power*, Vol. 38, No. 12, pp. 1389-1398 (in Japanese).

[4] C.Yasuda, R.Ito, C.Kita, and J.Masamori, (1989), "Diagnostic System Based on Vibration Measurement for Rotating Machines," *The Thermal and Nuclear Power*, Vol. 40, No. 2, pp. 167-180.

[5] S.Yamauchi, S.Fujii, Y.Sakamoto, and K.Terashita, (1992),"Development of Vibration Diagnosis Systems for Rotational Machineries," *Ishikawajima-Harima Engineering Review*, Vol.32, No.5, pp. 346-351 (in Japanese).

[6] K.Hashimoto, (1992), "Outline of Vibration Diagnosis System for Rotating Machinery in Thermal Power Plants," *The Thermal and Nuclear Power*, Vol. 43, No. 7, pp. 805-816 (in Japanese).

[7] The Iron and Steel Institute of Japan (ISIJ), (1986), Setsubi Shindan Gizyutsu Handbook, Maruzen. (in Japanese).

[8] I.E.Aleguindigue, A.Loskiewicz-Buczak, and R.E.Uhrig, (1993), "Monitoring and Diagnosis of Rolling Element Bearings Using Artificial Neural Networks," *IEEE Trans. on Industrial Electronics*, Vol. 40, No. 2, pp. 209-217.

[9] T.Iwatsubo, S.Kawamura, and A.Kanie, (1992), "The Diagnosis Method for Rotating Machines by Using Neural Network," *Proceedings of the 67th JSME Kansai Branch Spring Annual Meeting*, Vol. 3, No. 924, pp. 92-94 (in Japanese).

[10] M.Kotani, Y.Ueda, H.Matsumoto, and T.Kanagawa, (1995),"Acoustic Diagnosis for Compressor Using Neural Network," *Trans. of the Society of Instrument and Control Engineers*, Vol. 31, No. 3, pp. 382-390 (in Japanese).

[11] T.Hinami and K.Tsutsumi, (1991), "Fuzzy Diagnosis for Breakdown of the Bearing," Dynamics and Design Conf., Vol. 39, No. 910, Pt 1, pp. 30-33 (in Japanese).

[12] Y,Shao and K.Nezu, (1996), "An On-Line Monitoring and Diagnostic Method of Rolling Element Bearing with AI," *Trans. of the Society of Instrument and Control Engineers*, Vol. 32, No. 8, pp. 1287-1293.

[13] T.I.Liu, J.H.Singonahalli, and N.R.Iyer, (1996), "Detection of Roller Bearing Defects using Expert System and Fuzzy Logic," *Mechanical Systems and Signal Processing*, Vol.10, No. 5, pp. 595-614.

[14] L.C.Jain and R.K.Jain, Editors, (1997), Hybrid Intelligent Engineering Systems, World Scientific Publishing Company, Singapore.

[15] L.C.Jain, Editor, (1997), Soft Computing Techniques in Knowledge-based Intelligent Engineering Systems, Springer-Verlag, GmbH & Co., Germany.

[16] M.Yato, Y.Sato, and L.C.Jain, (1997), Fuzzy Clustering Models and Applications, Springer-Verlag, GmbH & Co., Germany.

[17] A.J.F.van Rooij, L.C.Jain, and R.P.Johnson, (1997), Neural Network Training Using Genetic Algorithms, World Scientific Publishing Company, Singapore.

[18] E.Vonk, L.C.Jain, and R.P.Johnson, (1997), Automatic Generation of Neural Network Architecture using Evolutionary Computing, World Scientific Publishing Company, Singapore.

Chapter 9:

Fuzzy Expert Systems in ATM Networks

FUZZY EXPERT SYSTEMS IN ATM NETWORKS*

C. Douligeris

Department of Electrical Computer Engineering
University of Miami
Coral Gables, FL 33124
U.S.A.

S. Palazzo

Istituto di Informatica e Telecomunicazioni
University of Catania
V.le A. Doria 6
95125 Catania
Italy

In this chapter we present the application of fuzzy expert systems in ATM networks. In particular, we show how fuzzy rule-based systems can be used effectively in admission control, policing, rate control, and buffer management. We provide extended examples of applying fuzzy rate control and fuzzy policing in the context of ATM control and we examine commonalities and differences between fuzzy-based and neural-based systems.

1. Introduction

The Asynchronous Transfer Mode (ATM) is emerging as the most attractive information transport technique within broadband integrated networks supporting multimedia services, because of its flexibility. ATM supports the full range of traffic characteristics and Quality of Service (QOS) requirements, efficient statistical multiplexing and cell switching [1].

---

* Portions reprinted, with permission, from:
C. Douligeris and G. Develekos, "Neuro-Fuzzy Control in ATM Networks," *IEEE Communications Magazine*, Vol. 35, No. 5, pp. 154-162, May 1997. © 1997 IEEE
V. Catania, G. Ficili, S. Palazzo and D. Panno, "A Comparative Analysis of Fuzzy Versus Conventional Policing Mechanisms for ATM Networks," *IEEE/ACM Transactions on Networking*, Vol. 4, No. 3, pp. 449-549. June 1996. © 1996 IEEE

0-8493-9804-5/99/$0.00+$.50

Multimedia connections carrying several different classes of traffic will have a wide variety of demands for quality service. These demands are negotiated during the call set-up procedure and, based on the ability of the network to satisfy these demands without jeopardizing the grade of service provided to already established connections, a call is accepted or rejected. On the other hand, network utilization is a primary concern to the network provider. Thus, effective control mechanisms are necessary to maintain a balance between QOS and network utilization, both at the time of call set-up and during progress of the calls across the ATM network. These include Connection Admission Control, Usage Parameter Control or Policing, and Network Resource Management [2].

Connection Admission Control (CAC) is defined as the set of actions taken by the network during the call set-up phase in order to determine whether a virtual channel/virtual path connection request can be accepted or rejected. A connection request is accepted only when sufficient resources are available to establish the call through the entire end-to-end path at its required QOS and maintain the agreed QOS of existing calls. To meet the above requirement, it is necessary to evaluate the degree of availability of the current network loading and the impact of adding a new connection.

Traffic is characterized by its Source Traffic Descriptor, Cell Delay Variation, and Conformance Definition based on one or more definitions of the Generic Cell Rate Algorithm [3]. Quality of service requirements may involve cell loss probability, end-to-end delay, and cell delay variation. Some researchers classify the traffic sources into many classes; the decision of CAC is based on the number of connections of each class. The problem is how to classify the calls. Even if it is possible to perform classification, the number of classes is often large and it is cumbersome to consider all combinations of different classes to make the accept/reject decision.

Another popular approach is to find the equivalent bandwidth (EBW) of individual sources and extend it to multiple sources. The new connection's anticipated traffic bandwidth requirement is estimated from the traffic parameters specified by the user. The problem of using this approach is that the expression to get the EBW of multiple sources is too complicated to be calculated in real time. In addition, the expressions for the EBW of single and multiple connections follow some particular arrival process model, an assumption that restricts us to only a subset of the known traffic types for which a source model can be established and verified, and may not hold at all for service types that will arise in the future.

After a call is accepted, there is a need for flow control to guarantee that sources behave as agreed upon during the call set-up phase. This procedure is called Policing or Usage Parameter Control (UPC). Policing is used to ensure that sources stay within their declared rate limits, so they do not adversely affect the performance of the network. Policing is done by the network provider at the Virtual Circuit or Virtual Path Level and action is taken if a source does not abide by its contract. The actions range from complete blocking of a source to selectively dropping packets to tagging packets so that they may be dropped at a later point, if necessary. Violations of the negotiated traffic requirements may result due to malfunctioning equipment, malicious users, or simply due to delay jitter for cells traveling through the network.

The UPC function is centered around a decision: to penalize or not penalize a cell when its arrival triggers an overflow of one of the leaky buckets that have been deployed for the policing of the cell stream. Decision-making is also evident in the Connection Admission Control phase of an incoming call: based on the call's traffic descriptors and QOS requirements, as well as the network's status, an accept-reject decision has to be made, as well as a determination of the bandwidth that needs to be allocated upon acceptance. It is this inherent decision-making nature of these procedures that has attracted the interest of a number of researchers that investigate pending control problems for plausible deployment of fuzzy logic and neuro-fuzzy principles.

The growing success of fuzzy logic in various fields of application, such as control, decision support, knowledge base systems, data base and information retrieval, and pattern recognition, is due to its inherent capacity to formalize control algorithms which can tolerate imprecision and uncertainty, emulating the cognitive processes that human beings use every day [4-7]. Fuzzy systems are, in fact, suitable for approximate reasoning, above all, in systems for which it is difficult, if not impossible, to derive an accurate mathematical model. Imprecision or uncertainty can, for instance, affect the input values or parameters of the system, as well as the inference rules which characterize the control algorithm. In such cases, fuzzy logic is a powerful tool which allows us to represent qualitatively expressed control rules quite naturally, often on the basis of a simple linguistic description. In addition, when applied to appropriate problems - especially in control systems - fuzzy systems have often shown a faster and smoother response than conventional systems, also thanks to the fact that fuzzy control rules are usually simpler and do not require great computational complexity. The latter aspect, along with the spread of VLSI hardware structures dedicated to fuzzy computation, makes fuzzy systems cost effective [8]. In the field of telecommunications fuzzy systems are also beginning to be used in areas such as network management and queueing theory [9-29].

In this chapter, we first concentrate on the use of fuzzy expert systems for control in ATM networks. We then concentrate on two particular examples of rate control and policing to show how the fuzzy expert system works and what are the main components of the controllers. A comparison between fuzzy-based and neural-based controllers is also discussed. Advantages and disadvantages of the proposed algorithms as well as areas of open research in the field are also presented.

2. Fuzzy Control

The wide range of service characteristics, bit rates, and burstiness factors that one encounters in broadband networks combined with the need for flexible control procedures makes the use of traditional control methods very difficult and often fragmentary in terms of the cases involved or the controls analyzed. It is apparent that it is impossible to analyze all the different situations that may arise in an ATM network and it is also difficult to update if new services are introduced [30]. Nontraditional control methods that may use adaptive learning and be flexible enough to support a variety of criteria and wide range of parameters include the use of neural networks and fuzzy logic.

While scanning the literature of ATM control one finds several arguments that reinforce our conviction that fuzzy control is appropriate for an ATM environment. With regard to the definition of the service characteristics of the sources. Rathgeb [31] states, "Another problem is caused by the inaccuracies and uncertainties in the knowledge about relevant parameters, like the mean bit rate, in the establishment of the call." These inaccuracies are amplified by the delay variation introduced in the network and significantly affect the instantaneous mean bit rate, used in most policing functions, as well as the peak bit rate.

Moreover, "it has to be recognized that the set of policing parameters proposed by CCITT in recommendation I.311, namely, average cell rate, peak cell rate, and duration of peak is not sufficient to completely describe the behavior of ATM traffic sources. Furthermore not all these characteristics may be known at call set-up with the required accuracy and some of them may be modified before the cells reach the policing function..." [31]. The difficulty lies in the fact that the sources to be characterized have different statistical properties as they range from video to data services, and it is necessary to define parameters that can be monitored during the call [32-34]. A traffic parameter contributing to a source traffic descriptor should be of significant use in resource allocation, enforceable by the network operator, and understandable by the user. The latter requirement is especially necessary to allow the user to estimate the value of the parameter in relation to the type of traffic that will be generated. This is still an open issue as, in the case of both average parameters such as long-term average cell rate, average burst duration, average inter-burst time, and in the case of upper-bound parameters such as the Sustainable Cell Rate [3, 35], it is difficult for the user to accurately estimate their value.

Several intuitive control rules are found in the ATM literature. In [36], where a thorough study is done with real traffic, it is concluded that there are some linguistic type of rules as to whether congestion has occurred or congestion has passed. Arguments like "if the network traffic load is heavier than usual but performance is acceptable, congestion has not occurred" [36], or "even if a buffer is more full than usual, the queue is not congested" can be expressed very accurately by using linguistic variables such as *rather full, heavier, more congested*, etc. It should be noted that using only three or even two arguments to characterize variables that take a large number of values is not a restriction since even in conventional control it has been advocated that a small number of classes be incorporated.

Linguistic arguments can also be found in the following statement discussing the relation between the loss curve and the magnitude deviation: "the loss curve is too drastic for small magnitude deviation, since even nominal sources may slightly exceed the exact negotiated mean rate from time to time" and "one should not be too severe on small magnitude deviation and should increase its severity as the magnitude of the deviation becomes more significant" [37]. As a matter of fact, the authors of [37] approximate a sharp loss curve with a smoother one based on the above linguistic arguments. If fuzzy logic control theory had been used, a precise justification of these arguments could have been provided.

Taking into consideration factors like the source traffic descriptor, the amount of current network congestion along the path of the incoming call, and Quality of Service requirements of the new and the pre-existing calls is a daunting task for any mathematical model. A number of publications have recently demonstrated the merit of fuzzy logic in dealing with such a complex setting. Four distinct areas of applications, namely Fuzzy Admission Control (e.g., [24]), Fuzzy Policing (e.g., [17]), Fuzzy Rate Control (e.g., [22]), and Fuzzy Buffer Management (e.g., [27]) have been investigated. Figure 1 shows the positions of a Fuzzy Logic Controller (FLC) that would perform any one of these controls. In this chapter we will present the use of fuzzy expert systems in detail in rate regulation and in policing.

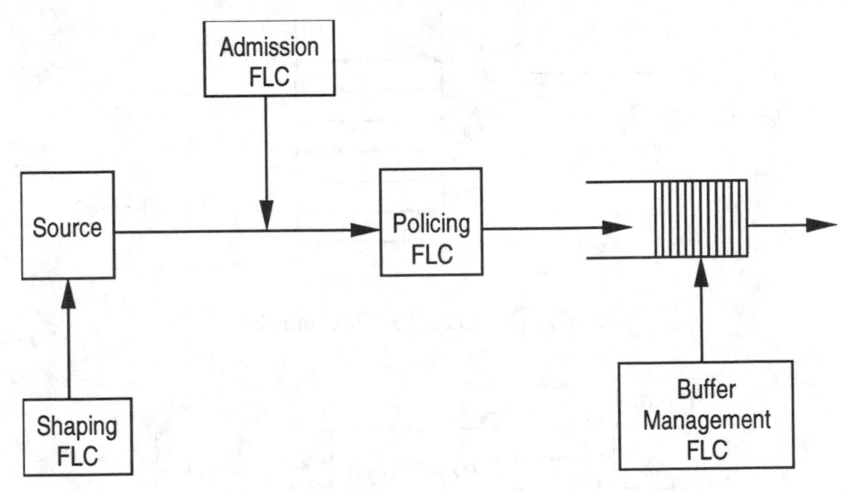

Figure 1. Fuzzy Expert System ATM Controllers

3. Fuzzy Feedback Rate Regulation in ATM Networks

Our objective in this part of the chapter is to propose a novel controller scheme that regulates the peak rate as well as reduces the Cell Loss Rate (CLR). Our proposed scheme works in connection with the Leaky Bucket (LB). The Leaky Bucket operates like a virtual queueing system, where each cell arrival increases the bucket size by one until a maximum value of S_{th}, while, at constant and regular time intervals, $1/D_r$, the bucket size is decremented (Figure 2). Cells arriving to find the bucket size equal to S_{th} are discarded or tagged. If only the LB is used, the CLR is too high if the threshold S_{th} is set too low. Meanwhile, the peak rate cannot be controlled if the threshold is set to a high value. As the traffic gets bursty, it is difficult to set the best choice to meet these two conditions. The proposed system considers the propagation delay time Δb to predict the possible cell loss in the near future. If cell discarding is imminent, the source transmission rate is reduced to a level that depends on the "strength" of the feedback signal. A correct prediction for the feedback signal is very important. If the backpressure is overdone, the additional delay time incurred on the cells whose transmission has been postponed will be intolerable, although the cell loss ratio is very low. On the other hand, if it is underestimated, the cell loss ratio may still be excessive. In this work, we propose a hybrid mechanism that uses the Leaky Bucket as the cell loss controller and generates a backpressure signal that is sent back to the transmitting source and is the outcome of fuzzy processing of the status of the LB pseudo-queue, of an indicator of the changing rate of the whole system, and of a variable that monitors the error of previous decisions. These three parameters are fuzzified to take linguistic values like HIGH, SMALL, MEDIUM, LARGE, POSITIVE, and NEGATIVE.

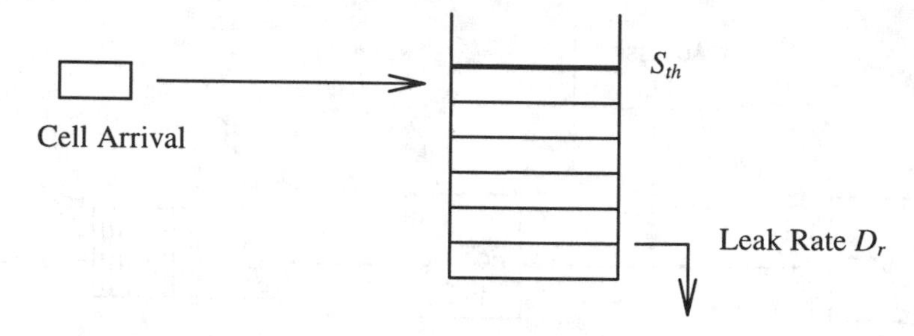

Figure 2: A Leaky Bucket Controller

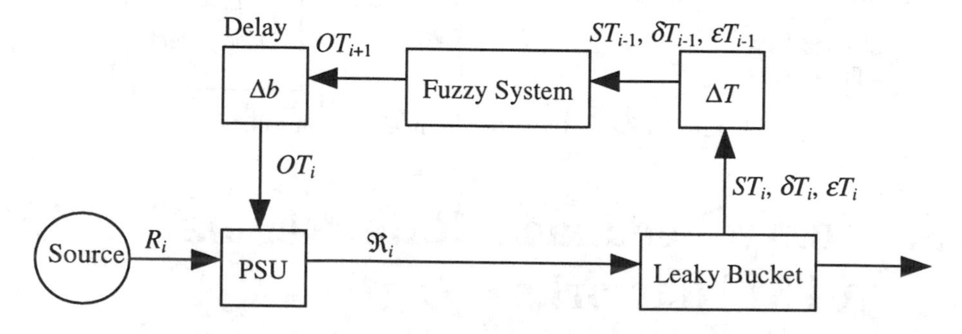

Figure 3. The Architecture of Feedback Rate Regulator

3.1 Fuzzy Feedback Control Model

The architecture of our proposed control model is shown in Figure 3. Assume that a traffic source declares its peak rate as B_p and mean rate as B_m. At time T_i the source generates data with a rate R_i. This rate is regulated to $\Re_i$ by a Pre-Shaping Unit (PSU) which reduces the data rate depending on the output OT_i of the fuzzy controller.

A Leaky Bucket is used as the policing function to shape the regulated traffic $\Re_i$. We define the depletion rate of LB as D_r, $D_r = 1/T$. The time interval to sample the status of the LB is equal to $\Delta T = T_i - T_{i-1}$, for all $i, j > 0$. Let DT_i be the number of cells discarded between time T_i and $T_i + \Delta T$ if cells arrive at a time when $S_i = S_{th}$, where S_i is the current value of the LB counter.

To monitor the expected number of cells that arrive if R_i is not regulated in transmitting during ΔT, we define $\delta T_i = (S_i - S_{i-1}) + DT_i + CT_i$, where CT_i is the estimated number of cells stopped by the PSU from time T_i to $T_i + \Delta T$.

In our model, we use $\Re_i = (1-OT_i)\, R_i$ to regulate the source rate. So we define $CT_i = (OT_i \times R_i \times \Delta T)/424$. A possible scenario works as follows: at time T_i, the source begins to transmit data at rate $\Re_i$. This status will be maintained for ΔT seconds. In the meantime, the fuzzy logic system receives the status of LB for the time interval $[T_{i-1}, T_i]$. The fuzzy logic system is able to predict possible cell discarding in time interval $[T_{i+1}, T_{i+2}]$. If ΔT is not less than Δb plus the processing time of the fuzzy logic system, then we can regulate the source rate in ΔT time intervals.

We assume that ρ_{max} is the number of cells generated by the source at its peak rate B_p during ΔT. From the system description, we get $\delta T_i < [(1 - D_r/B_p) \times \rho_{max}] = \delta_{max}$.

δ_{max} is an important parameter in our prediction, because it represents the maximum possible increment in the pseudo-queue of the Leaky Bucket during the period of ΔT, if the real peak rate doesn't violate the declared peak rate. We use it as a key parameter to decide the membership functions of the fuzzy controller.

Our Fuzzy Logic (FL) System contains two subsystems as shown in Figure 4. One of them is to process S_i and δT_i, then predicting possible cell loss. The second one is to monitor εT_i. Its purpose is to adjust the crispy output of the first one, when error in previous estimations is detected.

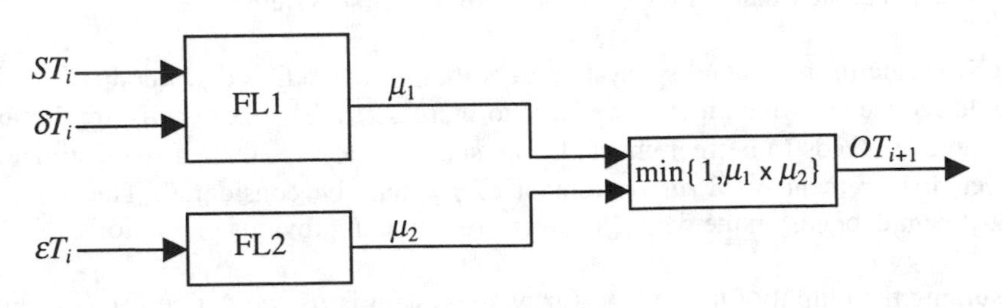

Figure 4. High-Level Diagram of the Fuzzy System Components

S_i and δT_i are fed into the first of our fuzzy logic subsystems. This subsystem generates a crispy output μ_1 as the possibility of cell discarding in the near future. The bigger the μ_1 is, the more cells are likely to be discarded. In order to make our prediction, we use another fuzzy logic subsystem which monitors previous estimation errors and then adjusts the value of μ_1. To do this, an error function is defined below as:

$$\varepsilon T_i = \begin{cases} -\dfrac{DT_i}{\delta_{max}} & if\ DT_i > 0 \\ OT_i & if\ DT_i = 0 and (S_{th} - S_i) > \delta_{max} \\ 0 & otherwise \end{cases}$$

The negative sign of εT_i indicates that OT_i was underestimated, which may have resulted in some cells being discarded. On the other hand, the positive sign represents an overestimation of OT_i. The first condition states that a good prediction should not cause any cell loss. Any cell loss must have been caused by a previous low estimation. The second condition states that, if the length of pseudo-queue is still long enough, no reduction of transmission rate should be taken. This error is processed by the second fuzzy logic subsystem. Our scheme considers the propagation delay time of the feedback signal; it makes the prediction more difficult, but it approaches the real world application.

Three fuzzy If-Then rules are used in the first subsystem:

- If S_i is HIGH and δT_i is Positive SMALL then do shaping LESS
- If S_i is HIGH and δT_i is Positive MEDIUM then do shaping MEDIUM
- If S_i is HIGH and δT_i is Positive LARGE then do shaping MORE

Two fuzzy rules are used in our second subsystem:

- If εT_i is smaller than 0 then increase output of first system
- If εT_i is greater than 0 then decrease output of first system.

The first rule in the second subsystem says that, if any cell is discarded ($\varepsilon T_i < 0$), we should reduce $\Re_i$, which means we have to increase μ_1. The more cells are discarded, the more μ_1 needs to be incremented. The second rule says that, if overestimation is detected, increment of $\Re_i$ or decrement of μ_1 must be considered. The amount of adjustment depends on the degree of overestimation in previous predictions.

Assuming the output of the second fuzzy subsystem is μ_2, we define $OT_{i+1} = min\ \{1, \mu_1\ x\ \mu_2\}$ as the output of our fuzzy system. The membership functions used in our system are shown in Figure 5.

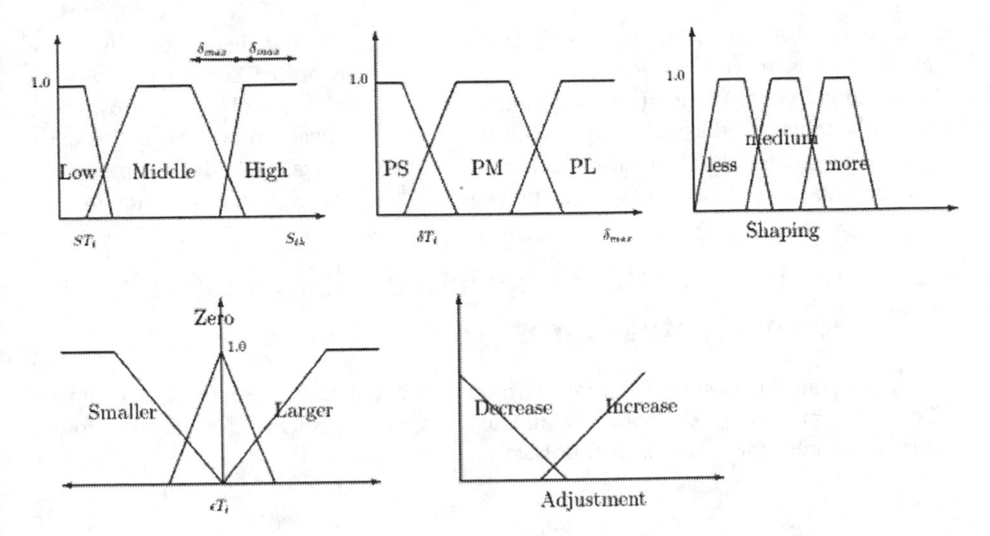

Figure 5. Membership Functions for Fuzzy Rate Regulator

3.2 Traffic Shaping

The Usage Parameter Control (UPC) mechanism discussed above has its intrinsic limitation in the ability to ensure that the negotiated connection parameters are respected, due to the stochastic behavior of the controlled source. An alternative approach is to pre-shape the cell generation process.

In the following subsections, we propose two alternatives which accept our fuzzy system's crisp output to perform traffic shaping, namely Rate Regulation (RR) and Rate Reduction with Rate Increment (RRRI).

In RR, we neither control the peak bit rate nor the burst length as long as LB still has enough "space" to accept cells. On the other hand, if possible overflow (cell discarding) in the LB is detected, the transmission rate is reduced to $\Re_i = (1\text{-}OT_i)\ R_i$. The transmission rate is updated every ΔT seconds. Its objective is to avoid cell discarding by LB. So when traffic is regulated, some delay time is usually needed to complete the requested transmission. For example, assume an unshaped traffic is to generate data at a constant rate R_c cells/second for T_c seconds. The source generates data at a constant rate of R_c cells/per second for $n\Delta T$ seconds at first, $n<m$. It is then regulated to a generation rate of $\Re_i$ cells/second for the rest of the transmission ($\Re_i < R_c$). It then takes $(m\text{-}n)\Delta T(R_c/\Re_i\text{-}1)$ seconds of delay to complete the transmission.

In RRRI, we assume that there is infinite buffer space for the source. The buffer is similar to the one discussed above and is able to serve at a rate equal to or lower than the declared peak bit rate [2]. All cells generated by the source pass through the buffer in FIFO sequence. The shaper is now located at the output of the buffer. So, even if $\Re_i$ is equal to 0, the source can still generate cells that enter the buffer at a rate R_i. The main difference between RR and RRRI is that the latter allows the transmission rate to increase to B_p when OT_{i-1} and OT_i are both equal to 0 and there are cells in the buffer. That is to say, if there are cells delayed in the buffer and no cell will be discarded in the near future (judging from the trend of OT_{i-1} and OT_i), we allow those cells to be transmitted as soon as possible at the declared peak bit rate. So in this case, $\Re_i$ may be greater than R_i. This mechanism may still have extra delay time if the transmission rates of the last few time units are high compared to the depletion rate.

3.3 Computational Experience with the Fuzzy Feedback Regulator

We use a continuous state AR Markov model to evaluate the proposed mechanism. The Autoregressive (AR) Markov model was described in [52]. A first order representation of the model is as follows:

$$y_t = a y_{t-1} + b w_t$$

where y_t is the data rate at time t, a and b are constants and w is Gaussian white noise. This model has been proposed to approximate a single video source. It is suitable for simulating the output bit rate of a VBR video source to a certain extent, when a = 0.87881, b = 0.1108, and w_t has a mean equal to 0.572 and a variance equal to 1. The lag time, T_b, is 1/30 seconds. The unit of y is 7.5 Mbits/sec. This model attempts to characterize the bit rate in each separate frame interval of video traffic. The speed at which the frame information is transmitted does not change over time.

We generated more than 3 million ATM cells in $10{,}000T_b$ = 333.3 seconds. The actual peak bit rate for the data set is 7.5 x 1.4532 = 10.899 Mbits/sec and the actual mean bit rate is 7.5 x 0.561 = 4.2075 Mbits/sec. We assume $\Delta T = \Delta b$ = 200 slot times. A slot time is defined as the unit time in simulation. For simplicity, we define it as a cell transmission time at the actual peak rate. So, ΔT and Δb are both about 7.8ms in our simulations. Since the best selections of threshold and depletion are difficult to get, we select different parameters for LB to do the simulations. In Figure 6, we select S_{th} = 1,000 and use different depletion rates to run our simulations. We find that the cell loss rate decreases when D_r increases. The cell loss rates of RR and RRRI have at least 10 times improvement compared with the unshaped traffic. Also, the delay time to complete the transmission decreases. The time overhead for RRRI is close to 0 and the worst case for RR is less than 5% (note that there is no time penalty for the unshaped traffic). Figure 7 presents the simulation results with fixed depletion rate, D_r = 0.6 x 7.5 Mbits/second and different threshold values are applied to the LB. We see the same qualitative properties as in Figure 6. Similar results have been observed with ON/OFF sources as well.

4. A Fuzzy Model for ATM Policing

In order to give an example of how fuzzy inference rules can be defined in a typical problem of ATM control, in this section we present the fuzzy model introduced in [25] for UPC purposes.

The fuzzy policer proposed in [25] is a window-based control mechanism in which the maximum number N_i of cells that can be accepted in the i-th window of length T is a threshold which is dynamically updated by inference rules based on fuzzy logic.

The target of this fuzzy policer is to make a generic source respect the average cell rate negotiated, λ_n, over the duration of the connection. According to what is the expected behavior of an ideal policing mechanism, it should allow for short-term fluctuations, as long as the long-term negotiated parameter is respected, and it should also be able to immediately recognize a violation. Since the duration of the connection is not known *a priori*, achieving this aim entails accurate choice of the control strategy as it is the latter which determines the tolerance the source is to be granted when it exhibits periods of high transmission rate. If, for example, a source is considered which, at a certain instant, starts to transmit at a higher cell rate than negotiated, it is a question of establishing whether and for how long the policer has to allow such behavior, seeing as it is or is not permissible according to the duration of the connection. If, in fact, excessive tolerance is chosen and the connection is about to end, there is a risk of failure to detect any violation which may have occurred; if, on the other hand, the control is too rigid, a certain amount of false alarms will eventually occur if the source considerably reduces its transmission rate for the rest of the connection. So, control is based on global evaluation of the behavior of the source from the beginning of the connection up to the instant in which the control is exercised. A period of high transmission rate is tolerated as long as the average rate calculated since the beginning of the connection does not exceed the negotiated value. In addition, in order not to increase the false alarm probability, an additional period of temporary violation is tolerated according to "credit" the source may have earned. More specifically, the control mechanism grants credit to a source, which in the past has respected the parameter negotiated, by increasing its control threshold N_i, as long as it perseveres with nonviolating behavior. Vice versa, if the behavior of the source is violating or risky, the mechanism reduces the credit by decreasing the threshold value.

The parameters describing the behavior of the source and the policing control variables are made up of linguistic variables and fuzzy sets, while control action is expressed by a set of fuzzy conditional rules which reflect the cognitive processes that an expert in the field would apply.

The source descriptor parameters used are the average number of cell arrivals per window since the start of the connection, A_{oi}, and the number of cell arrivals in the last window, A_i. The first gives an indication of the long-term trend of the source; the second indicates its current behavior. A third parameter, the value of N_i in the last window, indicates the current degree of tolerance the mechanism has over the source. These parameters are the three linguistic variables which make up the fuzzy policer input. The output chosen is the linguistic variable ΔN_{i+1} which represents the variation to be made to the threshold N_i in the next window.

The model of the fuzzy system, comprising the control rules and the term sets of the variables with their related fuzzy sets, was obtained through a tuning process which started from a set of initial insight considerations and progressively modified the parameters of the system until it reached a level of performance considered to be adequate. In particular, the term sets of input variables have the following fuzzy names: Low (L), Medium (M), and High (H). The term set of the output variable is composed of seven fuzzy sets with the following fuzzy names: Zero (Z), Positive Small (PS), Positive Medium (PM), Positive Big (PB), Negative Small (NS), Negative Medium (NM), Negative Big (NB).

The membership functions chosen for the fuzzy sets are shown in Figures 6, 7, and 8, where N is equal to the expected value of cells per window ($N = T\ \lambda_n$); MAX represents the maximum value between 1.5 N and the maximum number of cells that can arrive in a window (T/t_c), where t_c is the cell interarrival time during a burst; and N_{i_max} indicates the upper bound value for the N_i variable. Choice of this value is one of the main issues in sizing the mechanism. It has to take two conflicting requirements into account. The first requires a high N_{i_max} value to ensure that any greater tolerance the source is granted will cause an improvement in the false alarm probability. The second requires a low N_{i_max} value to prevent excessive inertia in the detection of violation from causing a degradation in responsiveness. The best trade-off between responsiveness and false alarm probability was obtained choosing $N_{i\_max} = 9N$. The same tuning process led to the choice of $N_1 = 3.5N$ as the value to be attributed to N_i at the beginning of the connection.

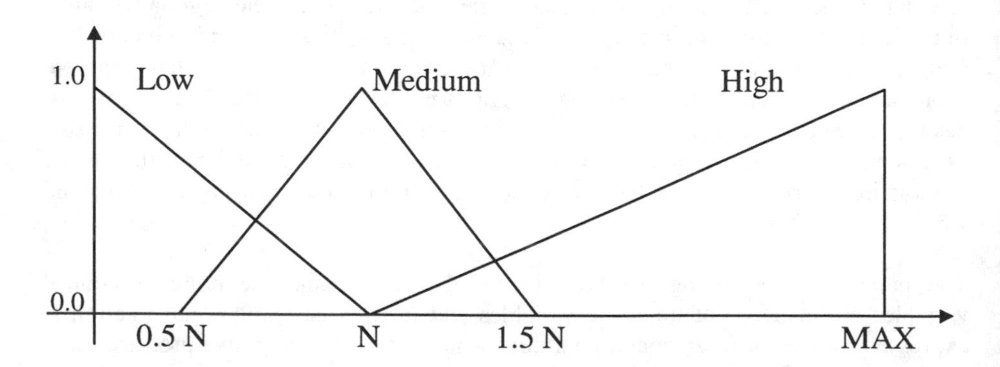

Figure 6. Membership functions for the A_{oi} and A_i input variables

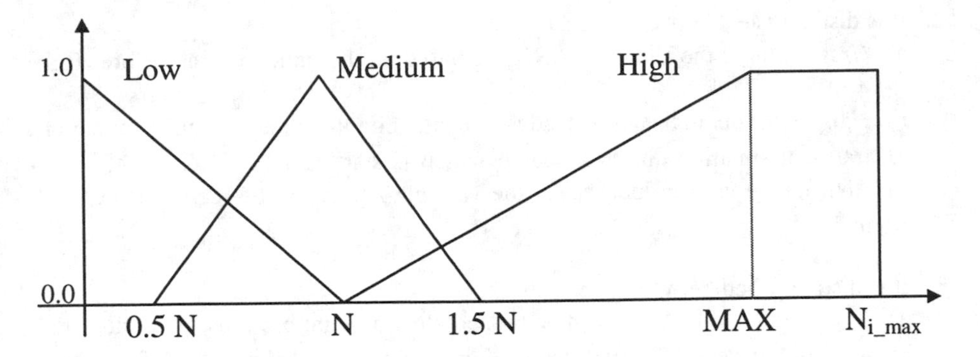

Figure 7. Membership functions for the N_i input variable

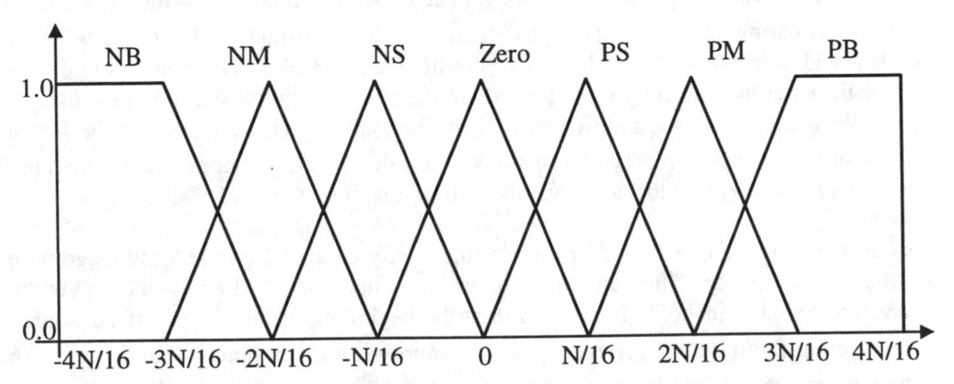

Figure 8. Membership functions for the ΔN_{i+1} output variable

Table 1 shows the fuzzy conditional rules for the policer. By way of illustration, Rule 1 in Table 1 has to be read as:

If (A_{oi} *is low) and* (N_i *is high) and* (A_i *is low) then* (ΔN_{i+1} *is positive big).*

To make the fuzzy policer's knowledge base easy to understand, the three cases in which the source is fully respectful (A_{oi} *is low*), moderately respectful (A_{oi} *is medium*), and violating (A_{oi} *is high*), respectively, are considered.

1. N_i is necessarily high due to the fact that the source has gained credit. Thus, if the number of cells which arrived in the last window is low or medium, that is, the source continues nonviolating behavior, its credit is increased (Rules 1, 2); vice versa, if A_i *is high*, a sign of a possible beginning of violation on the part of the source or an admissible short-term statistical fluctuation, the threshold value remains unchanged (Rule 3).

2. It is distinguished between two subcases:
   a) *N_i is medium*: the choice of ΔN_{i+1} is based on the same logic as before (Rules 4-6).
   b) *N_i is high*: this indicates a steady-state situation due to a respectful source or a transient situation due to a source which is starting to violate; the choice of ΔN_{i+1} is greatly influenced by the variable A_i, as can be seen in Rules 7, 8, and 9.

3. It is distinguished between three subcases:
   a) *N_i is low*: the threshold must be immediately brought back to values close to N to avoid excessively rigorous policing which would raise the false alarm probability (Rules 10-12).
   b) *N_i is medium*: without doubt this is a steady-state situation in which the source is violating and the threshold has therefore settled around N. Here again it may make sense to increase the source credit (Rules 13, 14) to be able to cope with the situation correctly if the arrivals in the current window are medium-low.
   c) *N_i is high*: this situation occurs when the source starts to violate in the initial stages of the connection. The threshold value has to be immediately lowered and the choice of the consequents is therefore clear (Rules 16-18).

As can be seen in Table 1, of 27 possible rules, only 18 appear in the knowledge base of the fuzzy policer. The remaining 9 are not included as they would never be activated. As the threshold N_1 is set *high* at the beginning of the control, if the source is respecting the negotiated rate, N_i can only grow up to its upper bound; so it can never happen that A_{oi} is *low* and at the same time N_i is *low* or N_i is *medium*; likewise, it cannot happen that A_{oi} is *medium* and N_i is *low*.

It should be noted that the fuzzy policer model is parametric with respect to the values MAX, N_{i_max}, and N_1 which are functions of N, t_c, and T. This allows the same model to be used for bursty sources with different statistical properties.

In [25], the performance of the fuzzy policer is evaluated in terms of selectivity and response time to violations and compared with that of the most popular policing mechanism, the leaky bucket (LB), and the most effective of the window mechanisms, the Exponential Weighted Moving Average window (EWMA). The policer is assessed against its capability to enforce mean cell rate in bursty sources, while the peak cell rate is considered as being separately controlled. The mean cell arrival rate negotiated is assumed to be $\lambda_n = 22$ cell/s. Both the traffic sources and the parameters characteristics of the LB and the EWMA are the same as those assumed in [31]. As mentioned previously, the dimension of the fuzzy policer depends upon the width, T, of the control window. This width is therefore fundamental for the performance of the fuzzy policer; it cannot be too wide as the policer would delay its control action on a violating source, nor can it be too small as a sufficiently long traffic estimation period is needed for the policer not to lose transparency, i.e., to prevent it from detecting permissible fluctuations in the source's bit rate as violating.

In [25] the fuzzy policer window size is chosen as equal to that of the EWMA, with which it is compared. Selectivity is measured as the probability, P_d, that the policing mechanism will detect a cell as excessive. The ideal behavior would be for P_d to be zero with the actual average cell rate up to the nominal one, that is, the mechanism is transparent toward a respectful source, and $P_d = (\sigma-1)/\sigma$ for $\sigma > 1$, where σ is the long-term actual mean cell rate of the source normalized to the negotiated mean cell rate.

Table 1 Fuzzy Policer Rules

	A_{oi}	N_I	A_i	ΔN_{i+1}
1	L	H	L	PB
2	L	H	M	PS
3	L	H	H	Z
4	M	M	L	PB
5	M	M	M	PS
6	M	M	H	Z
7	M	H	L	PB
8	M	H	M	Z
9	M	H	H	NB
10	H	L	L	PB
11	H	L	M	PM
12	H	L	H	PS
13	H	M	L	PB
14	H	M	M	PM
15	H	M	H	Z
16	H	H	L	NS
17	H	H	M	NM
18	H	H	H	NB

In Figure 9, the curve P_d versus σ is drawn. The fuzzy policer exhibits transparency for respectful sources ($\sigma \leq 1$) and, in the case of violating sources ($\sigma > 1$), a probability of detection of violation very close to ideal and certainly much greater than that of the other policing methods. In Figure 10, the dynamic behavior of the mechanisms is also compared in terms of the fraction of the violating cells detected versus the average number of cells emitted by a source with an actual cell rate 50% higher than the negotiated one, i.e., $\sigma = 1.5$.

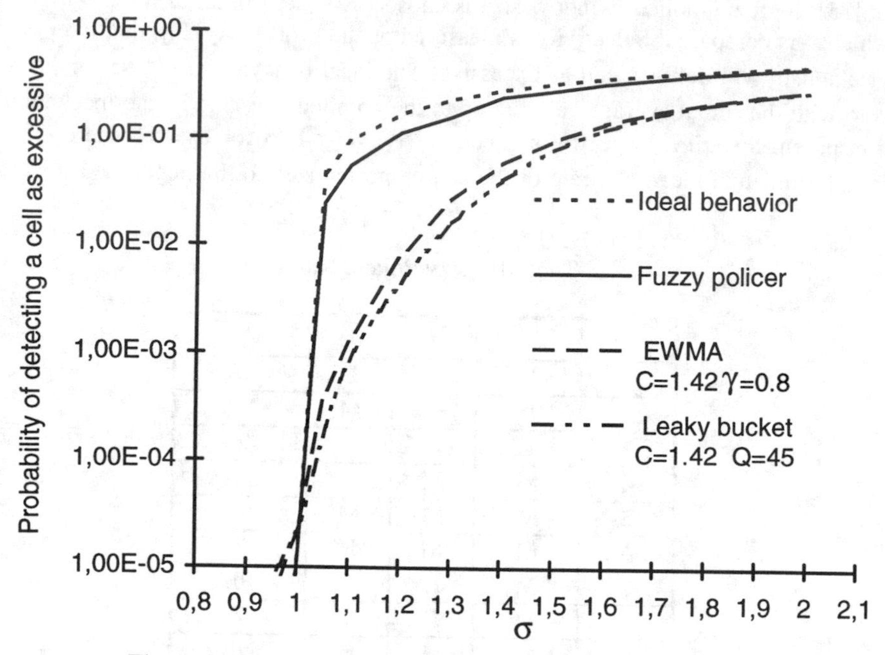

Figure 9. Selectivity performance of the fuzzy policer

From Figures 9 and 10, a comparison of traditional mechanisms shows that the mechanism which has the best behavior toward long-term violations (namely the EWMA), is the worst as far as response time is concerned; the opposite holds for the leaky bucket. This confirms the fact that traditional mechanisms are not able to cope efficiently with the conflicting requirements of ideal policing, that is, transparency and low response time. This problem is not encountered with the fuzzy policer. In fact, a trend very close to the ideal curve in the steady state corresponds to decidedly better dynamics than those of the other mechanisms. More specifically, although the leaky bucket starts detecting violation first, the percentage of cells detected as excessive is very low compared with an ideal detection probability of about 33%. Conversely, the detection probability of the fuzzy mechanism grows very fast thus showing a marked improvement over the other policing methods.

In [25], the behavior of the fuzzy policer is also tested in controlling sources featuring different types of violation. More precisely, the average cell arrival rate is violated either by increasing the average duration of the periods of silence E[Off] and keeping the average burst length E[On] constant, or vice versa. The performance results have shown that the fuzzy policer is robust and efficient, irrespective of the type of violation.

In conclusion, the fuzzy policer, when used to control bursty sources, offers performance levels which are decidedly better than those obtainable with conventional mechanisms; it is capable, in fact, of combining low response times to violations with a selectivity close to that of an ideal policer.

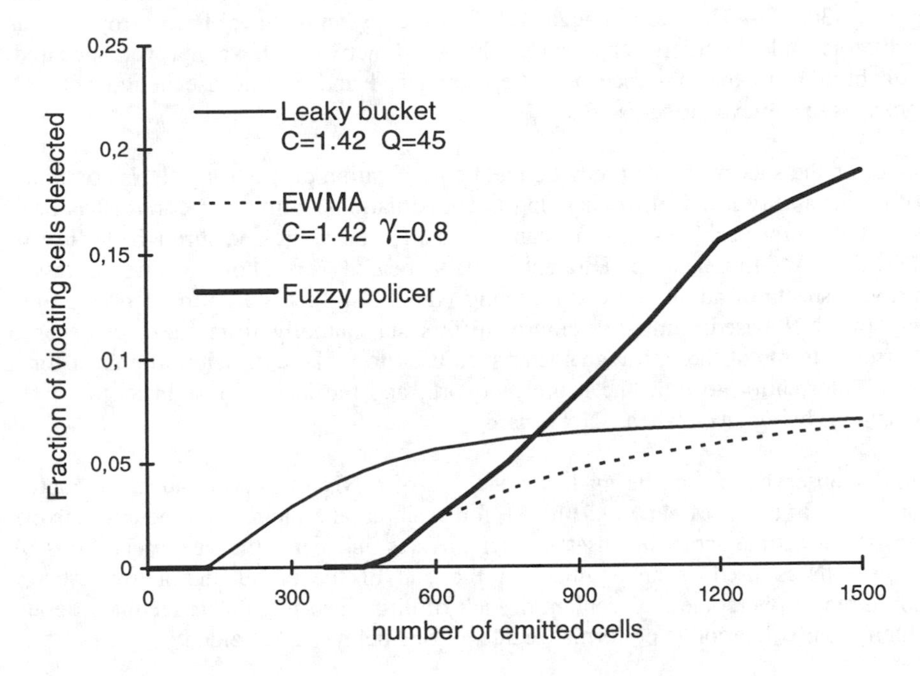

Figure 10. Dynamic behavior of the fuzzy policer

5. Relationship between Fuzzy and Neural Approaches

Another class of artificial intelligence techniques that has gained popularity in ATM network control is the use of Neural Networks (NNs). We can classify the use of Neural Network techniques in ATM control into four general categories: NN-based Admission Control, NN-based Policing, NN-based Traffic Characterization and Prediction, and NN-based Switch Routing Optimization. In the first two categories, the prediction and classification properties of NNs are deployed to either predict incoming traffic or classify incoming traffic, or combine prediction of the expected performance of the system with congestion notification. For such problems, feedforward NNs are used. The proposed networks differ in their number of layers, size, training technique used, and input/output representation. In the last category NNs are used as generic optimizers and thus Hopfield-type NNs are suggested.

Traffic prediction and classification is an inherent property of NNs. Thus, NNs used in Admission Control perform classification of acceptable and unacceptable traffic types. NNs used in Congestion Control need to first predict the rate of arrival so that they can suggest optimum control actions. Evaluation of the predictive and classification properties of NNs in an ATM environment without necessarily proposing or showing any direct application in control has been reported in several papers [30, 38-43]. Even though fuzzy logic systems have been proposed as predictors and classifiers, applications in ATM networks have not yet appeared, probably due to the existence of more established and easy-to-use neural network packages to achieve the same goal.

Based on the success of the prediction and classification properties of NNs combined with their ability to adapt to changing traffic situations, admission control has been one of the first problems in broadband control to be addressed through the use of NNs. Since the first paper by Hiramatsu [30] appeared, several others have addressed various aspects of admission control using NNs [44-47]. It needs to be pointed out here that NN based admission control differs substantially from fuzzy rule based control in terms of the traffic characteristics used to make a decision, the need for *a priori* information to train the neural network, and the lack of insight as to why a decision is being taken when a NN is used.

Similar observations can be made for the use of a NN to provide adaptive control congestion in an ATM network. In [48], for example, the input to the neural network consists of a time series of observed arrival rates and performance observations. A second NN is used as an emulator at the end of the broadband network to be monitored to overcome the problem of lack of direct learning. Thus feedback loops, which usually deteriorate performance due to long delays, are avoided.

Douligeris and Liu investigated the use of NN extensively and compared various traffic arrival streams and control methodologies. In [49] they use the NN as a device that observes the output of a leaky bucket and feedbacks to the source an optimum rate of transmission. MPEG traffic traces are used as traffic streams and the NN controller is compared with static feedback controllers. The performance of the NN-based control shows a considerable improvement in Cell Loss Rate and an excellent delay performance.

From the above it is evident that both Artificial Neural Networks and Fuzzy Logic based systems can play an important role in the control of ATM networks, since they can provide adaptive, model-free, real time control to the user.

Fuzzy Logic control provides the capabilities of simultaneously achieving several objectives (like mean and peak rate control), and avoiding the drawbacks of bang-bang controls by providing smoother changes in the call/reject regions in admission and congestion control. Robustness of the achieved performance with regard to the number of rules and the exact positioning of the membership functions allow easy implementations. Neural Network based implementations provide adaptive learning capabilities, high computation rate, generalization of learning, and a high degree of robustness and fault tolerance.

By comparison, Neural Networks provide a black box that performs as expected in situations where there may be no *a priori* knowledge or experience, while Fuzzy Logic based systems use expert knowledge and experience to control the network.

At present, there is no systematic procedure to design fuzzy logic systems. The most intuitive approach is to define membership functions and rules based on the knowledge of an experienced person and then perform adjustment if the design fails to produce the proper output. The distributed representation and learning capabilities of neural networks make them excellent candidates for integration with the fuzzy mechanism, introducing the new approach of using neural networks to find optimal input/output membership functions. In a typical architecture, given that a fuzzy rule based system can be represented by a neural system with the proper structure, fuzzy rules and membership functions [50, 51] are implemented using layers of neurons that carry out the fuzzification, inference, and defuzzification actions. Such a design obviously readily lends itself to a feed-forward error back-propagation learning procedure. As a more efficient alternative to the standard random weight initialization, the designer can use an expert's knowledge and experience to set the initial parameters and allow the neural network learning to carry out the fine-tuning. Such a structure is shown in Figure 11. The two inputs correspond to the fuzzy input variables and subsequent layers carry out the fuzzification, inference, and defuzzification operations in a NN-like manner that allows for on-line training of the membership function properties. The output corresponds to the output of the original fuzzy controller [23]. In this fashion, neuron layers cease to be black boxes with no intuitively apparent functionality, thereby adding transparency to the neural network, while fuzzy systems obtain self-adaptation properties. Such integrated methodologies will allow the Fuzzy Logic Systems to operate in areas where there is insufficient data or the data is completely unavailable in the beginning of the operation of the system but gradually becomes available.

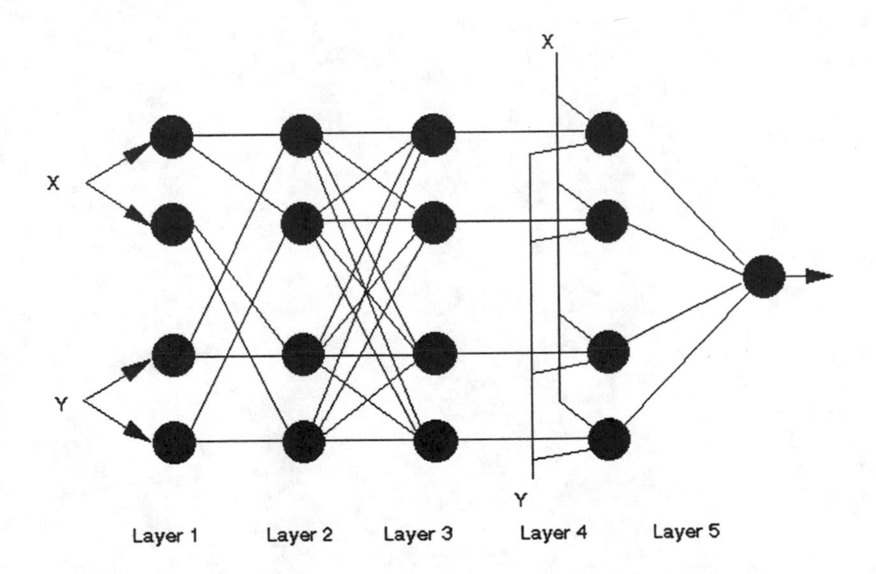

Figure 11. NN implementation of a fuzzy rule-based system

6. Conclusions

Most of the systems used for Neural Networks or Fuzzy Logic control use very standard methodologies from the respective literature. It seems that the Neural Networks and Fuzzy Logic literature, and, of course, that for Neuro-Fuzzy, have addressed a variety of issues in system design, stability, and convergence that have not been evaluated extensively in ATM traffic control. With current basic models showing their applicability to the problem at hand, it is time that researchers look for more elaborate models and techniques so that they can propose more efficient, robust, and fast algorithms.

Acknowledgments

The authors would like to thank all their colleagues in the University of Miami and the University of Catania, who provided the results of research projects for this chapter, and IEEE for permission to publish results that appeared in [53] and [54].

References

[1] R. O. Onvural, *Asynchronous Transfer Mode networks: performance issues*, Artech House, 1994.

[2] I. W. Habib and T. N. Saadawi, "Controlling flow and avoiding congestion in broadband networks," *IEEE Communications Magazine*, Vol. 29, No. 10, October 1991.

[3] ATM Forum, "Traffic management specification," Version 4.0, Draft Version, April 1995.

[4] L.A. Zadeh, "Fuzzy sets," *Inform. Contr.*, Vol. 8, pp. 338-353, 1965.

[5] M. Sugeno, *Industrial Applications of Fuzzy Control*, Amsterdam: North-Holland, 1985.

[6] M.M. Gupta and T. Yamakawa, *Fuzzy Computing*, Amsterdam: North-Holland, 1985.

[7] T. Munakata, and Y. Jani, "Fuzzy systems: an overview," *Communications of the ACM*, Vol. 37, No. 3, pp. 69-76, March 1994.

[8] V. Catania, A. Puliafito, M. Russo, and L. Vita, "A VLSI fuzzy inference processor based on a discrete analog approach," *IEEE Transactions on Fuzzy Systems*, Vol. 2, No. 2, May 1994.

[9] H.M. Prade, "An outline of Fuzzy or possibilistic models for queuing systems," *Proc. Symp. Policy Anal. Inform. Syst.*, Durham, 1980.

[10] R.-J. Li and E.S. Lee, "Analysis of Fuzzy queues," *Comput. Math. Appl.*, Vol. 17, No. 7, pp. 1143-1147, 1989.

[11] J.M. Holtzman, "Coping with broadband traffic uncertainties: statistical uncertainty, fuzziness, neural networks," *Proc. GLOBECOM '90*, San Diego, Dec. 1990.

[12] N. Millstrom, A. Bonde Jr., and M. Grimaldi, "A hybrid neural-fuzzy approach to VHF frequency management," *Proc. of the 1993 IEEE Workshop on Neural Networks for Signal Processing*, 1993.

[13] Y. Lirov, "Fuzzy Logic for distributed systems troubleshooting," *Proc. Second IEEE Int. Conf. Fuzzy Systems*, San Francisco, 1993.

[14] L. Lewis and G. Dreo, "Extending trouble ticket systems to fault diagnostics," *IEEE Network Magazine*, pp. 44-51, Nov. 1993.

[15] A. Bonde and S. Ghosh, "A comparative study of Fuzzy versus 'Fixed' thresholds for robust queue management in cell-switching networks," *IEEE/ACM Transactions on Networking*, Vol. 2, No. 4, pp. 337-344, August 1994.

[16] M. Abdul-Haleem, K.F. Cheung, and J. Chuang, "Fuzzy Logic based dynamic channel assignment," *Proc. ICCS '94*, Singapore, Nov. 1994.

[17] K.F. Cheung, D.H.K. Tsang, C.C.Cheng, and C.W.Liu, "Fuzzy Logic based ATM policing," *Proc. ICCS '94*, Singapore, Nov. 1994.

[18] T.D. Ndousse, "Fuzzy Neural control of voice cells in ATM networks," *IEEE Journal on Selected Areas in Communications*, Vol. 12, No. 9, pp. 1488-1494, Dec. 1994.

[19] D. Jensen, "B-ISDN network management by a fuzzy logic controller," *Proc. GLOBECOM '94*, San Francisco, Nov. 1994.

[20] A. Jennings and K. Au, "Fuzzy Resource Management in High Speed Networks," *Proc. 2nd Australian and New Zealand Conf. on Intelligent Information Systems*, Brisbane, Australia, 29 Nov. - 2 Dec. 1994.

[21] C. Douligeris and G. Develekos, "A Fuzzy Logic Approach to Congestion Control in ATM Networks," *Proc. ICC'95*, Seattle, June 1995.

[22] A. Pitsilides, Y.A. Sekercioglu, and G. Ramamurthy, "Fuzzy Backward Congestion Notification (FBCN) Congestion Control in Asynchronous Transfer Mode (ATM) Networks," *Proc. GLOBECOM'95*, Singapore, November 1995.

[23] Y. Liu, "Adaptive Intelligent Traffic Management in ATM Networks," Ph.D. Dissertation, University of Miami, Coral Gables, FL, December 1995.

[24] R.G. Cheng and C.J. Chang, "Design of a Fuzzy Traffic Controller for ATM Networks," *IEEE/ACM Transactions on Networking*, Vol. 4, No. 3, pp. 460-469, June 1996.

[25] V. Catania, G. Ficili, S. Palazzo, and D. Panno, "A Comparative Analysis of Fuzzy Versus Conventional Policing Mechanisms for ATM Networks," *IEEE/ACM Transactions on Networking*, Vol. 4, No. 3, pp. 449-459, June 1996.

[26] Q. Hu, D.W. Petr, and C. Brown, "Self-tuning Fuzzy Traffic Rate Control for ATM Networks," *Proc. ICC'96*, Dallas, June 1996.

[27] Y. Liu and C. Douligeris, "Nested Threshold Cell Discarding with Dedicated Buffers and Fuzzy Scheduling," *Proc. ICC'96*, Dallas, June 1996.

[28] V. Catania, G. Ficili, S. Palazzo, and D. Panno, "Using Fuzzy Logic in ATM Source Traffic Control: Lessons and Perspectives," *IEEE Communications Magazine*, Vol. 34, No. 11, pp. 70-81, Nov. 1996.

[29] K. Uehara and K. Hirota, "Fuzzy Connection-Admission-Control for ATM Networks Based on Possibility Distribution of Cell Loss Ratio," *IEEE Journal on Selected Areas in Communications*, Vol. 15, No. 2, pp. 179-190, February 1997.

[30] A. Hiramatsu, "ATM Communications Network Control by Neural Networks," *IEEE Transactions on Neural Networks*, Vol. 1, No. 1, pp. 122-130, March 1990.

[31] E. Rathgeb, "Modelling and performance comparison of policing mechanisms for ATM networks," *IEEE Journal on Selected Areas in Communications*, Vol. 9, No. 3, pp. 325-334, April 1991.

[32] A. Berger and A. Eckberg, "A B-ISDN/ATM traffic descriptor, and its use in traffic and congestion controls," *Proc. GLOBECOM '91*, Phoenix, Dec. 1991.

[33] A. Berger, "Desiderable properties of traffic descriptors for ATM connections in a Broadband ISDN," *Proc. of the 14th ITC*, Antibes, June 1994.

[34] J. Andrade, "ATM source traffic descriptor based on the peak, mean and second moment of the cell rate," *Proc. of the 14th ITC*, Antibes, June 1994

[35] ITU-TSS Recommendation I.371: "Traffic control and congestion control in B-ISDN," Frozen Issue, March 1995.

[36] H.J. Fowler and W.E. Leland, "Local Area Network Traffic Characteristics, with Implications for Broadband Network Congestion Management," *IEEE Journal on Selected Areas in Communications*, Vol. 9, No. 7, pp. 1139-1149, September 1991.

[37] B. Lague, C. Rosenberg, and F. Guillemin, "A Generalization of Some Policing Mechanisms," *Proc. INFOCOM 1992*, Florence, Italy, April 1992.

[38] A. Hiramatsu, "Integration of ATM Call Control and Link Capacity Control by Distributed Neural Networks," *IEEE Journal on Selected Areas in Communications*, Vol. 9, No. 7, pp. 1131-1138, September 1991.

[39] E.S. Yu and C.Y.R. Chen, "Traffic Prediction using Neural Networks," *Proc. GLOBECOM '93*, Houston, Nov. 1993.

[40] T. Okuda, M. Anthony, and Y. Tadokoro, "A Neural Approach to Performance Evaluation for Teletraffic System," *Proc. IEEE ICC '94*, New Orleans, May 1994.

[41] A.A. Tarraf, I.W. Habib, and T.N. Saadawi, "Congestion Control Mechanism for ATM Networks Using Neural Networks," *Proc. ICC '95*, Seattle, June 1995.

[42] H.R. Mehvar and T. Le-Ngoc, "NN Approach for Congestion Control in Packet Switch OBP Satellite," *Proc. ICC '95*, Seattle, June 1995.

[43] R.C. Lehr and J.W. Mark, "Traffic Classification using Neural Networks," *Proc. Communication Networks and Neural Networks: The Challenge of Network Intelligence*, Duke University, NC, March 1996.

[44] R.J.T. Morris and B. Samadi, "Neural Networks in Communications: Admission Control and Switch Control," *Proc. ICC '91*, Denver, June 1991.

[45] A.D. Estrella, E. Casilari, A. Jurado, and F. Sandoval, "ATM Traffic Neural Control: Multiservice Call Admission and Policing Function," *Proc. Int. Workshop on Applications of Neural Networks to Telecommunications (IWNNT)*, Princeton, NJ, October 1993.

[46] J.E. Neves, L. de Almeida, and M. J. Leitao, "ATM Control by Neural Networks," *Proc. Int. Workshop on Applications of Neural Networks to Telecommunications (IWNNT)*, Princeton, NJ, October 1993.

[47] S.A. Youssef, I.W. Habib, and T.N. Saadawi, "A Neural Network Control for Effective Admission Control in ATM Networks," *Proc. ICC '96*, Dallas, June 1996.

[48] X. Chen and I.M. Leslie, "Neural Adaptive Congestion Control for Broadband ATM Networks," *IEE Proceedings I*, Vol. 139, No. 3, June 1992.

[49] Y.C. Liu and C. Douligeris, "Rate Regulation with Feedback Controller in ATM Networks - A Neural Network Approach," *IEEE Journal on Selected Areas in Communications*, Vol. 15, No. 1, January 1997.

[50] H.R. Berenji and P. Khedkar, "Learning and Tuning Fuzzy Logic Controllers Through Reinforcements," *IEEE Transactions on Neural Networks*, Vol. 3, No. 5, pp. 724-740, September 1992.

[51] J.S. Jang, "ANFIS: Adaptive-Network-Based Fuzzy Inference System," *IEEE Transactions on Systems, Man, and Cybernetics*, Vol. 23, No. 3, pp. 665-685, May 1993.

[52] B. Maglaris, D. Anastassiou, P. Sen, G. Karlsson, and J. Robbins, "Performance Models of Statistical Multiplexing in Packet Video Communications," *IEEE Transactions on Communications,* vol. 6, No. 7, pp. 834-844, July 1988.

[53] C. Douligeris and G. Develekos, "Neuro-Fuzzy Control in ATM Networks," *IEEE Communications Magazine*, Vol. 35 , No. 5, pp. 154-162, May 1997.

[54] V. Catania, G. Ficili, S. Palazzo, and D. Panno, "A Comparative Analysis of Fuzzy Versus Conventional Policing Mechanisms for ATM Networks," *IEEE/ACM Transactions on Networking*, Vol. 4, No. 3, pp. 449-459, June 1996.

Chapter 10:

Multimedia Telephone for Hearing-Impaired People

MULTIMEDIA TELEPHONE FOR HEARING-IMPAIRED PEOPLE

F. Lavagetto
Dept. of Telecommunications, Computer and Systems Sciences
University of Genova
Italy

Human behaviors are expressed through different complementary modalities working in cooperation toward motorial/sensor goals actually guaranteed by successful coordination. Everyday experience provides a quantity of clear evidence of this phenomenon like sight-motor coordination in grasping, sewing, or walking. Speech production and perception are further examples of biological multimodal mechanisms in which different sensorial channels are used to convey information (production) and whose outputs are fused to decode information (perception).

The conversion of speech into visual information addresses the fascinating world of multimedia integration and multimodal communication. The possibility of converting the communication modality while preserving the conveyed information further highlights the foreseeable applications of related techniques and systems. The advent of a unified worldwide market will require the introduction of new standards in each of the many components of the service-integrated environment like terminal equipment, interfaces, and networks. The possibility of having only a few telecommunication companies in the world will encourage the need of a less fragmented market, cooperatively oriented to providing large-scale service.

Rehabilitation technology (RT) will definitely play a central role in the above depicted scenario since an increasing share of consumers will explicitly demand applications and services in the fields of interpersonal communications, man-machine interaction, tele-work and tele-education with special aids for overcoming impairments due to age, handicaps, and temporary or permanent diseases. In a future technological society based on integrated multimedia services at low cost, easy access, high capillarity and privacy, multimedia approaches to interpersonal [1-4] communication will definitely represent a means of formidable strength to overcome most of still existing social barriers.

As far as interpersonal communications are concerned, low cost and compact terminal equipment interfaced to mass communication lines, capable of converting in real-time, incoming messages from whatever source modalities to more suitable destination

0-8493-9804-5/99/$0.00+$.50

modalities would provide relevant hints toward the goal of social integration. Waiting for future revolutionary services provided by intelligent networks relying on large bandwidth connections, short-term applications must be primarily oriented to exploitation of the existing public networks and, first of all, the analogue telephone lines. The deployment of the telephone switched lines has many features including the utmost advantage of guaranteeing home-to-home connections at very affordable costs, sound network management and plant maintenance, facilities for international connections, progressive upgrading due to forthcoming technological improvements, interconnection to other digital communication services, and interface to ground cellular communication between mobile stations.

1. Introduction

Speech communication is considered as the richest means of human interaction. The service providers in telecommunications have always managed this worldwide business without paying enough attention to the needs of hearing-impaired consumers who are evidently unsuited to the acoustic medium and are, therefore, partially or totally excluded from this primary source of communication. Several attempts have recently been made to process the speech analogue signal in order to filter out noise, reduce the distortion, enhance the quality, and, finally, drive suitable electro-acoustic-visual devices [5-7]. Thanks to these sophisticated techniques, many communication barriers have been overcome and new relay and mediation services are offered in the field of social and cultural integration.

Moreover, the exploitation of *a priori* knowledge on the bimodal acoustic-visual nature of speech production and perception makes it possible to process the speech signal and extract suitable parameters, capable of driving the animation of a synthetic mouth where lips movements are faithfully reproduced [8-9]. The hardware required is very simple and basically located at the receiver (some suitable preprocessing can be optionally performed at the transmitter). It consists of some electronics for processing incoming speech and driving the animation of the lip icon on a small display device. "Intelligent" receiver equipment, with its low complexity and cost-effectiveness, is also reasonably compact and could be optionally plugged into any conventional telephone set.

Within the European TIDE initiative (Technology Initiative for Disabled and Elderly people) in the field of Rehabilitation Technology, the consortium SPLIT[1] has addressed directly the ambitious goal of converting speech into lip-readable visualization for the development and experimentation of a multimedia telephone for hearing-impaired people. A prototype version of this system is shown in Figure 1. Here a normal hearing caller is connected, through a conventional telephone (PTS)

---

[1] SPLIT, activated in 1994 with the participation of DIST as full member, was a 2-year pre-competitive project oriented to the development and experimentation of advanced multimedia technology in the field of speech rehabilitation of deaf persons.

line, to an Intelligent Network (IN) node equipped with processing and converting incoming speech into corresponding visual parameters which are then transmitted along a ISDN (Integrated Service Data Network) down-link to a PC terminal located at the hearing-impaired receiver.

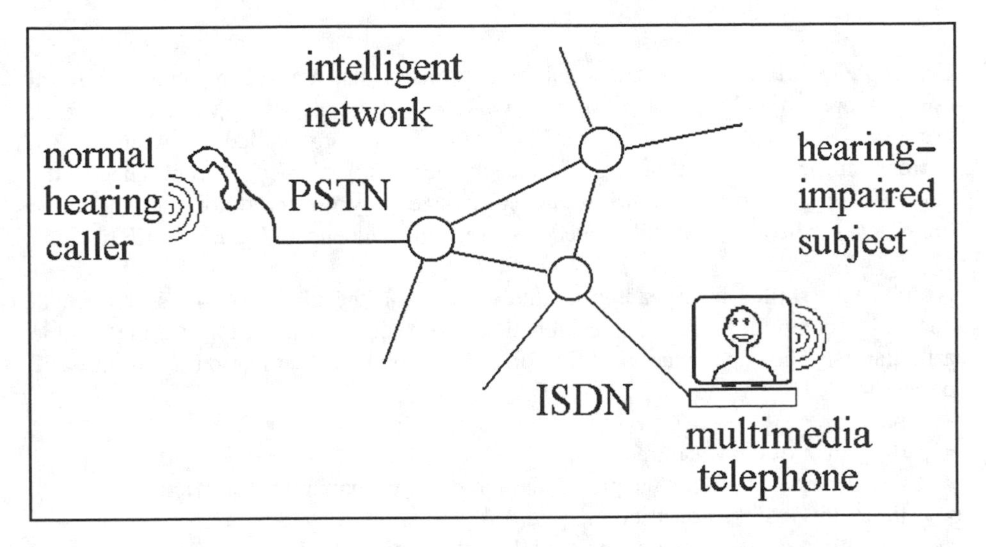

Figure 1. The multimedia telephone: a system for real-time conversion of analogue telephone calls into digital graphic animation (1063-6528/95$04.00 © 1995 IEEE).

Promising industrial applications are foreseen considering the large market of potential consumers interested not only in telephone interpersonal communication but also in a multitude of other unconventional applications, including better human interfaces to public offices and environments. Let us take simple examples like a school/university class, a conference room, an open-air talk, or a theater representation.

Many other examples of multimedia applications can be mentioned, where lip synchronization with speech represents a key issue, both short/medium term as cited above and medium/long term, like those oriented to more ambitious objectives: multimedia digital manipulation for content creation, virtual reality, augmented reality, and interactive gaming.

2. Bimodality in Speech Production and Perception

Speech is the concatenation of elementary units, phones, generally classified as vowels if they correspond to stable configurations of the vocal tract or, alternatively, as consonants if they correspond to transient articulatory movements. Each phone is then characterized by means of a few attributes (open/closed, front/back, oral/nasal, rounded/unrounded) which qualify the articulation manner (fricative like /f/,/s/, plosive like /b/, /p/, nasal like /n/, /m/, ...) and articulation place (labial, dental, alveolar, palatal, glottal).

Some phones, like vowels and a subset of consonants, are accompanied by vocal cords' vibration and are called "voiced" while other phones, like plosive consonants, are totally independent of cords' vibration and are called "unvoiced." In correspondence of voiced phones, the speech spectrum is shaped in accordance to the geometry of the vocal tract with characteristic energy concentrations around three main peaks called "formants," located at increasing frequencies $F1$, $F2$, and $F3$.

An observer skilled in lipreading is able to estimate the likely locations of formant peaks by computing the transfer function from the configuration of the visible articulators. This computation is performed through the estimation of four basic parameters:

- the length of the vocal tract L;
- the distance d between the glottis and the place of maximum constriction;
- the radius r of the constriction;
- the ratio A/L between the area A of the constriction and L.

While the length L can be estimated *a priori* taking into account the sex and age of the speaker, the other parameters can be inferred, roughly, from the visible configuration. If the maximum costriction is located in correspondence with the mouth, thus involving lips, tongue, and teeth as it happens for labial and dental phones, this estimate is usually reliable. In contrast, when the maximum costriction is nonvisible like in velar phones (/k/, /g/), the estimate is usually very poor.

2.1 The Task of Lipreading Performed by Humans

Lipreading represents the highest synthesis of human expertise in converting visual inputs into words and then into meanings. It consists of a personal database of knowledge and skills constructed and refined by training, capable of associating virtual sounds to specific mouth shapes, generally called "viseme," and, therefore, infer the underlying acoustic message. The lipreader's attention is basically focused on the mouth, including all its components like lips, teeth, and tongue, but significant help in comprehension comes also from the entire facial expression.

In lipreading, a significant amount of processing is performed by the lipreader himself/herself who is skilled in post-filtering the converted message to recover from errors and from communication lags. Through linguistic and semantic reasoning it is possible to exploit the message redundancy and understand by context; this kind of knowledge-based interpretation is performed by the lipreader in real time.

Audio-visual speech perception and lipreading rely on two perceptual systems working in cooperation so that, in case of hearing impairments, the visual modality can efficiently integrate or even substitute the auditory modality. It has been demonstrated experimentally that the exploitation of the visual information associated with the movements of the talker's lips improves the comprehension of speech; the Signal-to-Noise Ratio (SNR) is incremented up to 15 dB and the auditory failure is transformed into near-perfect visual comprehension. The visual analysis of the talker's face provides different levels of information to the observer improving the discrimination of signal from noise. The opening/closing of the lips is, in fact, strongly correlated to the signal power and provides useful indications on how the speech stream is segmented. While vowels, on one hand, can be recognized rather easily both through hearing and vision, consonants are, conversely, very sensitive to noise and the visual analysis often represents the only way for comprehension success. The acoustic cues associated with consonants are usually characterized by low intensity, a very short duration, and fine spectral patterning.

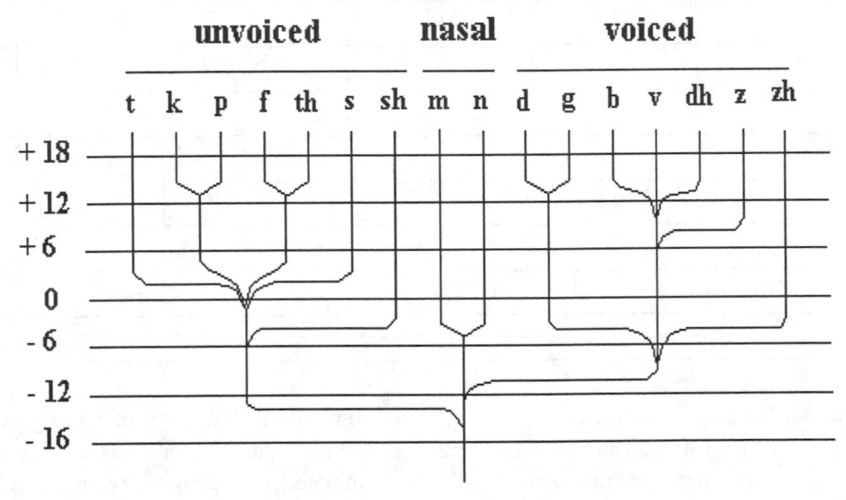

Figure 2. Auditory confusion of consonant transitions CV in white noise with decreasing Signal-to-Noise Ratio expressed in dB (From B.Dodd, R.Campbell, "Hearing by Eye: the Psychology of Lipreading," Lawrence Erlbaum Assoc. Publ.).

The auditory confusion graph reported in Figure 2 shows that cues of nasality and voicing are efficiently discriminated through acoustic analysis, different from place cues which are easily distorted by noise. The opposite situation occurs in the visual domain, as shown in Figure 3, where place is recognized far more easily than voicing and nasality.

Place cues are associated, in fact, to mid-high frequencies (above 1 KHz) which are usually scarcely discriminated in most hearing disorders, contrary to nasality and voicing, which reside in the lower part of the frequency spectrum. Cues of place, moreover, are characterized by short-time fine spectral structure requiring high frequency and temporal resolution, different from voicing and nasality cues, which are mostly associated to unstructured power distribution over several tens of milliseconds.

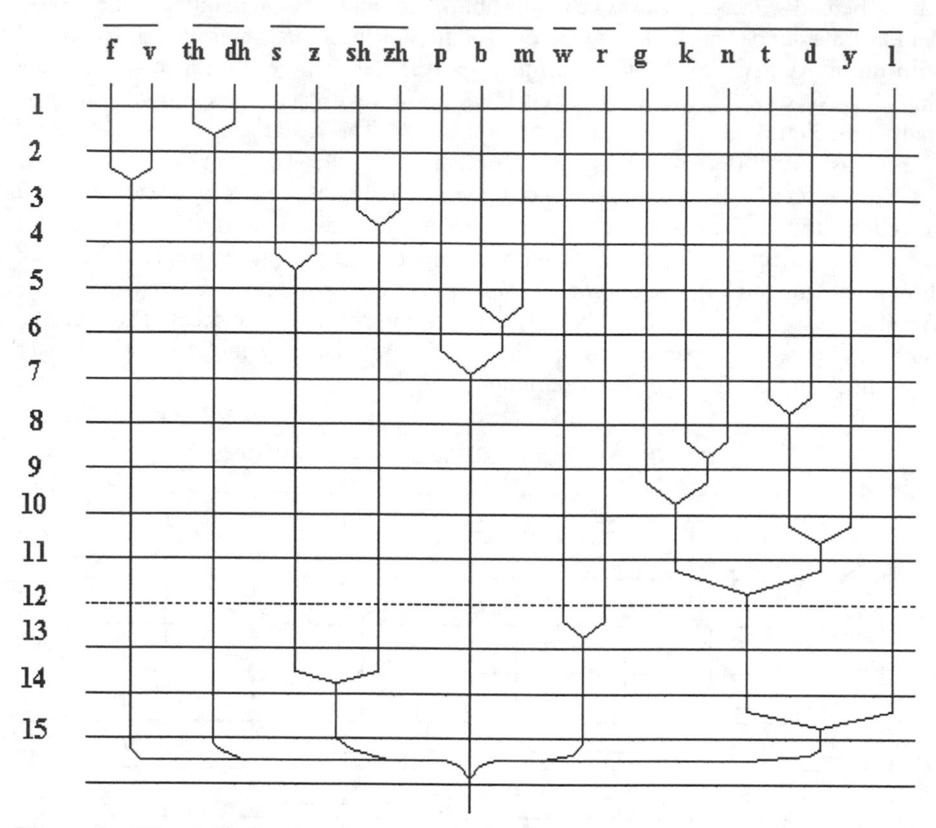

Figure 3 Visual confusion of consonant transitions CV in white noise among adult hearing-impaired persons, by decreasing the Signal-to-Noise Ratio. Consonants, which are initially discriminated, are progressively confused and clustered. When the 11th cluster is formed (dashed line), the resulting 9 groups of consonants can be considered distinct visemes (From B. Dodd, R. Campbell, "Hearing by Eye: the Psychology of Lipreading," Lawrence Erlbaum Assoc. Publ.).

In any case, seeing the face of the speaker is evidently of great advantage to speech comprehension and almost necessary in presence of noise or hearing impairments; vision directs the auditor attention, adds redundancy to the signal, and provides evidence of those cues which would be irreversibly masked by noise.

In normal verbal communication, the analysis and comprehension of the various articulation movements rely on a bimodal perceptive mechanism for the continuous integration of coherent visual and acoustic stimula. In case of impairments in the acoustic channel, due to distance, noisy environments, transparent barriers like a pane of glass, or to pathologies, the prevalent perceptive task is consequently performed through the visual modality. In this case, only the movements and the expressions of the visible articulatory organs are exploited for comprehension: vertical and horizontal lips opening, vertical jaw displacement, teeth visibility, tongue position, and other minor indicators like cheeks inflation and nose contractions.

Results from experimental phonetics show that hearing-impaired people behave differently from normal-hearing people in lipreading. In particular, visemes like bilabial /b, p, m/, fricative /f, v/, and occlusive consonants /t, d/ are recognized by each of them, while other visemes like /k, g/ are recognized only by hearing-impaired people. The occurrence of correct recognition for each viseme is also different between normal and hearing-impaired people: as an example, hearing-impaired people much more successfully recognize nasal consonants /m, n/ than normal hearing people. These two specific differences in phoneme recognition can be hardly explained since velum, which is the primary articulator involved in phonemes like /k, g/ or /m, n/, is not visible and its movements cannot be perceived in lipreading. A possible explanation, stemming from recent results in experimental phonetics, relies on the exploitation of secondary articulation indicators commonly unnoticed by the normal observer.

2.2 Speech Articulation and Coarticulation

When articulatory movements are correlated with their corresponding acoustic output, the task of associating each phonetic segment to a specific articulatory segment becomes a critical problem. Different from a pure spectral analysis of speech where phonetic units exhibit an intelligible structure and can be consequently segmented, the articulatory analysis does not provide, on its own, any unique indication on how to perform such segmentation.

A few fundamental aspects of speech bimodality have inspired interdisciplinary studies in neurology, physiology, psychology, and linguistics. Experimental phonetics have demonstrated that, in addition to speed and precision in reaching the phonetic target (that is, the articulatory configuration corresponding to a phoneme), speech exhibits high variability due to multiple factors such as

- psychological factors (emotions, attitudes);
- linguistic factors (style, speed, emphasis);
- articulatory compensation;
- intra-segmental factors;
- inter-segmental factors;
- intra-articulatory factors;
- inter-articulatory factors;
- coarticulatory factors.

To give an idea of the interaction complexity among the many speech components, it must be noticed that emotions with high psychological activity automatically increase the speed of speech production, and that high speed usually determines articulatory reduction (Hypo-speech) and a clear emphasized articulation is produced (Hyper-speech) in the case of particular communication needs.

Articulatory compensation takes effect when a phono-articulatory organ is working under unusual constraints; as an example when someone speaks while he is eating or with the cigarette between his lips.

Intra-segmental variability indicates the variety of articulatory configurations which correspond to the production of the same phonetic segment, in the same context, and by the same speaker. Inter-segmental variability, on the other hand, indicates the interaction between adjacent phonetic segments and can be expressed in "space," like a variation of the articulatory place, or in "time," meaning the extension of the characteristics of a phone.

Intra-articulatory effects are apparent when the same articulator is involved in the production of all the segments within the phonetic sequence. Inter-articulatory effects indicate the interdependencies between independent articulators involved in the production of adjacent segments within the same phonetic sequence.

Coarticulatory effects indicate the variation, in direction and extension, of the articulators movements during a phonetic transition. Forward coarticulation takes effect when the articulatory characteristics of a segment to follow are anticipated by previous segments, while backward coarticulation happens when the articulatory characteristics of a segment are maintained and extended to following segments. Coarticulation is considered "strong" when two adjacent segments correspond to a visible articulatory discontinuity or "smooth" when the articulatory activity proceeds smoothly between two adjacent segments.

The coarticulation phenomenon represents a major obstacle in lipreading as well as in artificial articulatory synthesis when the movements of the lips must be reconstructed from the acoustic analysis of speech, since there is no strict correspondence between phonemes and visemes. The basic characteristics of these phenomena is the nonlinearity between the semantics of the pronounced speech (despite the particular acoustic unit taken as reference) and the geometry of the vocal tract (representative of the status of each articulatory organ). Experimentation reveals that speech segmentation cannot be performed by means of only the articulatory analysis; articulators, in fact, start and complete their trajectories asynchronously exhibiting both forward and backward coarticulation with respect to the speech wave.

If a lip-readable visual synthetic output must be provided through the automatic analysis of continuous speech, much attention must be paid to the definition of suitable indicators, capable of describing the visually relevant articulation places (labial, dental, and alveolar) with the least residual ambiguity. This methodological consideration has been taken into account in the proposed technique by extending the analysis-synthesis region of interest to the region around the lips, including cheeks and nose.

2.3 Speech Synchronization in Multimedia Applications

In multimedia applications like video-phone and video-conferencing, this intrinsic bimodality of speech is definitely ignored so that audio and video signals are handled separately as independent channels. The highest delivery priority is typically given to audio to guarantee continuity and quality of decoded speech while far less concern is usually devoted to video. As a result, images are displayed as soon as they are decoded without taking into account their coherence with audio. Video, in fact, has no time reference when encoded and transmitted and, therefore, no synchronization can be guaranteed at the decoder.

As a consequence, the visual-acoustic bimodality of speech is lost and annoying artifacts are reproduced with perceivable incoherence between the movements of the speaker's lips and speech. This loss of quality imposes a severe impact on the human perceiver, especially if he has hearing impairments and his comprehension depends very much on the capability to correlate acoustic and visual cues of speech. Experimental results prove that a minimum of 15 synchronized video frames must be presented in one second to guarantee successful speech reading, meaning that acoustic cues extracted by the human hearing system from the speech input are not reliable for comprehension if they are not associated, over minimum time intervals of 60-70 ms, with coherent visual cues. In conclusion, it can be truly said that we hear not only by ear but also by eye.

To regain the lost synchronization between the acoustic and the visual modalities at the decoder, suitable post-processing must be applied with the usual constraint of real-time performances required by inter-personal communication applications like video-telephone and video-conferencing. In other applications based on virtual character animation, no real-time requirement is typically issued and a larger choice of possible solutions is offered.

Independently from the particular technical solution adopted, however, a correct audio/video re-synchronization mechanism must necessarily rely on *a priori* knowledge about the acoustic/visual correlations in speech whose evaluation, due the dependence on the language, on the speaker, on the linguistic and phonetic context, and on the speaker's emotional status, definitely represents a hard task.

In applications for very low bitrate video-phone coding, synthetic images of the speaker's mouth (generated from the articulatory estimates provided by speech analysis) can be interleaved with the actual images [34-35] thus increasing the frame rate at the decoder. Software and hardware demonstrators of speech-assisted video-phone are currently under development within the European ACTS project VIDAS.[2]

---

[2] VIDAS, activated in 1995 with the participation of DIST as coordinating member, is a 4-year project oriented to speech assisted video coding and representation.

In this scheme, a deformable wire-frame model is adapted to the mouth of an original frame and is then animated by means of the articulatory estimates provided by acoustic speech analysis. The animated textured mouth is therefore suitably pasted onto the original image thus generating interpolated/extrapolated synthetic images.

3. Lip Movements Estimation from Acoustic Speech Analysis

3.1 Corpus Acquisition

The first step is recording a valid corpus for training the system which is asked to correlate the two speech modalities and, afterward, be able to estimate visual cues from pure acoustic analysis of the speech signal. The main characteristics of a valid corpus are

- balance between phonemes (i.e., all the phonemes of the language must be present in almost similar occurrence);
- balance between frequencies (i.e., low, medium, and high frequency phones must be present in almost similar occurrence);
- balance between visemes (i.e., the different postures of the lips/tongue must be present with almost similar occurrence);
- coarticulation representativeness (i.e., phonemes sequences must be representative of any meaningful phone concatenation);
- intra-speaker representativeness (i.e., multiple utterances must be performed by the same speaker, possibly at different time intervals);
- inter-speaker representativeness (i.e., the corpus must include the largest possible number of speakers to be representative of different acoustic/articulatory characteristics in speech production).

In consideration of the huge amount of audio/video data to record and process, a reasonable approach would be that of first facing the problem from only a single-speaker point of view, thus decimating the corpus size. Since continuous speech analysis represents another main difficulty, an advisable suggestion would be that of focusing only on the analysis of single separate phones, diphones, and triphones, then passing on separate word analysis, and, only at the end, to continuous speech.

As pointed out before, a minimum rate of 15 Hz is necessary to perform any valid analysis of visual speech. A higher time resolution of video is recommended like 25/50 Hz (with video captured through PAL cameras) or 30/60 Hz (with NTSC cameras). Higher frame rates would help a lot, but the availability and cost of special professional cameras are often not affordable.

It may be noticed that, for the registration of the corpus, complete video information is definitely redundant since only the description of lips/tongue shape and position is needed while texture and color information are of no help in the analysis and represent 99% of the memory storage requirement.

For this reason, when the image acquisition system has enough computation power, it is usually more convenient to process video frames just after capturing and extracting the lips/tongue parameters on the flight without storing the entire image. This process is usually done through "Chroma-Key" techniques based on color segmentation. In this case both lips and tongue must be marked with special color make-up to aid in their extraction from the image. Getting rid of raw images just after capture, however, does not allow further parameter correction/integration as would be possible if images were stored.

As an alternative to articulatory measurement based on video camera acquisition (by far the more usual and easy method), infra-red cameras capable of detecting the 3D position of small markers (passive reflectors) placed along the lips contour could be used. In this case high time resolution can be reached (100 Hz), precise space localization and tracking of points is assured. Negligible memory storage is required, but the nature and size of markers typically constrain the spatial resolution and prevent locating them on the tongue. An effective solution would be to integrate the two acquisition procedures and record both video and markers at different rates. In case video is recorded for parameter extraction, the use of make-up is almost unavoidable for facilitating lip/tongue segmentation (see Figures 4 and 5). Lipstick can be used to enhance the lips/cheeks contrast with a suitable color, like blue-cyan, with significant chromatic difference from the typical "pink" hue of human skin. Tongue is usually difficult to be detected through image processing: appreciable advantage is obtained by pointing a light source frontally to the mouth and by using natural substances to color the mouth cavity (like blue-metylene or special paste used by TV actors and showmen to smooth-out the tongue color and enhance their contrast with teeth).

In order to evaluate the vertical aperture of the jaw (the jaw motion has three freedom degrees: vertical rotation, back-to-forward and, side translations), the distance between two rigid reference points must be measured. Typical points are chosen on the tip of the nose and on tip of the chin and must be marked in a way to facilitate their extraction and tracking.

In case of infra-red images with 3D localization, two reflectors can be placed for correspondence of these points. In case of video acquisition and processing, on the contrary, a suitable marker can be obtained by painting a small colored cross on the skin of nose and chin. Previous considerations of color still apply.

Since mouth articulation is properly three-dimensional and since some visemes are characterized by the protrusion of lips and by the position of the tongue tip with respect to teeth, a side view of the speakers mouth is almost necessary for integrating the frontal information in case of video-based acquisition. Stereo video acquisition of two orthogonal views (frontal and side) can be adopted by the use of a mirror placed on one side of the speaker's head and oriented at 45° degrees with respect to the camera.

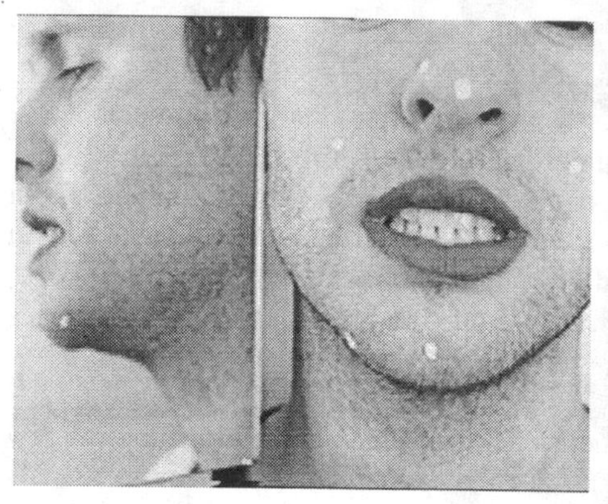

Figure 4. Make-up with coloured lipstick for chroma-key segmentation

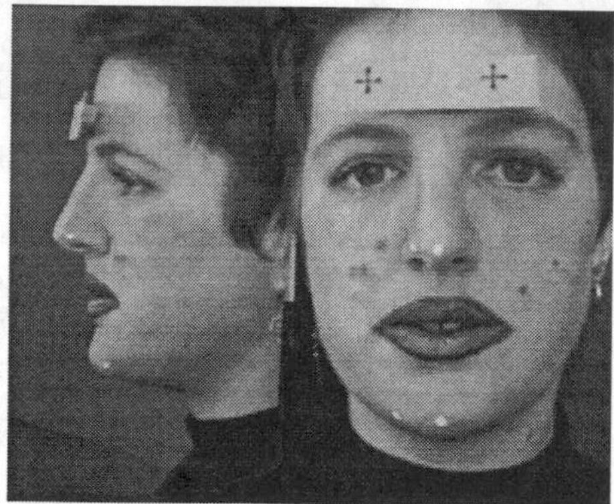

Figure 5. Make-up with lipstick and reference point on the forehead

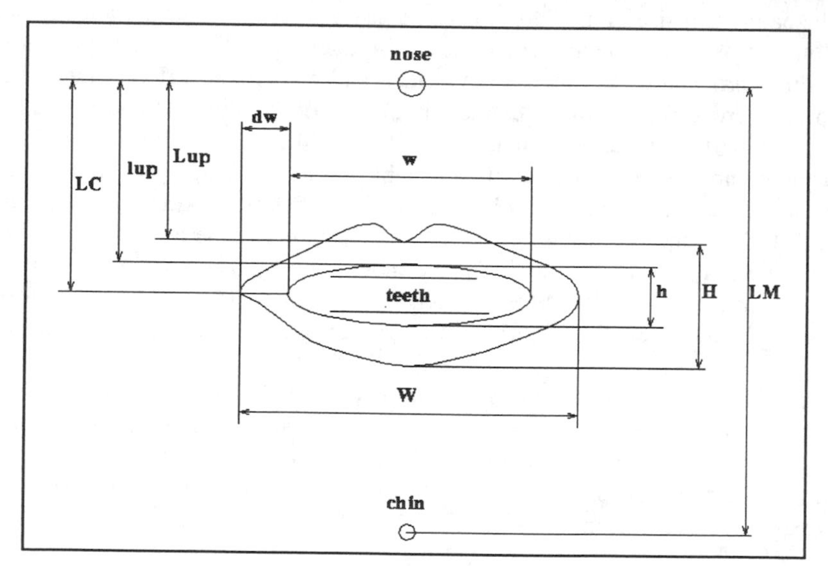

Figure 6. Model of the mouth with associated articulatory parameters (1063-6528/95$04.00 © 1995 IEEE).

H	external height of the mouth	***LC***	mouth-nose distance
h	internal height of the mouth	***Lup***	external lip-nose distance
W	external width of the mouth	***lup***	internal lip-nose distance
w	internal width of the mouth	***LM***	chin-nose distance (jaw aperture)
dw	segment of adjacency between the upper and the lower lips		

The mouth model, which has been employed in [8] and sketched in Figure 6, is defined by a vector of 10 parameters (*LC, lup, Lup, dw, w, W, LM, h, H, teeth*). The mouth articulatory parameters described in Figure 6 have been analyzed in order to evaluate their cross-correlation and provide a measure of their mutual dependence. In the following some examples of significant cross-correlation surfaces are reported in Figure 7, from which a basis of 5 almost noncorrelated parameters (*LM, H, W, dw, Lup*) has been defined.

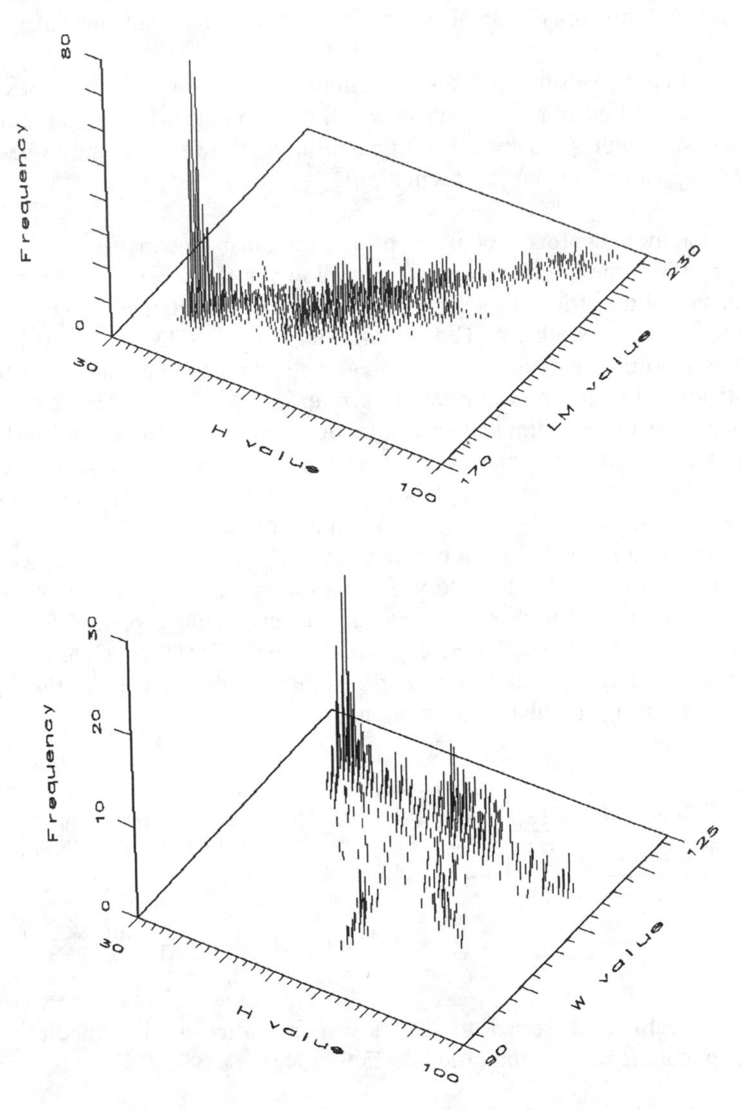

Figure 7. Analysis of the cross-correlation among *H-LM* and *H-W* pairs of articulatory parameters

3.2 Acoustic/Visual Speech Analysis

Extensive experimentation on normal hearing and hearing-impaired subjects [2-4] has clearly demonstrated that if, on one hand, phonemes can be associated rather easily to well defined mouth configurations (called "visemes"), the inverse association is usually troublesome since the same posture of the mouth can correspond to different phonemes. As an example, the "bilabial" viseme is associated to different phonemes like /m,p,b/, and the "velar" viseme is associated to different phonemes like /k,g/.

Moreover, intense investigations on the articulatory dynamics [5-11] stress the role played by the coarticulatory phenomena which describe the effects on articulation due to past acoustic outputs (backward coarticulation) and to future going-to-be-produced acoustic information (forward coarticulation).

A rather common approach consists of a preliminary phoneme recognition step followed by phoneme-to-viseme mapping as shown in the scheme of Figure 8. In this case estimates of the articulatory parameters are typically obtained by means of vector quantizers, neural networks, or Hidden Markov Models (HMM) [12-19], based on preliminary learning procedures for training the system to associate acoustic speech representations to coherent visual information. A very wide use of these methodologies is done in bimodal speech recognition for improving the performances of the system by adding visual cues to the conventional acoustic cues. It helps in exploiting the audio/video complementarity as it is usually done in speech comprehension performed by humans [20]. In all these methodologies, the phoneme-viseme association is performed in two separate and consecutive steps concerned with phoneme recognition and articulatory estimation, respectively. For this aspect, the approach is similar to that characterizing various algorithms proposed for converting text or phonetic transcriptions into audio/visual speech. Here the task of coarticulation modeling is performed based on *a priori* knowledge either during phoneme recognition or during articulatory estimation.

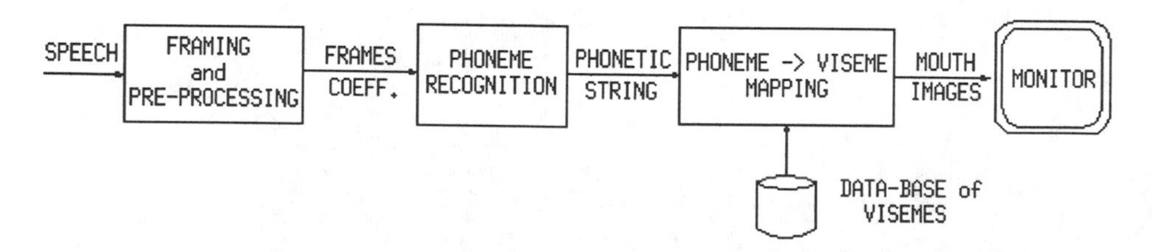

Figure 8. Speech is converted to lip movements after an intermediate stage of phoneme recognition (1063-6528/95$04.00 © 1995 IEEE).

A new approach to the problem, formulated in [8] and reported in the following section, is based on a well-established technology, i.e., Time-Delay Neural Network (TDNN). It has shown the possibility of merging the two computational steps of phoneme recognition and articulatory estimation, performed so far one after the other, into a single process embedding coarticulation modeling (see Figure 9). The advantage of using the TDNN solution for the direct estimation of the mouth articulation from acoustic speech is clearly due to the finite memory of its neurons. Since their output represents the response to the weighted sum of a variable number of past inputs, the system can base its estimation on a suitably sized noncausal speech registration. The supervised training of this kind of system, based on a large audio/video synchronous training set, has demonstrated appreciable performance in articulatory estimation without requiring any *a priori* knowledge.

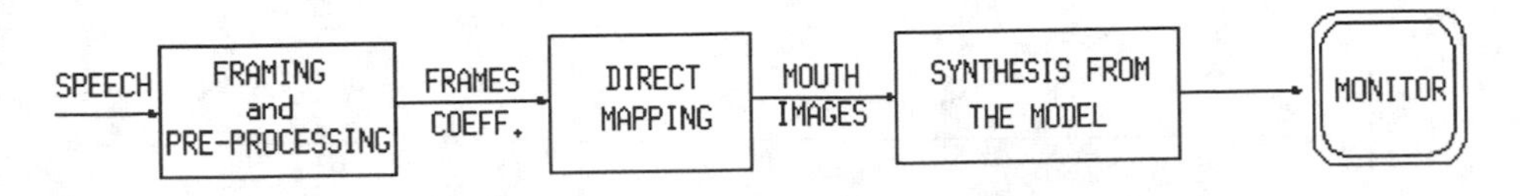

Figure 9. Speech is converted directly into lip movements without any intermediate stage of phoneme recognition (1063-6528/95$04.00 © 1995 IEEE).

4. The Use of Time-Delay Neural Networks for Estimating Lip Movements from Speech Analysis

4.1 The Implemented System

The speech conversion system has been implemented on a SGI workstation INDIGO 4000 XZ, 100 MHz, 48 bpp true color, z-buffer, double buffering, and graphic accelerator. The speech signal, after being sampled at 8 KHz and quantized linearly at 16 bits, undergoes multistage processing oriented to

- spectral preemphasis;
- segmentation into nonoverlapped frames of duration $T = 20$ ms (160 samples per frame);
- linear predictive analysis of 10-th order;
- power estimation and computation of the first 12 cepstrum coefficients;
- frame normalization.

Preemphasis is obtained through a FIR filter with transfer function $F(z) = 1 - a\, z$ ($a = 0.97$). The frame duration T has been chosen equal to 20 ms in order to associate two consecutive audio frames to the same video frame (25 video frames/sec). Each frame has been filtered by means of Hamming windowing to reduce the spectral distortion and analyzed through the Durbin procedure for the estimation of 10 LPC coefficients. By means of simple linear operations, the envelope of the cepstrum is estimated and its first 12 coefficients are computed. The frame power (obtained directly from the value R^(0) of the estimate R^(t) of the autocorrelation function) is normalized to the range [P_{min}, P_{max}], where P_{min} and P_{max} are known *a priori* and represent the noise power and the maximum expected signal power, respectively. The 12 cepstrum coefficients of the frame are then normalized to the range [−1, 1] and finally multiplied by the normalized power. From the example shown in Figure 10 it is apparent that the normalization procedure reshapes the cepstrum coefficients according to the power envelope.

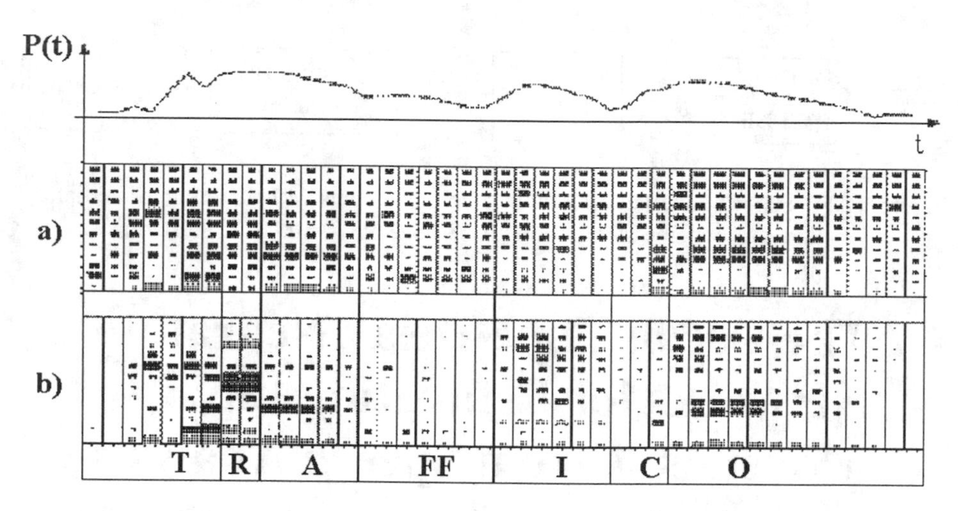

Figure 10. Sonogram of the Italian word "traffico" before (a) and after normalization of the ceptrum coefficients (b) (1063-6528/95$04.00 © 1995 IEEE).

The normalized 12-dimensional vector of cepstrum coefficients is then presented to the actual conversion system. As described in Figure 11, conversion is based on a bank of Time-Delay neural networks (TDNN), each of them trained to provide estimates of the corresponding articulatory parameters. The TDNN outputs are then smoothed and sub-sampled 1:4 in order to associate the same configuration of articulatory parameters to 4 consecutive frames. The smoothing filtering is applied to stabilize the estimates while sub-sampling is forced by hardware constraints; the visualization system used can, in fact, display video frames at a maximum frequency of 12.5 frames/sec. corresponding to 80 ms speech segments. A synthesis program finally employs the vector of articulatory parameters to modify the wire-frame structure which models the face.

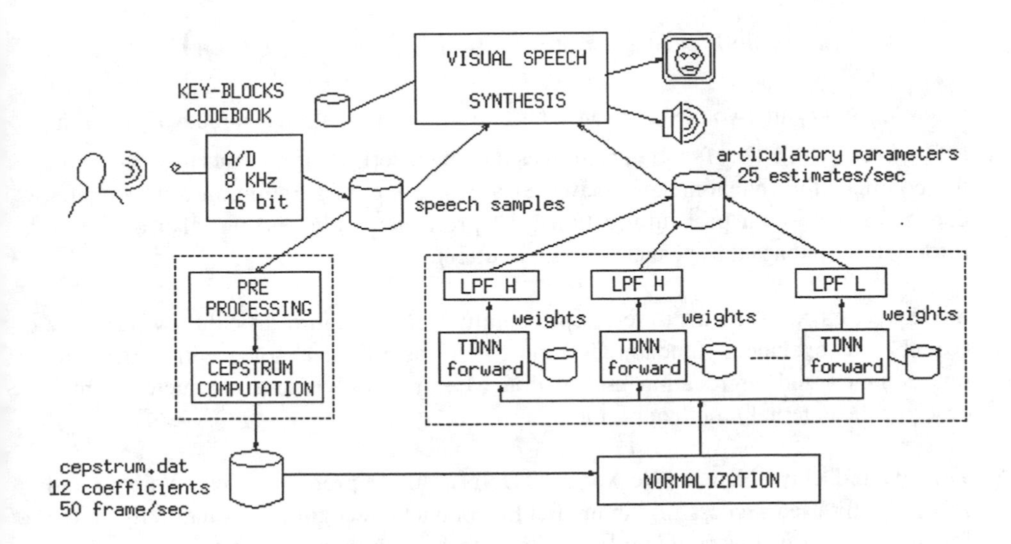

Figure 11. Scheme of the analysis-synthesis system implemented on the Silicon Graphic workstation (1063-6528/95$04.00 © 1995 IEEE).

4.2 The Time-Delay Neural Network

Classification and functional approximation are typically static tasks in which a unique output vector is associated to any possible input. Many natural processes have however an intrinsic time evolution like those related to the coordinated generation of the many body gestures involved in walking, dancing, writing, singing or, closer to the work reported in this chapter, in speaking. In these cases the recognition of a particular input configuration and the definition of suitable output values must be performed through the analysis of time correlated data implying the availability of suitable mechanisms for representing the time dependence between the network structure and its dynamics.

There are many different ways through which a neural network can represent the time information: recursive connections can be introduced, cost functions with memory can be employed or, alternatively, suitable time delays can be used as in the case of TDNNs. All these solutions exhibit peculiar characteristics which are more suited to handle some specific problems than others, so their appropriate choice is critical. The task of estimating the articulatory mouth parameters from the acoustic speech waveform can be formalized as follows: given a set of pairs $\{\boldsymbol{x}(k), \boldsymbol{d}(k)\}$ where $\boldsymbol{x}(k)$ represents the k-th input vector to the network (whose components can be samples of the short-time spectrum envelope estimated from the k-th acoustic frame) and $\boldsymbol{d}(k)$ the corresponding target vector (whose components correspond to the articulatory parameters of the mouth measured at the same instant), a function may be defined as follows:

$$\boldsymbol{u}(k) = \boldsymbol{d}\hat{}(k) = \mathrm{G}\left(\boldsymbol{x}(k-k_p), \boldsymbol{x}(k-k_p+1), \ldots, \boldsymbol{x}(k+k_f-1), \boldsymbol{x}(k+k_f)\right)$$

where k_p and k_f are two positive constants and $\boldsymbol{u}(k)$ represents an estimate $\boldsymbol{d}\hat{}(k)$ of the actual vector $\boldsymbol{d}(k)$. This function expresses the characteristic of noncausality typical of the coarticulatory phenomenon for which the mouth configuration, at a given time, depends not only on past information (just pronounced phones) but also on future information (phones the speaker is going to utter).

Since vectors $\mathbf{x}(k)$ are a discrete representation of the entire articulatory structure (vocal cords excluded), function G( ) must work as a filter charged with processing this complex and aggregating information to extract only those parameters which describe the external mouth appearance.

The Time-Delay Neural Network (TDNN), first proposed by Waibel and subsequently used successfully in the field of phonetic recognition, is naturally suited for the solution of this problem. In contrast to conventional neurons which provide their response to the weighted sum of the current inputs, the TDNN also extends the sum to a finite number of past inputs (neuron delay or memory). In this way the output provided by a given layer depends on the output of the previous layers computed over an extended domain (in our case the time domain) of input values. The particular structure of a TDNN also allows the extension of the classical back-propagation algorithm and its complexity optimization.

The elementary unit of such network is the classic perceptron, where the weighted sum includes not only the current input pattern but also a certain number D of past inputs, as shown in Figure 12. The implemented function is

$$u_{l,j}^{t} = \sum_{d=0}^{D(l)} \sum_{i=0}^{N(l)} \left(u_{l-1,i}^{t-d} w_{l,i,j}^{d} \right)$$

As shown in Figure 13, a multi-layer perceptron network is composed of a pyramid of these elementary units for providing enhanced temporal characteristics. The first hidden layer concentrates on the temporal information coming from $D(1)$ input patterns. The subsequent layers collect information from temporal windows of increasing size.

It is worth noting that, in this way, less weights are used with respect to those which would be necessary to cover a temporal window of the same extension with a classic multi-layer perceptron network.

Moreover, the lower layers correlate only information close in time, while information coming from an extended temporal domain is integrated by the higher layers of the network. At a given time, the network output (in our architecture, one single neuron) depends on the temporal pyramid developed in the previous layers.

Feeding the network with patterns representative of a dynamic process, the generated output patterns can be put in correspondence to a distinct dynamic process correlated to the input one. The difference between the sequence of patterns, generated by the TDNN and the "target" output sequence, is used to train the network through the back-propagation algorithm.

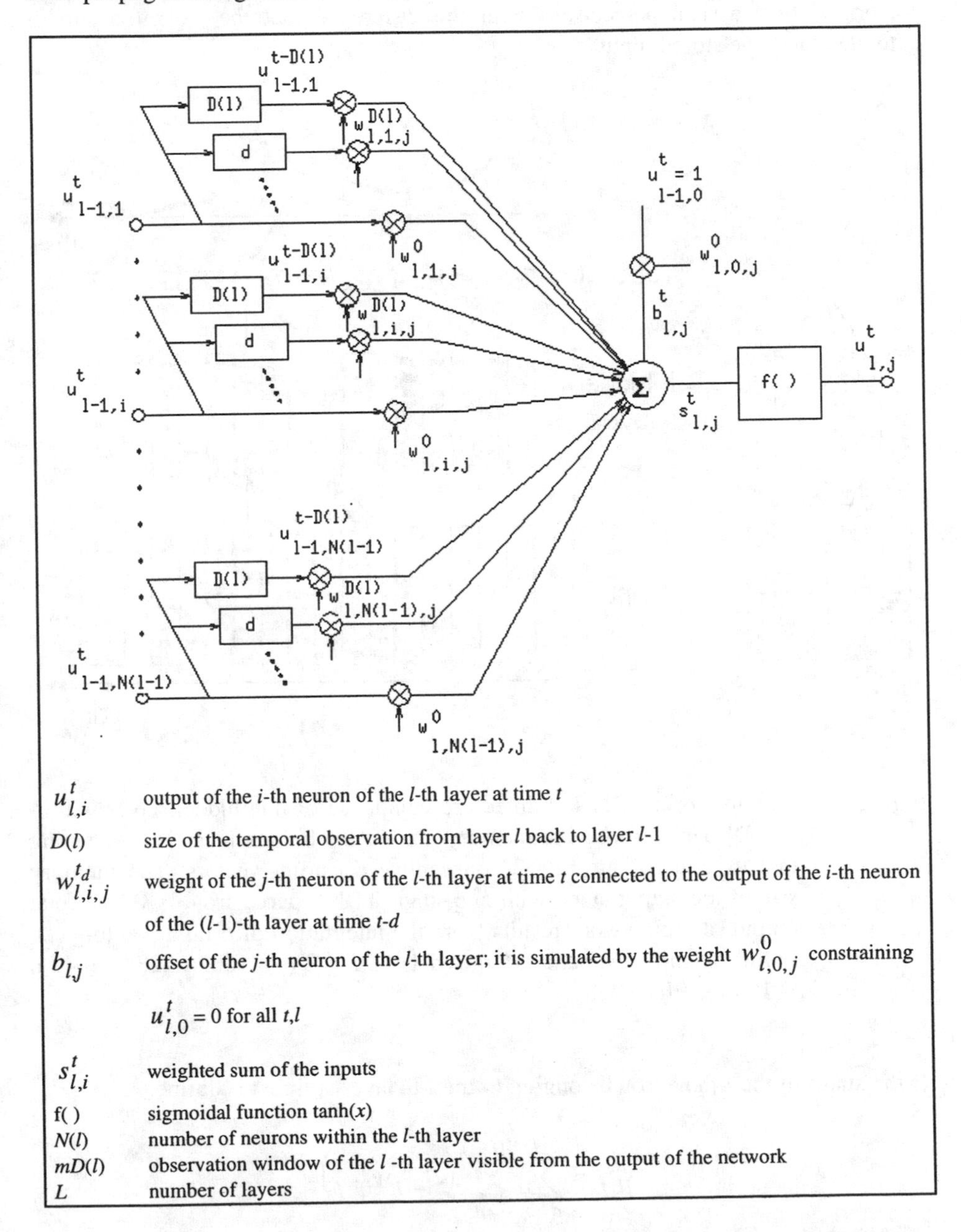

Figure 12. Scheme of the TDNN perceptron (1063-6528/95$04.00 © 1995 IEEE).

Counter l is referred to the layer which contains the currently examined neuron; in the following, $l = 0$ will represent the input to the network (which coincides with the pattern it must learn) and $l = 1, 2, ..., L$-1 the subsequent hidden layers with L being the total number of layers. Each neuron has a finite size buffer used to store the $D(l)$ past inputs; the nonlinear operation performed by the neuron is similar to that performed by classical perceptrons with the difference that the weighted sum is extended to all the stored inputs.

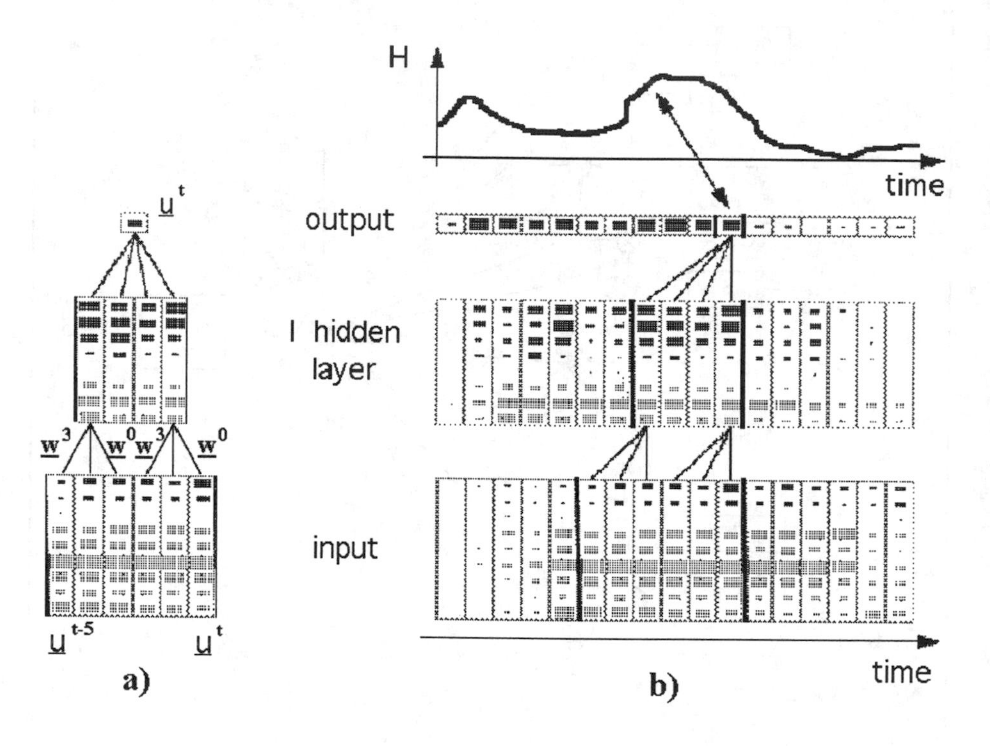

Figure 13. Example of a TDNN architecture composed of one hidden layer with 4 TDNN perceptrons. The memory size of the hidden layer is 3, meaning that its outputs integrate 3 consecutive cepstrum vectors. The memory size of the output layer, with one single TDNN perceptron, is 4. The final output of the network results from the integration of 6 input vectors (a). In this example the pattern-target delay DT is 3 (b) (1063-6528/95$04.00 © 1995 IEEE).

If the status of the j-th neuron belonging to the l-th layer at time t is defined as

$$s_{l,j}^{t} = \sum_{d=0}^{D(l)} \sum_{i=1}^{N(l-1)} \left(u_{l-1,i}^{t-d} w_{l,i,j}^{d} \right)$$

the neuron output $u_{l,j}^{t}$ can be expressed as

$$u_{l,j}^{t} = \mathrm{f}\left(s_{l,j}^{t}\right) = \mathrm{f}\left(\sum_{d=0}^{D(l)} \sum_{i=1}^{N(l-1)} \left(u_{l-1,i}^{t-d} w_{l,i,j}^{d}\right)\right)$$

Let us examine a simple TDNN with one single hidden layer and describe how the information introduced into the input layer is propagated up to the output layer.

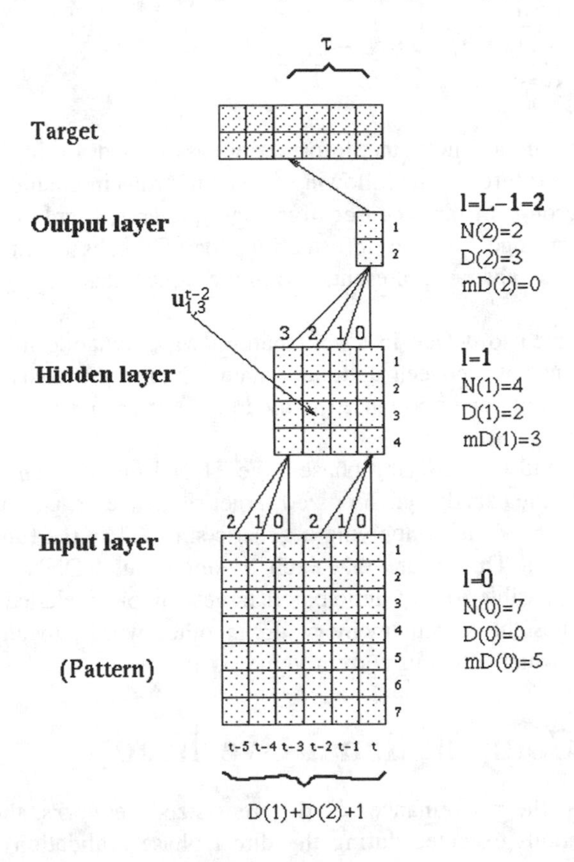

Figure 14. Information flow in a TDNN. The structure has been exploded in time to highlight the extension of the time window (memory) processed by the network (1051-8215/97$10.00 © 1997 IEEE).

Indicating with $D(1)$ the buffer size for neurons in the hidden layer, $D(1)+1$ input patterns are necessary to originate the first valid output value; in other words, the time window "seen" from the hidden layer includes $D(1)+1$ time instants. If $D(2)$ is the buffer size for neurons in the output layer, $D(2)+1$ vectors must be provided by the previous layer before the network is able to produce the first valid output. The time window "seen" from the output layer is therefore extended to $D(1)+D(2)+1$ time instants.

From these considerations we can imagine a time-exploded representation of the information flow through the network as shown in Figure 14, where the parameter $mD(l)$, indicating the number of past time instants for which the output from the l-th layer affects the network output at any given time, is defined as

$$mD(l) = \begin{cases} 0 & l = L-1 \\ \sum_{d=0}^{L-1} D(h) & l = L-2, \ldots, 0 \end{cases}$$

Further on, we can notice how the information associated to input patterns is subsequently integrated through the following layers: neurons in the hidden layer sum data coming from groups of three consecutive input patterns, highly correlated with one another, and arrange suitable information for the subsequent layer which therefore operates on an abstract representation of the same data.

The network is inclined to detect, in a very natural way, dynamic information like that associated to transients between stationary status or to sequence dynamics which actually represent cues of key importance in any articulatory analysis.

During the learning and the verifying phases, the TDNN outputs can be compared, sample by sample, to the exactly synchronized target sequence or, alternatively, to its anticipated version obtained after applying a generic shift DT in the future limited to the interval $[1, mD(L)]$. This means that each instantaneous TDNN output can be made as similar as possible to any anticipated target sample included in the time-window which defines the system memory or, in other words, to any past target sample among those which affect its current value.

4.3 TDNN Computational Overhead

In order to compare the performances of different sized networks, the number of floating-point operations executed during the direct phase (validation) represents a valid and homogeneous yardstick. This number, in fact, is directly correlated to the proficiency of the network since it is in proportion to the network size (number of free variables) and, consequently, to the complexity of the solvable problems.

The number of operations can be computed in two steps. Let us define the number of input patterns which are necessary for computing each output of the network, that is, the time-window $maxW$ which is "seen" from the input layer

$$maxW = 1 + mD(0)$$

The number of operations N_{op1} necessary to initialize the network or, in other terms, to fill all the neuron buffers with data derived from the first $maxW$ input patterns and determine the first valid output value, is equal to

$$N_{op1} = \sum_{l=1}^{L-1} N(l)[N(l-1)+1][D(l)+1][maxW - \sum_{k=1}^{l} D(k)]$$

while the number of operations N_{opS} corresponding to each new input vector, necessary for producing subsequent output values, is equal to

$$N_{opS} = \sum_{l=1}^{L-1} N(l)[N(l-1)+1][D(l)+1]$$

If N_{pat} defines the number of input patterns, the number of floating-point operations (sums and multiplications) necessary for the feed-forward phase is

$$N_{opT} = N_{op1} + N_{opS} \ldots\ldots (N_{pat} - maxW)$$

The value of N_{opT} provides a measure of the computational speed of the network and, therefore, of its performances. Let us also notice that, since N_{opS} is equivalent to the number N_w of network weights, it actually represents an estimate of the TDNN memory requirements.

4.4 Learning Criteria for TDNN Training

The quality of the TDNN learning can be assessed through different methods depending on the kind of performances required. In order to optimize the network size and parameters, the cost curves defined in the previous section have been analyzed. This analysis, however, loses validity when the performances of structurally different networks must be compared. In this case it is necessary to determine suitable performance figures for validating the estimates provided by the network which are independent from the network structure itself. Moreover, secondary figures must be considered like the computational speed (number of operations executed in the direct phase), the storage required and, optionally, a threshold on the maximum acceptable error.

Indicating with u_t and d_t the TDNN output and target vectors at time t, the following costs can be defined:

$$\text{mean square error } \mathbf{MSE} = \frac{1}{N_{pat}} = \sum_{l=1}^{N_{pat}} [u_t - d_t]^2$$

This figure, despite being the most commonly used error, turns out to be unreliable for the specific estimates of the network. What is basically required from the network is, in fact, the ability to follow the parameters modes without necessarily tracking their trajectories point by point. A linear phase distortion, as an example, producing a constant time delay between the pattern and the target sequences, yields a high **MSE** value but can often correspond to acceptable or even very good estimates.

$$\text{maximum error } \mathbf{MAX} = \max \left(| u_t - d_t | \right) |_{t \text{ in } [1, N_{pat}]}$$

Even if the network output values are limited to the interval [−1.0, +1.0], this kind of error does not provide reliable indications since it depends on the distribution of d_t. The analysis of **MAX** during learning indicates that, after an initial decreasing phase, it increases in contrast to the **MSE** value which is progressively reduced (corresponding to an improvement of the global network performances).

$$\text{pattern-target cross-correlation coefficient } \boldsymbol{r} = \mathrm{E}\left\{ (u_t - \mathrm{E}\{u_t\})(d_t - \mathrm{E}\{d_t\}) \right\} / \sigma_u \sigma_d$$

In contrast to the previous figures, the cross-correlation coefficient measures the similarity between the two sequence with invariance to translations and to the particular distribution of the sequence samples.

$$\text{cross-correlation } \mathbf{R}_{ud}[\tau] = \mathrm{E}\{u_t d(i+\tau)\} = \sum_{t=0}^{N_{pat}-\tau} u_t d_{i+\tau}$$

It provides the same advantage of the cross-correlation coefficient; the occurrence of time translations is detected in case the maximum of $\mathbf{R}_{ud}[\tau]$ is shifted with respect to $\tau = 0$. For evaluating the reliability of the estimate, the curve $\mathbf{R}_{ud}[\tau]$ can be compared to the self-correlation $\mathbf{R}_{dd}[\tau]$.

4.5 Multi-Output vs. Single-Output Architecture

Keeping constant the global computational overhead of each examined system configuration, the performance figures previously described have been adopted to check the convenience of using a single multi-output TDNN in alternative to multiple single-output TDNNs. For this reason a parametric optimization has been applied to find out the most suitable configuration for a 5-output TDNN charged with estimating a vector of 5 mouth articulatory parameters, in particular, as shown in Figure 10, the vertical offset of the upper lip *Lup* and the contact segment between the upper and lower lips *dw*.

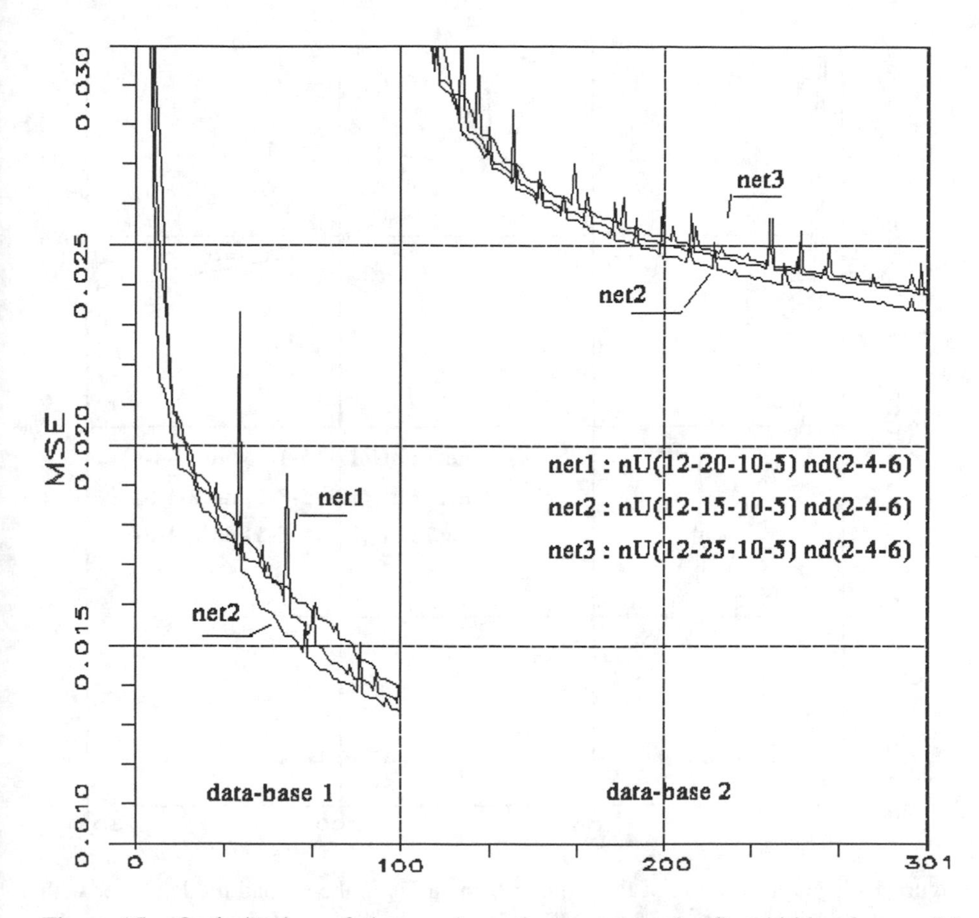

Figure 15. Optimization of the number of neurons in the first hidden layer with minimization of the **MSE**. In this case, either in case of data-base 1 (vowels) or data-base 2 (V/C/V transitions), the configuration net2 with 15 neurons in the first hidden layer provides the least **MSE** (1051-8215/97$10.00 © 1997 IEEE).

Constraining the investigated TDNN configurations to have constant complexity, 2 hidden layers and a pattern-target delay $DT = 8$, the cost descending speed (correlated to the level of learning) has been analyzed by varying the number of neurons in the first and in the second hidden layer and by changing the size of the neuron memory buffer. Although these tests have been carried out separately for minimizing the **MSE**, minimizing **MAX**, or maximizing **r**, the optimal TDNN configuration has proven to be almost independent of the specific cost functional adopted. Figures 15 and 16 provide an example of the optimization procedure followed in case of **MSE** minimization.

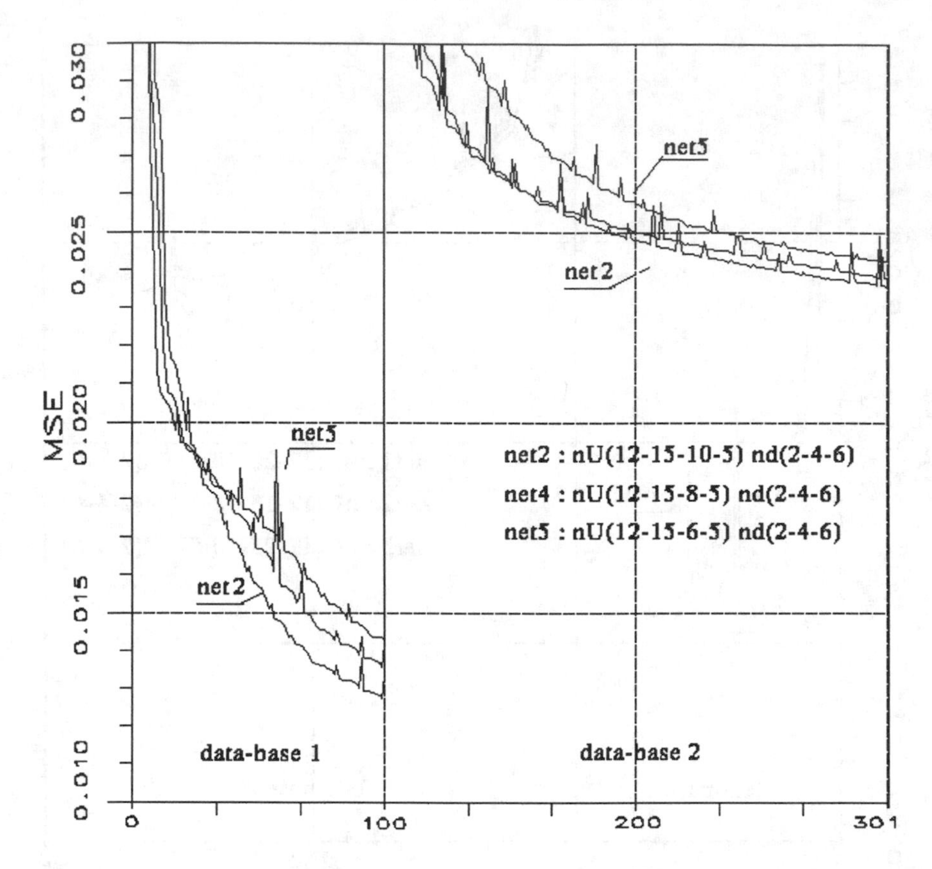

Figure 16 Optimization of the number of neurons in the second hidden layer with minimization of the **MSE**. In this case the number of neurons in the first hidden layer has been fixed to 15 for all the configurations. Either in case of data-base 1 (vowels) or data-base 2 (V/C/V transitions), the configuration net2 with 10 neurons in the first hidden layer provides the least **MSE** (1051-8215/97$10.00 © 1997 IEEE).

The TDNN has been trained on the first 2 sections of the corpus (vowels and V/C/V transitions) for a total of 1000 and 2000 iterations "by epoch," respectively. In order to compare the system based on one single 5-output TDNN with the system based on a bank of 5 single-output TDNNs, parameters have been suitably configured to maintain a constant computational overhead.

The two most performant 5-output configurations **NetA** and **NetB** have been chosen for comparing the results. The former has 15 neurons with 3-order memory in the first hidden layer, 10 neurons with 6-order memory in the second, and 9-order memory neurons in the output layer. The latter has 25 neurons with 2-order memory in the first hidden layer, 10 neurons with 4-order memory in the second, and 6-order memory neurons in the output layer. The optimal single-output configuration is indicated as **NetO**. *nU* indicates the number of neurons contained in each layer; *nD*, the memory size (delay) of each layer; N_{op1}, the number of operations required to initialize the network; and N_{opS}, the number of operations executed by the TDNN for each input vector, with obtained data reported in Table 1 have been obtained. It can be noticed that the number of operations executed by **Net0** has been multiplied by 5 to take into consideration the presence of 5 TDNNs in parallel.

Table 1. Comparison of the various TDNN configurations with respect to the computational overhead they require. **NetA** and **NetB** are the two most performant 5-output configurations while **Net0** indicates the optimal single-output configuration (in this case, the multiplication by 5 takes this parallelism into account to make the evaluation homogeneous with the other two (1051-8215/97$10.00 © 1997 IEEE).

	nU	*nD*	N_{op1}	N_{opS}
Net0x5	12-8-3-1	2-4-6	22025	2375
NetA	12-15-10-5	3-6-9	24230	2450
NetB	12-25-10-5	2-4-6	20210	2660

Networks **NetA** and **NetB** have been trained according to two different procedures defined as T1 and T2, respectively.

T1 First learning on a corpus of vowels (DB1) with 1000 iterations, followed by a second learning on a corpus of V/C/V transitions (DB2) with 2000 iterations. A corpus of isolated words (DB3) has been used as a testing set.

T2 Learning on an extended corpus (DB1+DB2+DB3) with 1000+2000+2000 iterations. DB3 was also used as a testing set.

The performances of these three TDNN configurations have been evaluated in terms of **MSE**, **MAX**, and ***r*** for each of the 5 mouth articulatory parameters (*LM, H, W, Lup, dw*). The experimental results obtained by applying the two different training procedures **T1** and **T2** are reported in Tables 2 and 3, respectively.

Results reported in Table 2 confirm the hypothesis that configuration **Net0** independently minimizes the cost figure for each parameter outperforming **NetA** and **NetB**. In the case of **Net0**, in fact, the weights of each TDNN are tuned independently toward a unique objective in contrast to **NetA** and **NetB** where 5 uncorrelated objectives must be met by the same, although larger structure.

Table 2. Performance comparison evaluated for each mouth articulatory parameter for the configurations **Net0**, **NetA**, and **NetB** by applying the training procedure **T1** (1051-8215/97$10.00 © 1997 IEEE).

T1	PARAMETER								
	LM			H			W		
Cost	net0	netA	netB	net0	netA	netB	net0	netA	netB
MSE	2.761	3.665	3.358	3.380	4.691	4.150	3.762	3.460	2.312
MAX	0.863	1.072	0.762	0.783	1.103	0.773	0.653	0.732	0.771
r	0.9435	0.9186	0.9269	0.9386	0.9060	0.9188	0.8586	0.7855	0.7777
T1	PARAMETER								
	Lup			DW			Average		
Cost	net0	netA	netB	net0	netA	netB	net0	netA	netB
MSE	1.994	3.717	3.687	5.531	8.933	8.969	3.015	4.646	4.495
MAX	0.704	0.911	0.871	1.020	1.078	1.148	0.804	0.972	0.865
r	0.8090	0.5966	0.6031	0.7067	0.4448	0.4387	0.8513	0.7303	0.7330

Table 3. Performance comparison evaluated for each mouth articulatory parameter for the configurations **Net0**, **NetA**, and **NetB** after applying the training procedure **T2** (1051-8215/97$10.00 © 1997 IEEE).

T2	PARAMETER								
	LM			H			W		
Cost	net0	netA	netB	net0	netA	netB	net0	netA	netB
MSE	5.782	6.622	6.148	7.091	8.008	7.309	3.762	3.460	3.700
MAX	0.988	1.006	0.957	1.095	1.066	1.106	1.132	1.071	1.097
r	0.8599	0.8342	0.8489	0.8474	0.8260	0.8420	0.5619	0.5887	0.5655
T2	PARAMETER								
	Lup			DW			Average		
Cost	net0	netA	netB	net0	netA	netB	net0	netA	netB
MSE	16.18	18.27	18.75	10.93	10.12	10.03	8.569	9.296	9.187
MAX	1.384	1.430	1.490	1.354	1.090	1.210	1.190	1.132	1.172
r	0.3645	0.2937	0.2858	0.2701	0.2928	0.3184	0.5807	0.5671	0.5721

NetB yields average performances higher than **NetA** confirming that the optimal size for the time-window on which the articulatory estimates are based equals 260 msec. while 380 msec time-windows (19 x 20 msec) are used in the case of **NetA**.

4.6 MSE Minimization vs. Cross-Correlation Maximization

Different from the MSE-based learning where the TDNN output values try to track exactly the target sequence, when the cross-correlation is minimized, the network output is similar to the target sequence but usually different in amplitude.

Noticing that two coherent sinusoids always have unitary cross-correlation despite possible differences in amplitude and mean value, the output produced by the TDNN can be easily adapted to the specific target sequence by means of suitable scale and shift factors. The main advantage with this kind of learning is that, since there is no constraint on the output absolute value, the TDNN neurons operate in the linear interval of the activation function thus leading to fast convergence.

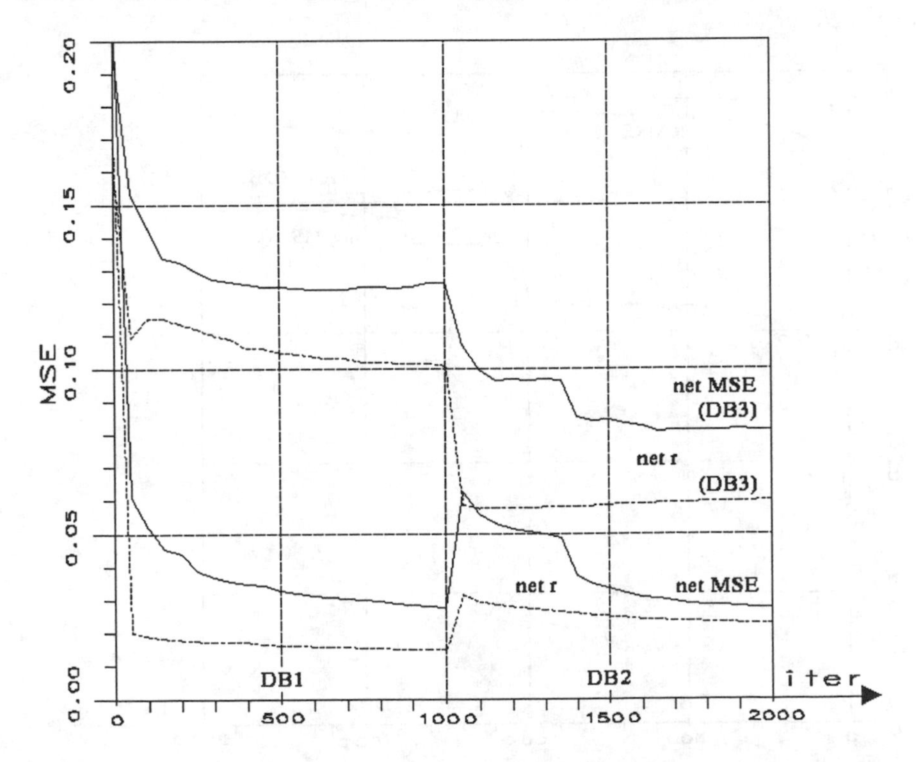

Figure 17. The lower curves provide a comparison between the **MSE**-based and ***r***-based performances expressed in terms of pattern-target MSE. The first 1000 iterations have been done using DB1 as training/testing set while DB2 was used in the second 1000 iterations. The upper curves have been computed using DB3 as cross-validation set for the entire (2000 iterations) learning phase (1051-8215/97$10.00 © 1997 IEEE).

The size of the TDNN can be indicated as nU(12-8-3-1), meaning 12-dimensional input vectors, two hidden layers with 8 and 4 neurons, respectively, and a single output neuron. Delays have been sized as nD(2-4-6), indicating a delay of 2, 4, and 6 time instants at the first hidden, second hidden, and output layer, respectively. The training of the network has been performed in two steps, first on a simple audio-video database DB1 (vowels) and then on a more complex database DB2 (V/C/V transitions). In Figures 17 and 18 a comparison between the **MSE** and the ***r*** learning curves is reported with reference to 2000 iterations "by epoch," 1000 for each database. The comparison is done with reference to the training set (DB1-DB2) and also to a testing set (DB3) containing isolated words and used for cross-validation. Comparison has been done in terms of patter-target **MSE** (minimization) as well as of patter-target cross-correlation (maximization). The experimental curves show that, for this specific estimation problem, learning based on cross-correlation outperforms the conventional **MSE**-based as far as both convergence speed and distortion are concerned.

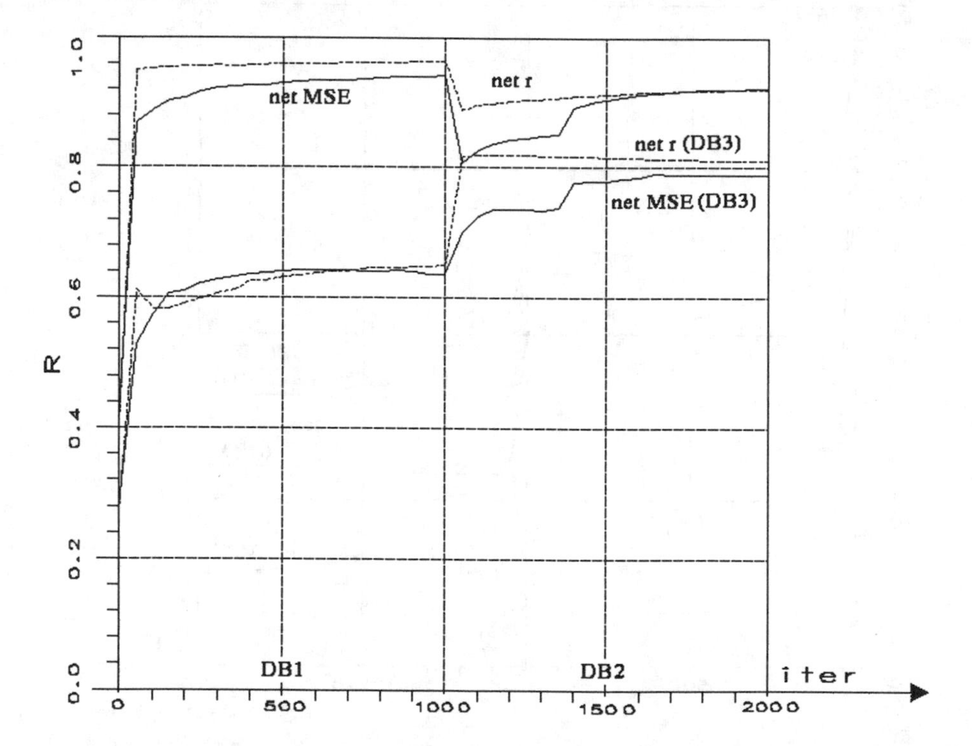

Figure 18. The lower curves provide a comparison between the **MSE**-based and ***r***-based performances expressed in terms of pattern-target cross-correlation. The first 1000 iterations have been done using DB1 as training/testing set while DB2 was used in the second 1000 iterations. The upper curves have been computed using DB3 as cross-validation set for the entire (2000 iterations) learning phase (1051-8215/97$10.00 © 1997 IEEE).

5. Speech Visualization and Experimental Results

Our experimental results are satisfactory using networks with two hidden layers, composed of 8 and 3 units each, with $D(1) = 2$, $D(2) = 3$ and $D(3) = 4$, so that each output pattern depends directly on the previous 9 input patterns.

In Figures 19 and 20, the articulatory parameter *H*, estimated through the network with reference to a test sequence, is compared to the actual parameter values.

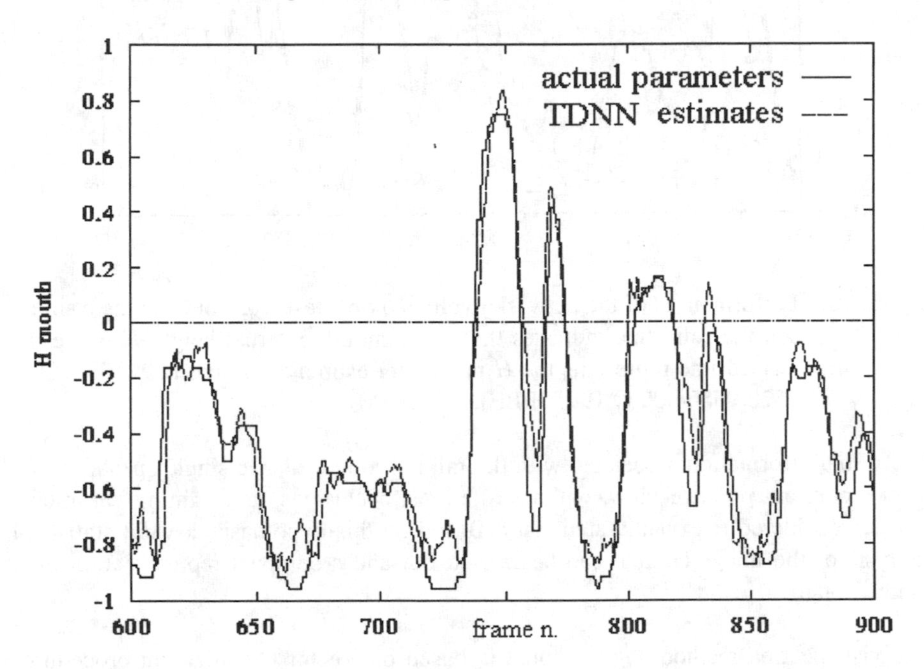

Figure 19. Performances of the network evaluated on a test word extracted from the training set: the solid line indicates the actual mouth external height *H* while the dashed line represents the estimated *H* parameter (1063-6528/95$04.00 © 1995 IEEE).

Any effective synthesis of visual speech cues should reproduce on the screen all the necessary articulatory information with usually associated a talking mouth. The articulatory estimates derived from the analysis of acoustic speech are usually very coarse and basically limited to the lips horizontal and vertical aperture. Important visible cues like teeth visibility and tongue position are generally characterized by acoustic signitures, too weak to be discriminated in noise; this fact reflects in poor visualization and consequent confusion in speech reading.

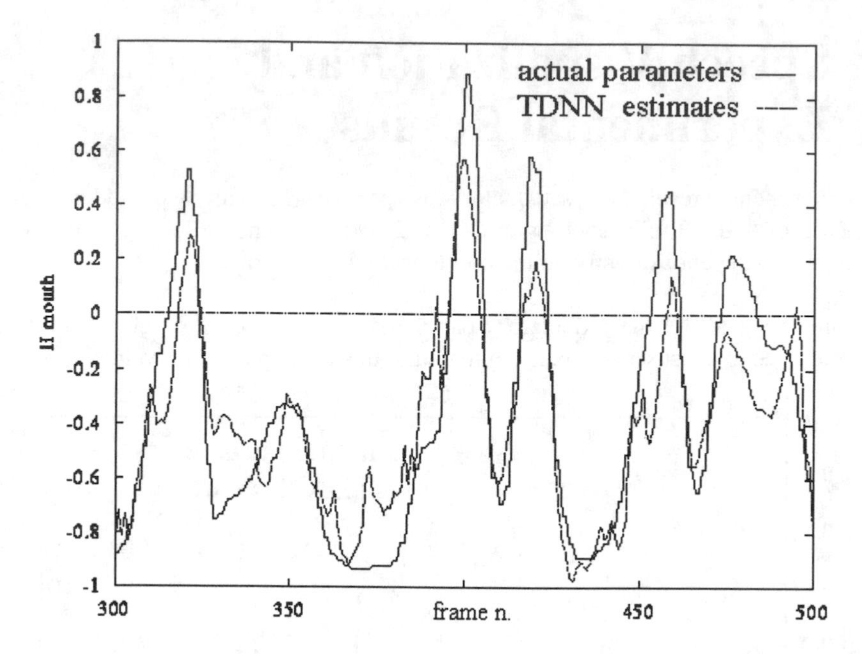

Figure 20. Performances of the network evaluated on a test word outside the training set: the solid line indicates the actual mouth external height H while the dashed line represents the H parameter estimated by the network (1063-6528/95$04.00 © 1995 IEEE).

The visual information associated with the talking mouth of one single speaker, seen from a constant point of view, without rotations, occlusions, and lighting variations, can be considered reasonably stationary. Based on this hypothesis, a valid statistical analysis of the image content can be carried out and a compact representation of it can be obtained.

The visualization methodology adopted is based on Vector Quantization procedures applied to clusterize visems in articulatory spaces of increasing dimensionality and then, to represent them in the pixel domain as a combination of elementary blocks. The audio-video Italian corpus we have recorded includes more than 30,000 images synchronized with speech where all the necessary details of the mouth visible articulation are reproduced. Each image has been automatically classified in terms of articulatory descriptors of varying complexity, ranging from the plain pixel coordinates of some specific feature to more sophisticated shape description of the lips' contour. Since each image contains both front and side views of the speaker's face, the articulatory description is expressed by orthogonal parameters. The articulatory vectors, which characterize the corpus images, have been clustered in spaces of increasing dimensionality yielding more and more precise quantized descriptions of the mouth configuration.

The resulting vector distribution is clustered in small subregions of the articulatory space which identify special configurations of the mouth representative of the various Italian visemes, thanks to the good properties of the employed clustering algorithms which allow the identification of small clusters. The main articulatory trajectories between the Italian visemes have been tracked and quantized into a pre-defined number of clusters (128 and 256), each of them associated to a corresponding image selected from the corpus, whose articulatory vector is closest to its centroid. These images have been taken as "key-frame."

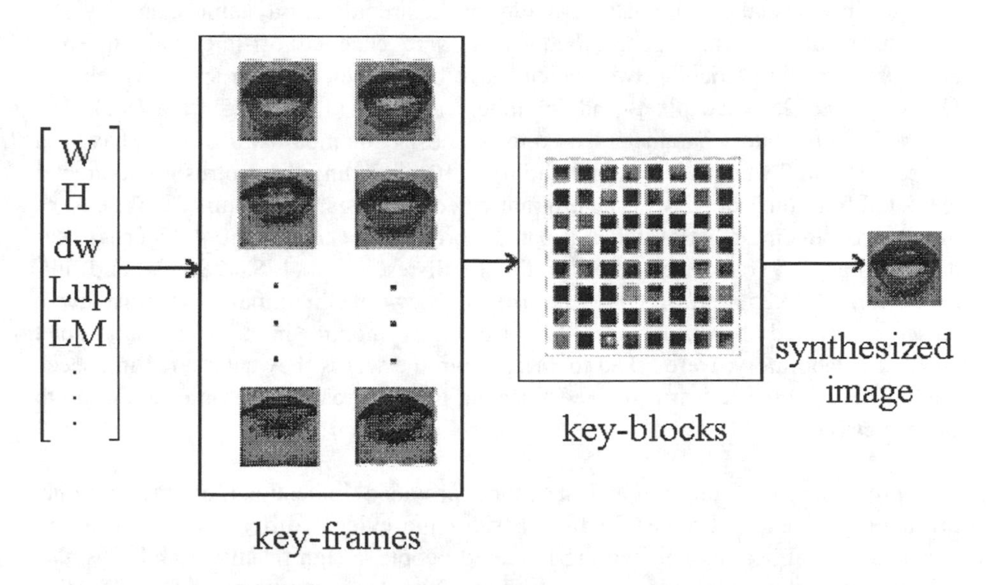

Figure 21. Description of the speech visualization procedure.

To reduce the memory requirements further, "key-frames" have been subdivided into 4x4 pixel blocks and the resultant vectors have been clusterized a second time in the 16-dimensional space of pixels providing a predefined number of "key-blocks" (128 and 256), which form the viseme reconstruction codebook (see Figure 21). Each of the 128 "key-frames" has been finally associated to a list of 7-bit indexes for addressing the suitable blocks in the reconstruction codebook. Experiments have been performed using 256x256 and 128x128 pel image formats composed of 4096 and 1024 "key-blocks," respectively.

The visual synthesis of speech has been evaluated by computing the Mean Square Error (**MSE**) between the original images of the corpus and those reconstructed by means of the "key-blocks." In this evaluation, the viseme reconstruction codebook has been addressed either by using the actual articulatory vectors measured from the images, or the estimates derived through speech analysis. Various dimensionalities of the articulatory space and of the reconstruction code-book have been used. The estimation **MSE** for each articulatory parameter has also been evaluated. The objective **MSE** evaluation alone, however, cannot provide enough indications on the performances since relevant components in speech reading depend on the quality with which the coarticulatory dynamics are rendered and on the level of coherence with acoustic speech.

Because of this, a set of subjective experiments has been carried out with both normal hearing and hearing-impaired subjects. Experiments consist of showing some sample sequences of visualized speech to persons asked to express their evaluation in terms of readibility, visual discrimination, quality of the articulatory dynamics, and level of coherence with acoustic speech. Sequences were encoded off-line with different configurations; in particular, two choices have been taken both for spatial resolution (128x128 and 256x256 pixel) and for time resolution (15 and 25 frame/sec). The number of articulatory parameters, used to synthesize the mouth articulation, has been increased from 2 (mouth height and width) to 10 (including the protrusion parameter extracted from the side view). The original video sequence, representing the speaker's mouth pronouncing a list of Italian isolated words (from a corpus of 400 words), was displayed at 12.5 frame/sec without audio at half-resolution (128x128 pels) and full-resolution (256x256 pels). Only the frontal view of the mouth was displayed. Observers, seated at a distance of 30 cm from a 21' monitor in an obscure and quiet site of the laboratory, were asked to write down the words they succeeded in speech reading. The sequence was displayed a second time, increasing the time resolution to 25 frame/sec.

The presentation of the original sequence provided indications on the personal proficiency of each observer. In fact, besides the evident difference of sensitivity between normal hearing and hearing-impaired people, a significant variability is also present within the same class of subjects. Therefore, the computation of the subjective evaluation score has been normalized for each individual on the basis of his/her speech reading perception threshold. For each observer, with reference to each of the two possible image formats (128x128 or 256x256), the minimum time resolution (12.5 or 25 Hz) allowing successful speech reading was found. Success was measured on the basis of a restricted set of "articulatory easy" words for which 90% correct recognition was required.

Further experimentation with synthetic images was then performed, observer by observer, using the exact time frequency which had allowed his/her successful speech reading of the original sequences. The whole test was repeated replacing the original images with synthetic ones, reconstructed by "key-blocks" addressed by means of actual (no estimation error) articulatory parameters. Finally, a third repetition of the test was performed addressing the reconstruction code-book by means of parameters estimated from speech through the TDNNs (with estimation error). Before each test repetition, the order of the words in the sequence were randomly shuffled to avoid expectations by the observer.

The number of people involved in the tests is still too small, especially as far as pathological subjects are concerned, but on-going experimentation aims at enlarging this number significantly. A total number of 15 observers were involved in the evaluations and only 2 of them are hearing impaired with 70 dB loss. The preliminary results reported in Tables 4 and 5 take into account, for each observer, only the words that he/she has correctly recognized in the original sequence. This has been done according to our particular interest in evaluating the subjective "similarity" of the reconstructed images to the original, as far as the possibility of correct speech reading was concerned.

Table 4. Results reported in each column express the average percentage of correctly recognized words. The test is subjective since, as explained in the text, both the video time resolution and the set of test words have been calibrated for each observer. In this case images are synthesized by selecting each time one of 128 possible key-frames (visemes) and by approximating it as a puzzle of 4x4 blocks extracted from a code-book of either 128 or 256 elements. The format of the images was 128x128 and 256x256.

128 key-frames	128 key-blocks				256 key-blocks			
Articulatory parameters used for speech visualization	Image format 128x128		Image format 256x256		Image format 128x128		Image format 256x256	
	true parameters	estimated parameters	true parameters	estimated parameters	true parameters	estimated parameters	true parameters	estimated parameters
W, H, dw	5%	-	5%	-	10%	5%	15%	8%
W, H, dw, LM	28%	10%	29%	12%	28%	10%	29%	12%
W, H, dw, LM, Lup	37%	25%	38%	25%	37%	25%	40%	28%
W, H, dw, LM, Lup, h	40%	25%	40%	25%	40%	25%	45%	28%
W, H, dw, LM, Lup, h, w	40%	27%	40%	27%	40%	27%	45%	28%
W, H, dw, LM, Lup, h, w, LC	42%	27%	42%	27%	42%	27%	45%	29%
W, H, dw, LM, Lup, h, w, LC, lup	42%	32%	42%	32%	42%	32%	47%	34%
W, H, dw, LM, Lup, h, w, LC, lup, teeth	50%	35%	52%	37%	52%	38%	55%	40%
W, H, dw, LM, Lup, h, w, LC, lup, teeth, protr	60%	42%	64%	45%	64%	45%	68%	49%

The use of parameters *W* and *H* alone did not allow speech reading; since observers were more inclined to guess than recognize, the test outcomes have been considered unreliable and have been omitted. From the results in both tables, it is evident that the progressive introduction of parameters *W*, *H*, *dw*, *LM*, and *Lup* raises significantly the recognition rate while slight improvement is gained when parameters *h*, *w*, *LC*, and *lup* are added. This can be easily explained by the fact that the former set of parameters represent a basis (mutual independence) while the latter set of parameters is strongly correlated to the previous one and can supply only marginal information. The information associated to teeth (supplied manually since it could not be estimated through the TDNN) has proved to be of great importance for improving the quality of speech visualization since it directly concerns the dental articulatory place and provides information on the tongue position.

Table 5. Results of the same subjective tests reported in Table 4, except that in this case 256 possible key-frames (visemes) were used.

256 key-frames	**128 key-blocks**				**256 key-blocks**			
Articulatory parameters used for speech visualization	Image format 128x128		Image format 256x256		Image format 128x128		Image format 256x256	
	true parameters	estimated parameters	true parameters	estimated parameters	true parameters	estimated parameters	true parameters	estimated parameters
W, H, dw	17%	12%	18%	12%	28%	20%	32%	24%
W, H, dw, LM	33%	20%	34%	22%	45%	32%	48%	38%
W, H, dw, LM, Lup	41%	28%	42%	30%	55%	44%	56%	38%
W, H, dw, LM, Lup, h	44%	30%	46%	32%	55%	44%	57%	42%
W, H, dw, LM, Lup, h, w	46%	32%	46%	34%	57%	44%	57%	44%
W, H, dw, LM, Lup, h, w, LC	48%	35%	48%	36%	58%	45%	60%	45%
W, H, dw, LM, Lup, h, w, LC, lup	51%	39%	53%	41%	58%	46%	60%	47%
W, H, dw, LM, Lup, h, w, LC, lup, teeth	62%	45%	65%	48%	68%	55%	80%	58%
W, H, dw, LM, Lup, h, w, LC, lup, teeth, protr	74%	51%	78%	54%	86%	62%	88%	64%

The "protr" parameter (last row of the tables) was measured from the side view of the speaker's face and expresses the lips protrusion; its use provided a very significant improvement in comprehension despite the fact that only the frontal view of the mouth is synthesized. This happens because different visemes, which were previously confused in the domain of frontal articulatory parameters, are now discriminated by the lips' protrusion. It can be noticed that the larger format (256x256) adds details which are relevant when few parameters are used (first three rows of the table) and when the "teeth" and "protr" parameters are added (last two rows).

By doubling the code-book size, higher texture resolution in viseme reconstruction is obtained while, by doubling the number of key-frames, higher viseme discrimination is achieved. If both the number of key-frames and the code-book size are increased, significant quality improvement is achieved. In this last case, the visemes reconstruction from true articulatory parameters is almost indistinguishable from the original (88% correct recognition) while the estimation error introduced by the TDNNs lowers this score to 64% as a consequence of artifacts like parameter amplitude distortion and parameter trajectories smoothing. These impairments are particularly severe when high articulatory dynamics occur like in plosives: the sudden closure of the lips usually affects one single 25 Hz frame whose parameters are too coarsely estimated by the TDNNs. Other severe impairments concern the reproduction of glottal stops which must be discriminated from generic silence intervals in order to be associated to suitable context-dependent visemes.

References

[1] Curinga, S., Grattarola, A.A., and Lavagetto, F. (1993), Synthesis and Animation of Human Faces: Artificial Reality in Interpersonal Video Communication, pp. 397-408, Proceedings of IFIP TC 5/WG 5.10 Conference on Modeling and Computer Graphics, Genova, Italy.

[2] Aizawa, K., Harashima, H., and Saito, T. (1989), Model-Based Analysis-Synthesis Image Coding (MBASIC) System for Person's Face, *Image Communication*, 1, pp. 139-152.

[3] Nakaya, Y., Chuah Y.C., and Harashima, H. (1991), Model-based/waveform hybrid coding for videotelephone images, pp. 2741-2744, Proceedings ICASSP-91, San Francisco, CA.

[4] Morishima, S. and Harashima, H. (1991), A Media Conversion from Speech to Facial Image for Intelligent Man-Machine Interface, *IEEE Journal on Sel. Areas in Comm.*, 9, 4, pp. 594-600.

[5] Bothe, H., Lindner, G., and Rieger, F. (1993), The Development of a Computer Animation Program for the Teaching of Lipreading, pp. 45-49, Proceedings of 1st TIDE Conference, Brussels, Belgium.

[6] Murakami, S. and Kumazaki, M. (1993), Lip Reading by 3D Features for Model Based Image Coding, paper No. 2.5, Proceedings of PCS-93, Lausanne, Switzerland.

[7] Hill, D.R., Pearce, A., and Wyvill, B., (1988), Animating Speech: an Automated Approach Using Speech Synthesised by Rules, *The Visual Computer*, 3, pp. 176-186.

[8] Lavagetto, F. (1995), Converting Speech into Lip Movements: A Multimedia Telephone for Hard of Hearing People, *IEEE Trans. on RE*, 3, 1, pp. 90-102.

[9] Morishima, S., Aizawa K., and H. Harashima (1988), Model-Based Facial Image Coding Controlled by the Speech Parameter, paper No. 4.4, Proceedings of PCS-88, Turin, Italy.

[10] Summerfield, A.Q. (1979), Use of Visual Information for Phonetic Perception, *Phonetica*, 36, pp. 314-331.

[11] Erber, N.P. (1972), Auditory, Visual and Auditory-Visual Recognition of Consonants by Children with Normal and Impaired Hearing, *Journal of Speech and Hearing Research*, 15, pp. 413-422.

[12] Wakita, H. (1973), Direct Estimation of the Vocal Tract Shape by Inverse Filtering of Acoustic Speech Waveforms, *IEEE Trans. on Audio Electroacoust.*, 21, pp. 417-427.

[13] Yuhas, B.P., Goldstein, M.H. Jr., and Sejnowski, T.J. (1989), Integration of Acoustic and Visual Speech Signal Using Neural Networks, *IEEE Communications Magazine*, 27, 11, pp. 65-71.

[14] Welsh, W.J., Simons, A.D., Hutchinson, R.A., and Searby, S. (1990), A Speech-Driven 'Talking-Head' in Real Time, paper 7.6, Proceedings of PCS-90, Cambridge, MA, 1990.

[15] Chen, T., Graf, H.P., and Wang, K. (1994), Speech-assisted Video Processing: Interpolation and Low Bitrate Coding, in Procedings of 28th Asilomar Conference on Signals, Systems and Computers, Pacific Grove, CA.

[16] Chen, T., Graf, H.P., and Wang, K. (1995), Lip Synchronization Using Speech-Assisted Video Processing, *IEEE Signal Processing Letters*, 2, 4, pp. 57-59.

[17] Chen, T., Graf, H.P., Haskell, B., Petajan, E., Wang, Y., Chen, H., and Chou, W. (1995), Speech-Assisted Lip Synchronization in Audio-Visual Communications, pp. 579-582, Proceedings of IEEE ICIP-95.

[18] Sokol, R. and Mercier, G. (1996), Neural-fuzzy Networks and Phonetic Feature Recognition as a Help for Speechreading, *Speechreading by Humans and Machines*, edited by D.G. Stork and M.E. Hennecke, Springer, pp. 497-504.

[19] Goldschen, A.J., Garcia, O.N., and Petajan, E.D. (1996), Rationale for Phoneme-Viseme Mapping and Feature Selection in Visual Speech Recognition, *Speechreading by Humans and Machines*, edited by D.G. Stork and M.E. Hennecke., Springer, pp. 505-515.

[20] Stork, D.G. and Hennecke M.E. (1996), Speechreading by Humans and Machines, Springer.

Chapter 11:

Multi-Objective Evolutionary Algorithms in Gas Turbine Aero-Engine Control

MULTI-OBJECTIVE EVOLUTIONARY ALGORITHMS IN GAS TURBINE AERO-ENGINE CONTROL

Andrew Chipperfield and **Peter Fleming**
Department of Automatic Control and Systems Engineering
The University of Sheffield
Sheffield S1 3JD England

Hugh Betteridge
Advanced Controls
Rolls-Royce Military Aero Engines Ltd.
Bristol BS12 7QE England

This chapter describes a novel approach to the problem of control mode analysis for gas turbine aero-engines. Using a multi-objective evolutionary algorithm, candidate control modes are selected and tested on models of the propulsion system to assess performance, safety, stability, and other important design criteria. An example controller design problem, considering some of the problems likely to be associated with new variable cycle engine concepts, is presented to demonstrate how the proposed approach may be employed to examine many design objectives, from different disciplines, in parallel. Potential control schemes are evaluated and compared with one another within an optimization framework. This comparative analysis of different control modes may be used to make more informed decisions regarding the nature of the control to be employed, acceptable performance margins, and elements of the engine design.

1. Introduction

Future concepts in aero-engine design, such as the variable cycle engine (VCE), will be expected to operate with greater fuel economy, over a wider flight envelope, and have extended mission capabilities. This type of engine will have more active internal variable geometry devices and sensors to meet the perceived operational requirements. Efficient operation will thus be achieved only through the accurate control of the variable geometry components to ensure that compressors and turbines run at the

0-8493-9804-5/99/$0.00+$.50

design conditions for airflow and pressure ratios. As the design conditions vary with operating point, e.g., cruise, combat, etc., it appears that the approaches adopted for conventional engine control configuration selection and design will be unsuited to the VCE [1].

The application of advanced control techniques, such as multivariable control, to VCEs is likely to provide the extra degrees of freedom necessary to enable cost-effective variation of the engine cycle according to the mission requirements [2]. Thus, a wider range of mission capability should be achievable with increased aircraft agility. However, the projected type of VCE design for both conventional and short take-off and vertical landing aircraft applications, and for compound helicopter power plants, shows a significant increase in the number and type of variable geometry devices. The task of selection of a control configuration is therefore complicated by the number of possible, but perhaps undesirable, configurations.

In this chapter, we address the problem of control configuration design for gas turbine aero-engines. The design of such systems is a complex process involving the collaboration of specialists from many different disciplines and yields a truly multi-objective design problem. An approach to control configuration design based on evolutionary algorithms (EAs) [3] is therefore proposed and developed. After a review of the fundamentals of aero-engine control, a short introduction to EAs and the concepts of multi-objective optimization is presented. The differences between a standard EA and a multi-objective one are then considered in some detail. An example aircraft engine control problem is described and an 'integration model,' combining design objectives from the different disciplines involved in the selection of suitable control modes, is constructed. The 'integration model' is then manipulated by a multi-objective genetic algorithm (MOGA), guided by the control engineer, in the search for appropriate control configurations.

While the control of conventional propulsion systems poses few problems to the control engineer, the example given in this chapter demonstrates that there are many candidate options available to the control designer, each offering different performance characteristics. Techniques such as the one proposed here will be essential if the true potential of future propulsion system designs are to be fully realized. Finally, while this chapter considers the use of multi-objective EAs in aircraft engine controller design, the techniques presented here should prove applicable to a wider range of control and engineering design problems, such as building energy management, controlled structures, and systems integration.

2. Gas Turbine Engine Control

The mechanical layout of a typical twin spool gas turbine engine is shown in Figure 1. Each spool comprizes a number of compressor and turbine stages aero-thermodynamically coupled to each other. Air is drawn into the fan (or LP compressor) through the inlet guide vanes which are used to match the airflow to the fan characteristics, and compressed. The air is then further compressed by the HP compressor before being mixed with fuel, combusted, and expelled through the HP and LP turbines. A portion of the air from the fan exit may bypass the HP compressor and turbines and be mixed with the combusted air/fuel mixture before being ejected through the jet pipe and nozzle to produce thrust.

The characteristics of operation of a fixed cycle gas turbine engine, such as specific thrust and specific fuel consumption, are fundamental to the engine design. The design thus becomes a compromise between meeting the conflicting requirements for performance at different points in the flight envelope and the achievement of low life-cycle costs, while maintaining structural integrity. However, variable geometry components, such as the inlet guide vanes (IGV) and nozzle area (NOZZ), may be used to optimize the engine cycle over a range of flight conditions with regard to thrust, specific fuel consumption, and engine life, assisting in the reduction of life-cycle costs [1].

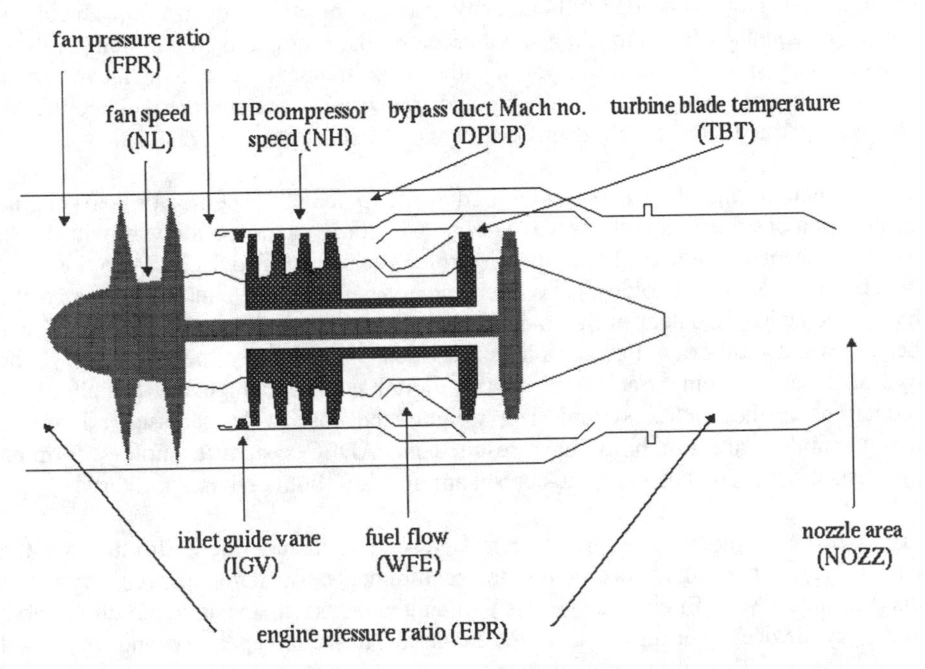

Figure 1. A gas turbine engine.

Traditionally, control systems for aircraft propulsion systems have been developed out of necessity to meet a particular need or to solve a certain problem. It is interesting to note a few of these developments to glean an understanding of the structure and operation of control systems that are commonplace today, and the likely requirements for future generations of aero-engines.

As early as 1946, the first electrical control, a throttle positioner on the Theseus engine in the Brabazon aircraft, was necessary. The simple reason in this case was that the wingspan was so great that the standard mechanical linkages employed up until that time could not position the engine throttle with sufficient accuracy due to flexing.

First generation gas turbines, with open-loop throttles, were prone to damage due to over-temperatures and, in some cases, engine fires. This led to the adoption of thermocouples mounted in the jet pipe to monitor and limit the exhaust gas temperature. Similarly, physical limitations, such as the maximum spool speed before blade separation or disk burst in the rotating turbo machinery, drove the development of controls to protect the engine from reaching over-speed conditions. By 1956, the first full-authority analogue electrical controller was introduced on the Proteus engine in the Britannia aircraft, followed shortly thereafter by a comparable controller on the Gnome engine to power helicopters.

Progress was considerably forced in the early 1960s by the stringent requirements of the world's first supersonic passenger transport aircraft, the Concorde. Its four Olympus 593 engines needed full-authority analogue electronic controllers capable of governing spool speed, controlling spool acceleration, limiting maximum temperature, pressure and spool speeds, and varying the exhaust nozzle area to achieve reheat operation (afterburning) during take-off and trans-sonic acceleration. This controller also incorporated hybrid fault identification and monitoring circuitry.

The advent of digital technology offered many potential benefits for aero-engine control demonstrated as early as 1976, when Concorde again led the way with flight trials of the first full-authority digital electronic controller (FADEC). Since then, the benefits of digital technology have been mostly gained from simplifications to the hydromechanichal content of the main engine and reheat fuel systems. The trend has been toward authority for complex functions, traditionally performed by the hydromechanics, being transferred to the digital control computers. Simultaneous reductions in the control system size, weight, and cost, with increased reliability, maintainability, and testability, have resulted in FADEC system technology forming the basis of all current aero-engine control applications, both civil and military.

Proposed development of the gas-turbine aero-engine include the ability to vary the engine cycle according to specific mission requirements. The complex architectures of the Variable Cycle Engines show a significant increase in the number of variable geometry devices, permitting a wider range of mission capability and increased aircraft agility. Operation of such engines to achieve their design performance is totally dependent on the precision and flexibility of simultaneous control of many engine parameters and stability margins. It has become clear that the engine control design methodology must evolve to incorporate techniques to take account of the multiple and, in some cases, conflicting requirements for the control of such complex engines.

Today, dry-engine control of a conventional engine is normally based on a single closed-loop control of fuel flow for thrust rating, engine idle and maximum limiting, and acceleration control. The closed-loop concept provides accuracy and repeatability of control of defined engine parameters under all operating conditions, and automatically compensates for the effects of engine and fuel system aging.

In these engines it is usual for any variable geometry to be positioned according to commands scheduled against appropriate engine and/or aircraft parameters. These schedules are often complex functions of several parameters, and adjustments may be frequently required to achieve the desired performance. Clearly, success of this open-loop mode of control is reliant on the positional accuracy achievable as there is no self-trimming to account for ageing as occurs in closed-loop modes. This results in penalties of reduced engine life and higher maintenance costs but allows a simple and reliable control structure to be employed. Advanced control concepts, such as multivariable control, are likely to offer advantages in terms of reducing fuel burn and life-cycle costs while maximizing available thrust without compromising safety and stability margins [4].

The control law definition task includes selection of the measurable engine parameters to be controlled. These control parameters should be suitable for representation of the essential, but not directly measurable, control parameters such as thrust, surge margin, and efficiency. The task of selecting a suitable control configuration is thus further complicated by the number of possible, but perhaps undesirable, configurations. There is, of course, the further design objectives of selecting controller configurations that satisfy robustness and disturbance rejection requirements.

Additionally, the safety implications of the alternative control strategies must also be recognized during the selection process. This will involve the examination and analysis of potential failure modes, assessment of their effects, and identification of suitable compensatory action including reconfiguration and fault accommodation. While the design of a control system for a conventional propulsion system poses few hard problems for the control engineer, there may be many candidate solutions available and the choice of the 'correct' system is paramount. Furthermore, new concepts in aircraft engines, such as the variable cycle engine, are likely to include more controllable elements and will almost certainly require the application of advanced control techniques if they are to realize their full potential benefits [2].

The multiplicity of interactive and potentially conflicting objectives, outlined above, to be considered during the controller selection, design, and integration has illuminated the desire for a truly multiobjective search and optimization technique to assist the control engineer in this task. One such technique under investigation and presented herein is based on evolutionary algorithms. An 'integration model,' combining design objectives across disciplines facilitates the use of complementary analysis methods and permits varying complexity in the discipline-dependent submodels. Such an integration model may be manipulated by the evolutionary algorithm, under the guidance of the control engineer, to help locate suitable control modes.

3. Evolutionary Algorithms

Evolutionary algorithms are based on computational models of fundamental evolutionary processes such as selection, recombination, and mutation, as shown in Figure 2. Individuals, or current approximations, are encoded as strings composed over some alphabet(s), e.g., binary, integer, real-valued, etc., and an initial population is produced by randomly sampling these strings. Once a population has been produced it may be evaluated using an objective function or functions that characterize an individual's performance in the problem domain. The objective function(s) is also used as the basis for selection and determines how well an individual performs in its environment. A fitness value is then derived from the raw performance measure given by the objective function(s) and is used to bias the selection process toward promising areas of the search space. Highly fit individuals will be assigned a higher probability of being selected for reproduction than individuals with a lower fitness value. Therefore, the average performance of individuals can be expected to increase as the fitter individuals are more likely to be selected for reproduction and the lower fitness individuals get discarded. Note that individuals may be selected more than once at any generation (iteration) of the EA.

```
procedure EA {
      t = 0;
      initialize P(t);
      evaluate P(t);
      while not finished do {
            t = t + 1;
            select P(t) from P(t-1);
            reproduce pairs in P(t);
            mutate P(t);
            evaluate P(t);
      }
}
```

Figure 2. An evolutionary algorithm.

Selected individuals are then reproduced, usually in pairs, through the application of genetic operators. These operators are applied to pairs of individuals with a given probability and result in new offspring that contain material exchanged from their parents. The offspring from reproduction are then further perturbed by mutation. These new individuals then make up the next generation. These processes of selection, reproduction, and evaluation are then repeated until some termination criteria are satisfied, e.g., a certain number of generations completed, a mean deviation in the performance of individuals in the population, or when a particular point in the search space is reached.

Although similar at the highest level, many variations exist in EAs. A comprehensive discussion of the differences between the various EAs can be found in [7].

4. Multi-Objective Optimization

The use of multi-objective optimization (MO) in control, and engineering design in general, recognizes that most practical problems require a number of design criteria to be satisfied simultaneously, viz:

$$\min_{x \in \Omega} F(x)$$

where $x = [x_1, x_2, \ldots, x_n]$ and Ω define the set of free variables, x, subject to any constraints and $F(x) = [f_1(x), f_2(x), \ldots, f_n(x)]$ are the design objectives to be minimized.

Clearly, for this set of functions, $F(x)$, it can be seen that there is no one ideal 'optimal' solution, rather a set of Pareto-optimal solutions for which an improvement in one of the design objectives will lead to a degradation in one or more of the remaining objectives. Such solutions are also known as noninferior or nondominated solutions to the multi-objective optimization problem.

Conventionally, members of the Pareto-optimal solution set are sought through solution of an appropriately formulated nonlinear programming problem. A number of approaches are currently employed including the e-constraint, weighted sum, and goal attainment methods [8]. However, such approaches require precise expression of a, usually not well understood, set of weights and goals. If the trade-off surface between the design objectives is to be better understood, repeated application of such methods will be necessary. In addition, nonlinear programming methods cannot handle multimodality and discontinuities in function space well and can thus be expected to produce only local solutions.

Evolutionary algorithms, on the other hand, do not require derivative information or a formal initial estimate of the solution region. Because of the stochastic nature of the search mechanism, GAs are capable of searching the entire solution space with more likelihood of finding the global optimum than conventional optimization methods. Indeed, conventional methods usually require the objective function to be well behaved, whereas the generational nature of GAs can tolerate noisy, discontinuous, and time-varying function evaluations [9]. Moreover, EAs allow the use of mixed decision variables (binary, n-ary and real-values) permitting a parameterization that more closely matches the nature of the design problem. Single objective GAs, however, do still require some combination of the design objectives although the relative importance of individual objectives may be changed during the course of the search process.

In previous work [10], we demonstrated an approach to the design of a multivariable control system for a gas turbine engine using multi-objective genetic algorithms. A structured chromosome representation [11], described later, was employed that allowed the level of complexity of the individual pre-compensators to be searched along with the parameters of the controller thus simultaneously permitting a search of both controller complexity and controller parameter space. Further, EAs have already been studied to some extent as a possible search engine in multidisciplinary optimization and preliminary design in aerospace applications [12] and have been shown to offer significant advantages over conventional techniques in this area and the related field of performance seeking control [13].

5. Multi-Objective Genetic Algorithms

The notion of fitness of an individual solution estimate and the associated objective function value are closely related in the single objective GA described earlier. Indeed, the objective value is often referred to as fitness although they are not, in fact, the same. The objective function characterizes the problem domain and cannot therefore be changed at will. Fitness, however, is an assigned measure of an individual's ability to reproduce and, as such, may be treated as an integral part of the GA search strategy.

As Fonseca and Fleming describe [14], this distinction becomes important when performance is measured as a vector of objective function values as the fitness must necessarily remain scalar. In such cases, the scalarization of the objective vector may be treated as a multicriterion decision-making process over a finite number of candidates – the individuals in a population at a given generation. Individuals are therefore assigned a measure of utility depending on whether they perform better, worse, or similar to others in the population and, possibly, by how much. The remainder of this section describes the main differences between the simple EA outlined earlier and MOGAs.

5.1 Decision Strategies

In the absence of any information regarding the relative importance of design objectives, Pareto-dominance is the only method of determining the relative performance of solution estimates. Nondominated individuals are all therefore considered to be 'best' performers and are thus assigned the same fitness [9], e.g., zero. However, determining a fitness value for dominated individuals is a more subjective matter. The approach adopted here is to assign a cost proportional to how may individuals in a population dominate a given individual, Figure 3. In this case, nondominated individuals are all treated as desirable.

If goal and/or priority information is available for the design objectives, then it may be possible to differentiate between some nondominated solutions. For example, if degradations in individual objectives still allow those goals to be satisfied while also allowing improvements in other objectives that do not already satisfy their design goals, then these degradations should be accepted. In cases where different levels of priority may be assigned to the objectives then, in general, it is important to improve only the high priority objectives, such as hard constraints, until the corresponding design goals are met, after which improvements may be sought in the lower priority objectives.

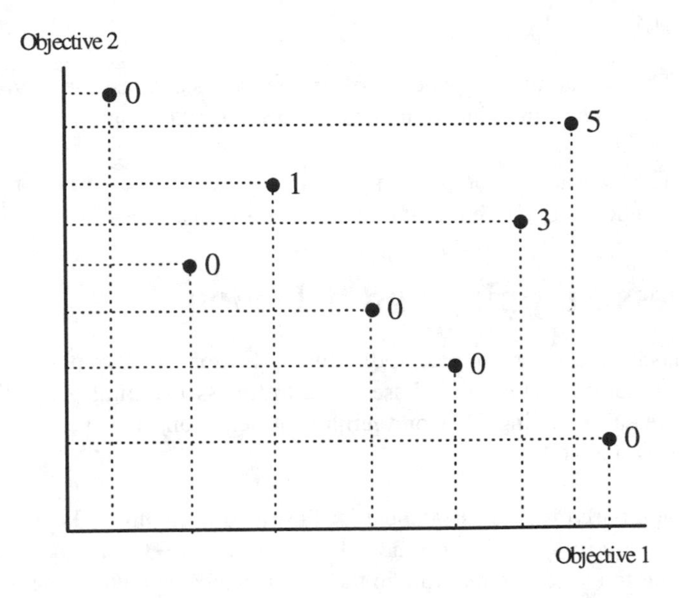

Figure 3. Pareto ranking.

These considerations have been formalized in terms of a transitive relational operator, *preferability*, based on Pareto-dominance, which selectively excludes objectives according to their priority and whether or not the corresponding goals are met [15]. For simplicity only one level of priority is considered here. Consider two objective vectors **u** and **v** and the corresponding set of design goals, **g**. Let the smile $\breve{\mathbf{u}}$ denote the components of **u** that meet their goals and the frown $\hat{\mathbf{u}}$ those that do not. Assuming minimization, one may then write

$$\mathbf{u}^{\breve{\mathbf{u}}} \le \mathbf{g}^{\breve{\mathbf{u}}} \quad \wedge \quad \mathbf{u}^{\hat{\mathbf{u}}} > \mathbf{g}^{\hat{\mathbf{u}}}$$

where the inequalities apply component wise. This is equivalent to,

$$\forall i \in \breve{\mathbf{u}}, u_i \le g_i \quad \wedge \quad \forall i \in \hat{\mathbf{u}}, u_i > g_i$$

where u_i and g_i represent the components of **u** and **g**, respectively. Then, **u** is said to be preferable to **v** given **g** if and only if

$$(\mathbf{u}^{\hat{\mathbf{u}}} \prec \mathbf{v}^{\hat{\mathbf{u}}}) \vee \left\{(\mathbf{u}^{\hat{\mathbf{u}}} = \mathbf{v}^{\hat{\mathbf{u}}}) \wedge \left[(\mathbf{v}^{\breve{\mathbf{u}}} \not\le \mathbf{g}^{\breve{\mathbf{u}}}) \vee (\mathbf{u}^{\breve{\mathbf{u}}} \prec \mathbf{v}^{\breve{\mathbf{u}}})\right]\right\}$$

where $\mathbf{a} \prec \mathbf{b}$ is used to denote that **a** dominates **b**. Hence **u** will be preferable to **v** if and only if one of the following is true:

1. The violating components of **u** dominate the corresponding components of **v**.

2. The violating components of **u** are the same as the corresponding components in **v**, but **v** violates at least one other goal.

3. The violating components of **u** are equal to the corresponding components of **v**, but **u** dominates **v** as a whole.

5.2 Fitness Mapping and Selection

After a cost has been assigned to each individual, selection can take place in the usual way. Suitable schemes include rank-based cost-to-fitness mapping [16] followed by stochastic universal sampling [17] or tournament selection, also based on cost, as described by Ritzel et al. [18].

Exponential rank-based fitness assignment is illustrated in Figure 4. Here, individuals are sorted by their cost – in this case the values from Figure 3 – and assigned fitness values according to an exponential rule in the first instance, shown by the narrow bars in Figure 4. A single fitness value is then derived for each group of individuals sharing the same cost, through averaging, and is shown in the figure by the wider bars.

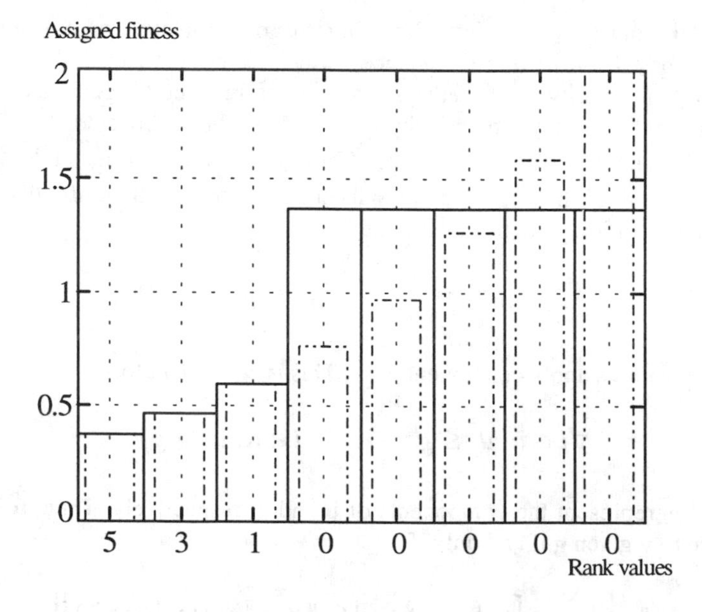

Figure 4. Rank-based fitness assignment.

5.3 Fitness Sharing

Even though all preferred individuals in the population are assigned the same level fitness, the number of offspring that they will produce, which must obviously be integer, may differ due to stochastic nature of EAs. Over generations, these imbalances may accumulate resulting in the population focusing on an arbitrary area of the trade-off surface, known as genetic drift [19]. Additionally, recombination and mutation may be less likely to produce individuals at certain areas of the trade-off surface, e.g., the extremes, giving only a partial coverage of the trade-off surface.

Originally introduced as an approach to sampling multiple fitness peaks, fitness sharing [20] helps counteract the effects of genetic drift by penalizing individuals according to the number of other individuals in their neighborhood. Each individual is assigned a niche count, initially set to zero, which is incremented by a certain amount for every individual in the population, including itself. A sharing function determines the contribution of other individuals to the niche count as a function of their mutual distance in genotypic, phenotypic, or objective space. Raw fitness values are then weighted by the inverse of the niche count and normalized by the sum of the weights prior to selection. The total fitness in the population is redistributed, and thus shared, by the population. However, a problem with the use of fitness sharing is the difficulty in determining the niche size, σ_{share}, i.e., how close together individuals may be before degradation occurs.

An alternative, but analogous, approach to niche count computations are kernel density estimation methods [21] as used by statisticians. Instead of a niche size, a smoothing parameter, h, whose value is also ultimately subjective, is used. However, guidelines for the selection of suitable values for h have been developed for certain kernels, such as the standard normal probability density function and Epanechnikov kernels. The Epanechnikov kernel may be written as [21]

$$K_e(d/h) = \begin{cases} \frac{1}{2} c_n^{-1}(n+2)\left[1-(d/h)^2\right] & \text{if} \quad d/h < 1 \\ 0 & \text{otherwise} \end{cases}$$

where n is the number of decision variables, c_n is the volume of the unit n-dimensional sphere, and d/h is the normalized Euclidean distance between individuals. Apart from the constant factor, , this kernel is a particular case of the family of power law sharing functions proposed by Goldberg and Richardson [20].

Silverman [21] gives a smoothing factor that is approximately optimal in the least mean integrated squared error sense when the population follows a multivariate normal distribution for the Epanechnikov kernel $K_e(d)$ as

$$h = \left[8c_n^{-1}(n+4)\left(2\sqrt{\pi}\right)^n / N \right]^{1/(n+4)}$$

for a population with N individuals and identity covariance matrix. Where populations have an arbitrary sample covariance matrix, S, this may simply be 'sphered,' or normalized, by multiplying each individual by a matrix R such that $RR^T = S^{-1}$. This means that the niche size, which depends on S and h, may be automatically and constantly updated, regardless of the cost function, to suit the population at each generation.

5.4 Mating Restriction

Mating restrictions are employed to bias the way in which individuals are paired for reproduction [22]. Recombining arbitrary individuals from along the trade-off surface may lead to the production of a large number of unfit offspring, called *lethals*, that could adversely affect the performance of the search. To alleviate this potential problem, mating can be restricted, where feasible, to individuals from within a given distance of each other, σ_{mate}. A common practise is to set $\sigma_{mate} = \sigma_{share}$ so that individuals are allowed to mate with one another only if they lie within a distance h from each other in the 'sphered' space used for sharing [14].

5.5 Interactive Search and Optimization

As the population of the MOGA evolves, trade-off information will be acquired. In response to the optimization so far, the control engineer may wish to investigate a smaller region of the search space or even move on to a totally new region. This can be achieved by resetting the goals supplied to the MOGA which, in turn, affects the ranking of the population and modifies the fitness landscape concentrating the population on a different area of the search space. The priority of design objectives may also be changed interactively using this scheme.

The introduction of a small number of random individuals at each generation, say 10-20%, has been shown to make the EA more responsive to sudden changes in the fitness landscape, as occurs when the optimization is changed interactively [23]. This technique may also be employed by a MOGA and is used in the example presented in the next section.

6. Gas Turbine Aero-Engine Controller Design

From the preceding sections, it is clear that the GA is substantially different from conventional enumerative and calculus-based search and optimization techniques. In this section, an example is presented that demonstrates how the GA may be used to address a problem that is not amenable to efficient solution via these conventional methods. The problem is to find a set of control loops and associated controller parameters for an aircraft gas turbine engine control system to meet a number of conflicting design criteria [24].

6.1 Problem Specification

The design example here illustrates how the proposed approach can be applied to the design of a control system for a gas turbine engine [24]. The object of the design problem is to select a set of sensors and design a suitable controller for a maneuver about a particular operating point while meeting a set of strict design criteria including stability, sensitivity, and the accommodation of degradation with engine aging.

Figure 5 shows the configuration of the basic simulation model used for this example. A linearized model of the Rolls-Royce Spey engine, with inputs for fuel flow (WFE), exhaust nozzle area (NOZZ), and HP inlet guide vane (IGV) angle, is used to simulate the dynamic behavior of the engine. Although the Spey engine is no longer in service, as far as controls are concerned the architecture is similar to that of modern engines such as the EJ200, the Eurofighter propulsion system. Sensors provided from outputs of the engine model are high and low pressure spool speed (NH and NL), engine and fan pressure ratios (EPR and FPR), and bypass duct mach number (DPUP). These sensed variables can be used to provide closed-loop control of WFE and NOZZ. Other engine parameters, such as the fan surge margin, LPSM, net thrust, XNN, and jet pipe (exhaust) temperature, JPT, are measured directly from the engine model.

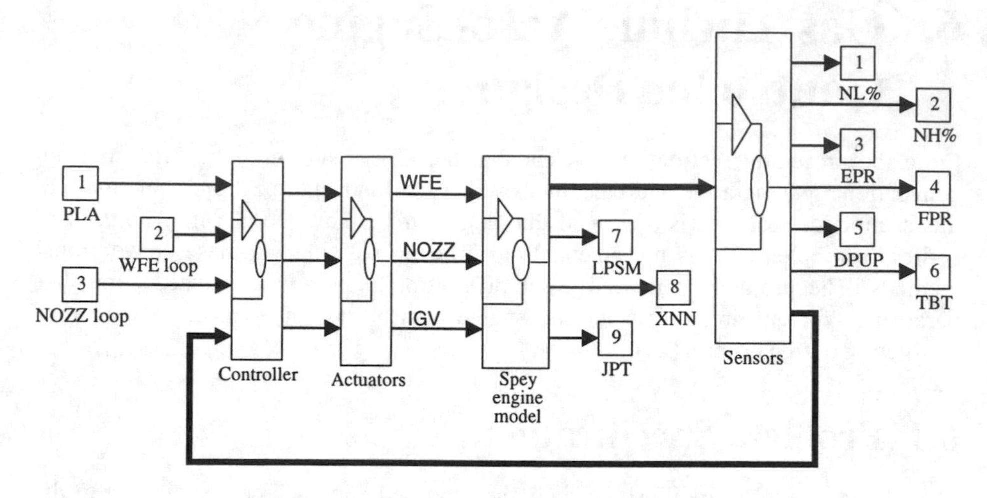

Figure 5. Basic gas turbine engine model.

Three inputs to the model are provided; input one is the demand reference signal and is translated from power lever angle (PLA) by the controller to provide the reference signal for the fuel loop, inputs two and three determine the measured parameters used to provide closed loop control. In this example, the nozzle area demand signal is derived from the fan working line and positioning of the HP IGVs are directly scheduled against the HP spool speed.

The possible control loops are

WFE	NL
	NH
	EPR
NOZZ	open-loop schedule
	FPR
	DPUP

For simplicity, a single 50% thrust-rating operating point is considered at sea level static conditions. The control options allow the use of PI control for SISO control of either the WFE loop alone, or for the WFE and NOZZ loops, or for multivariable control of both loops.

The system is required to meet the following design objectives:

(1) 70% rise-time $\leq$ 1.0s

(2) 10% settling time $\leq$ 1.4s

(3) XNN $\geq$ 40KN

(4) TBT $\leq$ 1540 °K

(5) LPSM $\geq$ 10%

(6) WFE sensitivity $\leq$ 2%

(7) δTR $\leq$ 0.25s

(8) δSFC $\leq$ 3.0%

(9) $\gamma \leq 1.0$

where objectives (1) and (2) are in response to a change in thrust demand of 33.33% to 66.66% and represent typical dynamic performance requirements for a military engine. XNN is the engine net thrust and is employed here as a measure of the accuracy of the mapping between the nominal and controlled engine performance. TBT is the maximum turbine blade temperature; a lower value indicates less thermodynamic stress and therefore longer engine life. LPSM is the fan surge margin representing aerodynamic safety margins and the WFE sensitivity, as a result of a 1% error in the sensed control parameter, is a measure of control tolerance. To ensure that candidate control schemes will provide adequate control over the lifetime of an engine, objectives (7) and (8) measure the difference in rise-time and thrust specific fuel consumption between a nominal engine and one with degraded compressor and turbine components. The degradations represent a typical variation in an engine over 6000 flight hours with a normal service schedule and are derived from previous deterioration studies [25]. Finally, the system should also be closed-loop stable, objective (9).

6.2 EA Implementation

The basic integration model was developed in SIMULINK and the associated performance measurements determined by simulation. Additionally, further models were constructed for the sensitivity and stability objectives. Actuators were modeled as first order systems with appropriate time constants and sensor parameters derived from the linear engine model outputs. To include realistic acceleration protection, input demands were rate limited.

A structured chromosome representation [11] was employed to allow the controller parameters for all possible control loops to reside in all individuals, Figure 6. Here, high-level genes, labeled WFE and NOZZ, are used to determine which loops are to be employed for control. Associated with each loop are the parameters of the corresponding controller, where $\{Pi_j, Ii_j\}$ may also contain the cross-coupling PI controller when the values of WFE and NOZZ dictate that a multivariable controller is to be employed. Note that as the NOZZ loop may be open-loop scheduled, there are no $P2_0$ and $I2_0$ parameters. In this manner, the chromosome may simultaneously contain a number of good representations, although only the set defined by the high-level genes will be active.

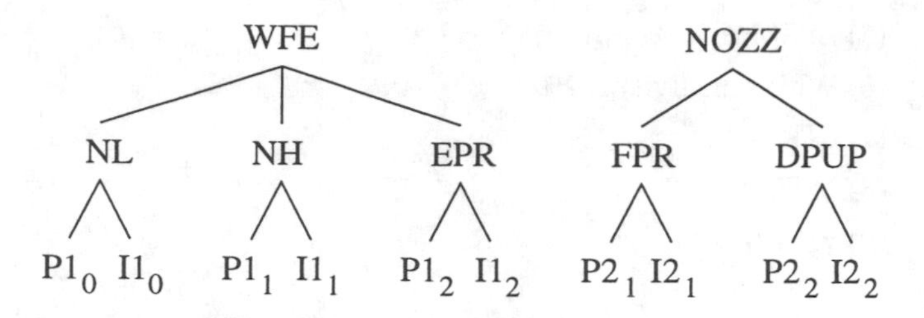

Figure 6. Structured chromosome representation.

The MATLAB Genetic Algorithm Toolbox [26] was used to implement the EA with additional routines to accommodate multi-objective ranking, fitness sharing, and mating restriction. Multi-objective ranking is based upon the dominance of an individual and how many individuals outperform it in objective space, combined with goal and priority information. In this example, the goals were set to the values given in the previous section and all objectives were assigned the same priority. In cases where objectives are assigned different priorities, higher priority objectives are optimized in a Pareto fashion until their goals are met, at which point the remaining objectives are optimized. Hard constraints may also be incorporated in the MOGA in this manner. Fitness sharing, implemented in the objective domain, favors sparsely populated regions of the trade-off surface and may be combined with mating restrictions to reduce the production of low performance individuals by encouraging the mating of individuals similar to one another.

In the example presented here, a binary MOGA with a population of 70 individuals was employed. The integer variables for WFE and NOZZ loop selection were encoded with eight bits and each controller parameter with 16 bits. Finally, lists of nondominated solutions for each controller configuration were maintained throughout the execution of the MOGA.

6.3 Results

Figure 7 illustrates a typical trade-off graph for Spey engine controller designs and the associated preference articulation window. In the trade-off graph, each line represents a nondominated solution found by the MOGA for the preferences shown. The *x*-axis shows the design objectives as described in the previous section, the *y*-axis the performance of controllers in each objective domain, and the cross-marks in the figure show the design goals. The preference articulation window demonstrates how the design goals may be varied (compared with the design goals presented earlier) and the use of different levels of goal priority (constraint, objective, and ignored) as discussed in the previous section.

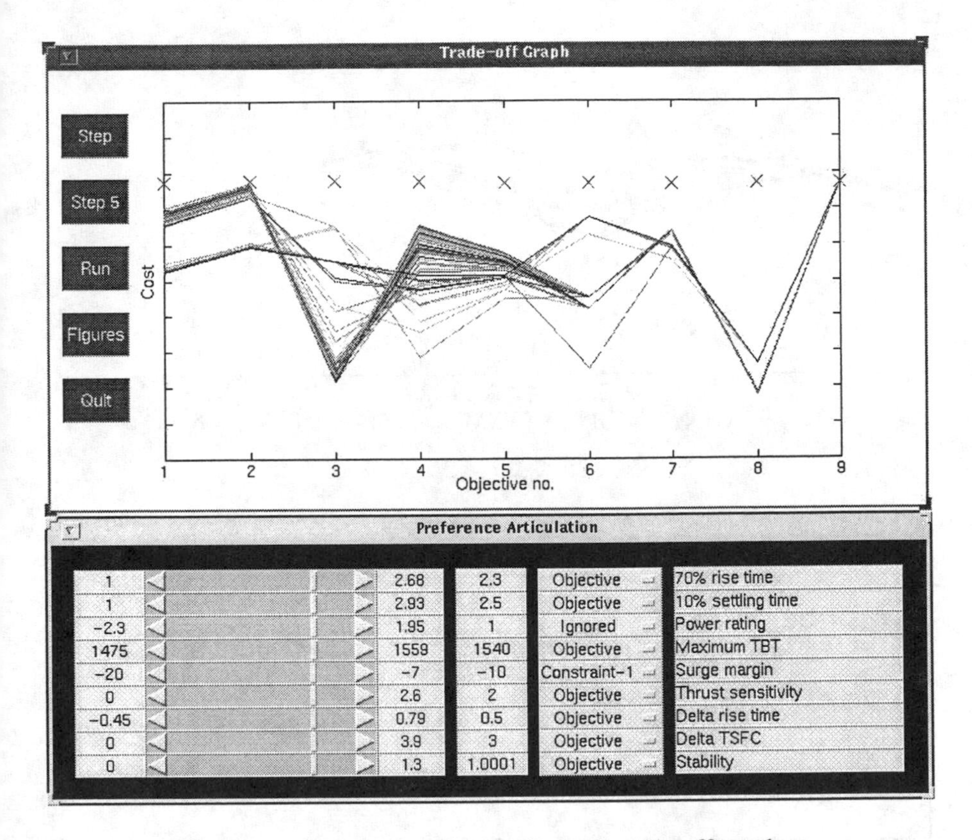

Figure 7. User interface and sample trade-off graph.

In Figure 7, only the preferred individuals, those that satisfy the design goals, are shown. When no individuals satisfy all the design goals, the nondominated or Pareto optimal solutions are displayed. Trade-offs between adjacent objectives result in the crossing of the lines between them, whereas concurrent lines indicate that the objectives do not compete with one another. For example, the power rating and TBT (objectives (3) and (4)) appear to compete quite heavily while the rise-time and settling-time (objectives (1) and (2)) do not exhibit the same level of competition. Note the appearance of distinct bands in the objective values, for example, objectives (6), (7), and (8), that occur for particular control configurations. An additional feature of the user interface is the ability to move the position of design objectives on the x-axis. This offers a convenient mechanism by which the engineer may examine the trade-offs between nonadjacent design objectives.

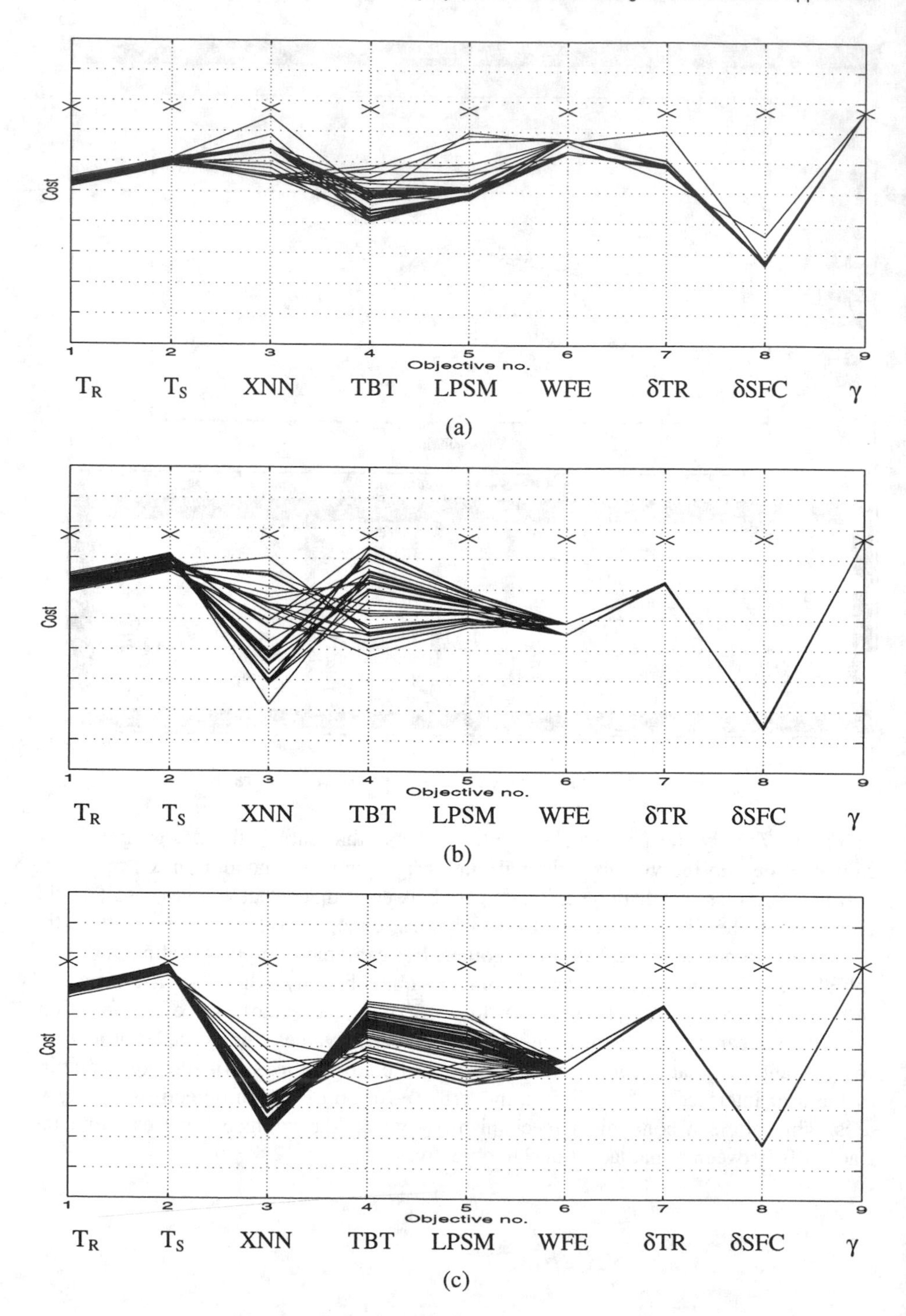

Figure 8. Trade-off for fuel control loops. (a) All NL controllers, (b) all NH controllers, and (c) all EPR Controllers

Examination of the results presented in such trade-off graphs may be used to gain insights into the nature of the system to be controlled and the trade-offs to be made between design objectives. Consider the family of trade-off graphs, grouped for the main fuel control loop, shown in Figure 8. It can clearly be seen that NL control provides the best transient response characteristics of all the controller types but suffers from a relatively high degree of sensitivity to sensor error. The sensitivity margin may be improved, for example, by selecting a better type of sensor for NL. The lowest power controllers were also found indicating that although a good thrust mapping to PLA can be achieved, there is less scope for compensation on a nonideal engine. This is confirmed by examining objectives (8) and (9) which indicate that although the transient performance margins may be easily accommodated, this mode of control suffers from the greatest increase in SFC with engine aging.

On the other hand, the NH controllers minimize the TBT at the expense of a slower response to changes in demand. This may allow the use of cheaper turbine materials, for example, if the other performance criteria are satisfactory or indicate a longer engine life expectancy and a reduced cost of ownership. Sensitivity to sensor error is better than with NL control and higher thrust ratings may also be achieved. The degraded engine also offers more consistent control in this mode than NL and only slightly less than EPR control.

EPR control results in the least sensitivity to sensor error while allowing a larger LPSM to be maintained. Step response times are generally the slowest while better TBT control could be achieved over NL control. Improvements in the step response could be achieved, but at the expense of reduced LPSM and increased TBT.

Clearly, competition exists to some extent between most of the design objectives and any satisfactory controller will be a compromise over the design criteria. As the MOGA approach encourages diversity in the population, direct comparisons may be drawn between different control schemes. For example, Figure 9 shows the trade-offs between two different controllers: a single SISO control loop for WFE against NL and an open-loop schedule for NOZZ against NH (labelled SISO); and a 2×2 multivariable control of WFE and NOZZ against EPR and FPR (labelled 2×2 MVC).

Figure 9(a) shows the trade-offs for all of the design objectives for each of the selected controller configurations and parameter sets. As expected, the SISO control of NL results in a faster rise and settling times compared with EPR-based control as well as a lower TBT. However, in this particular case, NL-based control also offers an increased LPSM, contrary to the general trends of Figure 6. The EPR-based multivariable controller offers better sensitivity to sensor error and smaller deviations in specific fuel consumption with engine aging. The increase in rise-time for this controller with a degraded engine and the slower performance with the nominal engine may, however, mean that this form of control cannot guarantee acceptable levels of performance over the engine life.

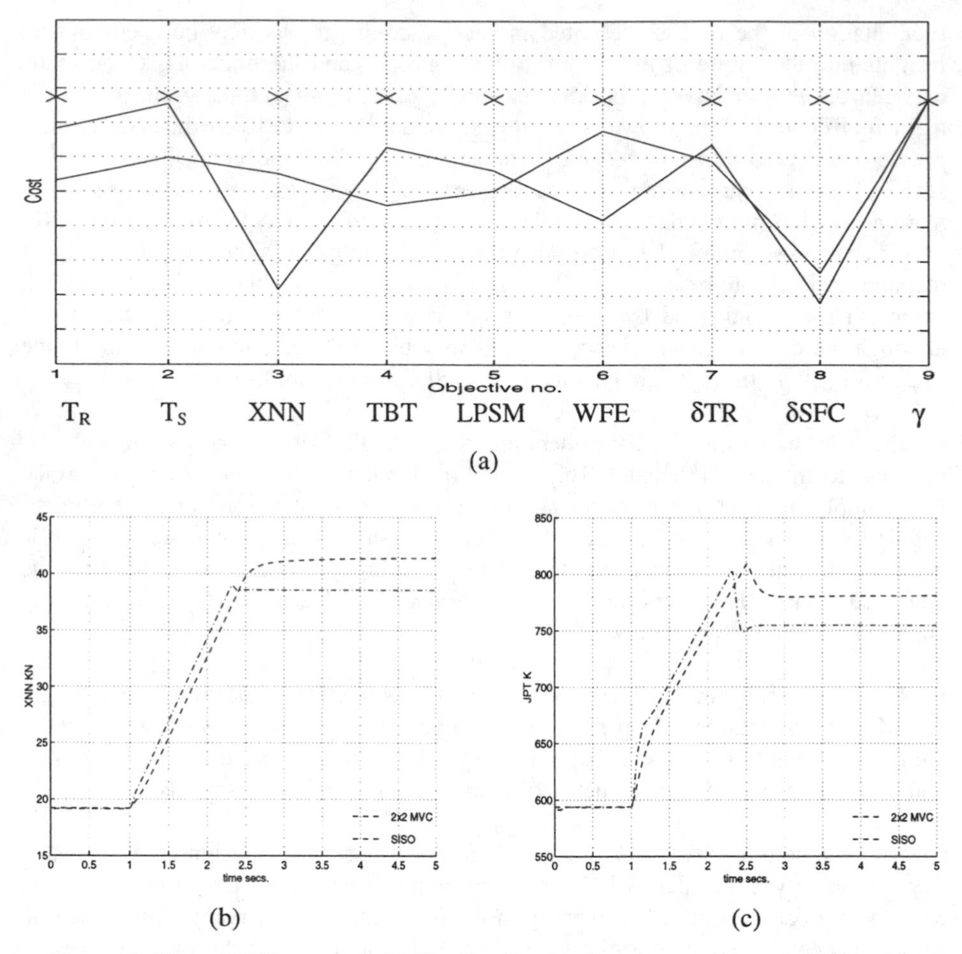

Figure 9. NL vs. EPR control. (a) Trade-offs, (b) thrust response, and (c) jet pipe temperature.

Figures 9(b) and (c) show the thrust response to the input demand and the jet pipe temperature over this maneuver. Figure 9(b) confirms that the NL-based control offers a faster rise than the EPR-based control, but also shows that the EPR-based controller does not overshoot the thrust demand. The multivariable controller offers a higher thrust rating as the nozzle area, NOZZ, is adjusted to maximize XNN while maintaining LPSM rather than following a fixed open-loop schedule as occurs with the SISO control. Similarly, the jet pipe temperature, Figure 9(c), settles to a higher level with the multivariable control, but does not suffer from the same transients as the SISO controller.

To consider the relative merits of the different controllers in this way using conventional search and optimization techniques would require the formulation of a number of separate optimization problems. Additionally, each optimization problem would have to be solved with a spread of weights and goals for the design objectives if the general characteristics of individual control configurations are to be identified relative to other available configurations. On the other hand, the method described with this example has shown that the MOGA-based approach allows us to determine the general control characteristics and compare individual controllers (configuration and parameter set) with one another within a single design framework in a relatively efficient manner. The range of design criteria considered, and the comparison of different control options, means that the engineer's choice for the final controller may be made on the basis of an informed selection over all of the design requirements. This should help match the controller to the plant and the desired design characteristics more closely, leading to a more integrated system.

6.4 Discussion

This example has considered the design of the control system at a single operating point. Within the proposed framework for evolutionary control mode analysis, it is feasible to perform a preliminary design of the control systems across a number of operating points and thus assess the suitability of each proposed control scheme more thoroughly. Indeed, where considerations are made for performance margins, models of degraded engine components have been included in the system to allow a more accurate assessment of design requirements and acceptable margins. The range of controllers may be further extended to include other controllable inputs and other control configurations. However, as many loops may not offer satisfactory control, the EA-based approach should be able to remove them from the search space once they have been determined to be infeasible. Similarly, the active set of design objectives may vary during the course of a search allowing more detailed analysis of conflicts between particular design objectives. This is an important consideration for more modern engine designs, such as ASTOVL and future variable cycle engines, which will require very precise control of more variable geometry components if their potential benefits are to be fully realized.

With the wider availability of sensor and actuator components, and more engine parameters becoming measurable and controllable, more detailed models may be included to cover such aspects as reliability, fault detection and isolation, weight, and economic considerations. Similarly, nonlinear models offer greater scope for realistic simulation and analysis and should be included within the architecture. However, given the potentially large search space and the nature of the interactions between the design parameters and objectives, this example has demonstrated that the MOGA would benefit from the addition of decision support, database and knowledge capture systems. Work is currently under way on a number of these aspects.

7. Concluding Remarks

This chapter has considered how multiobjective evolutionary algorithms may be employed in the search and optimization of suitable control modes for an aircraft gas turbine engine. The approach proposed differs from currently available techniques in that it affords the ability to simultaneously identify and examine a number of potential control configurations. This allows the control engineer the opportunity to examine the relative benefits of each control mode, highlighting both the positive and negative aspects of each individual scheme in a single framework, hopefully realizing a more informed design and efficient design process.

The example presented in this chapter has also shown how design points encountered during the search may be used to set performance specifications or determine areas of the system where components or subsystems may require redesign. As the specification and design of aircraft engines becomes more complex and the mission requirements for both civil and military engines get more stringent, the need for design tools such as those described in this chapter will increase.

Acknowledgments

The authors gratefully acknowledge the support of this research by UK EPSRC grants on "Evolutionary Algorithms in Systems Integration and Performance Optimization" (GR/ K 36591) and "Multi-objective Genetic Algorithms" (GR/J 70857). The authors also wish to thank Dr. Carlos Fonseca for his multi-objective extensions to the genetic algorithm toolbox.

References

[1] Garwood, K. R. and Baldwin, D. R., 1992. "The emerging requirements for dual and variable cycle engines," *10th Int. Symp. on Air Breathing Engines*, RAE-TMP-1220.

[2] Shutler, A. G. and Betteridge, H. C., 1994. "Definition, design and implementation of control laws for variable cycle gas turbine aircraft engines," *The Design & Control of the Next Generation of Civil & Military Engines: Do Variable Cycle Engines Have a Role?*, RAe, London, U.K.

[3] Chipperfield, A. J. and Fleming, P. J., 1996. "Genetic algorithms in control systems engineering," *IASTED Journal of Computers and Control*, Vol. 24, No.1, 88-94.

[4] Dadd, G. J., Sutton, A. E., and Greig, A. W. M., 1995. "Multivariable control of military engines," *Advanced Aero-Engine Concepts and Controls, AGARD Conf. Proc 572*, pp. 28/1 - 28/12.

[5] Hoskin, R. F., Net, C. N., and Reeves, D. E., 1991. "Control configuration design for a mixed vectored thrust ASTOVL aircraft in hover," *Proc. AIAA Guidance, Navigation and Control Conf.*, Vol. 3, 1625-1642, AIAA 91-2793.

[6] Shutler, A. G., 1995. "Control configuration design for the aircraft gas turbine engine," *IEE Computing & Control Engineering Journal*, 22 - 28.

[7] Spears, W. M., De Jong, K. A, Bäck, T., Fogel, D. B., and De Garis, H., 1993. "An overview of evolutionary computation," *Machine Learning: ECML-93 European Conference on Machine Learning, Lecture Notes in Artificial Intelligence*, No. 667, 442-459.

[8] Hwang, C. -L. and Masud, A. S. M., 1979. *Multiple Objective Decision Making - Methods and Applications: A State of the Art Survey*, Springer-Verlag, Germany.

[9] Goldberg, D. E., 1989. *Genetic Algorithms in Search, Optimization and Machine Learning*, Addison-Wesley.

[10] Chipperfield, A. J. and Fleming, P. J., 1996. "Multiobjective gas turbine engine controller design using genetic algorithms," *IEEE Trans. Ind. Electronics*, Vol. 45, No. 5, 583-587.

[11] Dasgupta, D. and McGregor, D. R., 1992. "Nonstationary function optimization using the structured genetic algorithm," *Parallel Problem Solving from Nature 2*, R. Manner and B. Manderick (Eds.), 145-154.

[12] Kroo, I., Altus, S., Braum, R., Gage, P., and Sobieski, I., 1994. "Multidisciplinary optimization methods for aircraft preliminary design," *5th AIAA/USAF/NASA/ ISSMO Symposium on Multidisciplinary Analysis and Optimization*, AIAA 94-4325.

[13] Gage, P. and Kroo, I., 1993. "A role for genetic algorithms in a preliminary design environment," AIAA 93-3933, AIAA Aircraft Design, Systems and Operations Meeting, Monterey, CA.

[14] Fonseca, C. M. and Fleming, P. J., 1995. "An overview of evolutionary algorithms in multiobjective optimization," *Evolutionary Computing*, Vol. 3, No. 1, 1-16.

[15] Fonseca, C. M., 1995. *Multiobjective Genetic Algorithms with Application to Control Engineering Problems*, Ph.D. Thesis, University of Sheffield.

[16] Baker, J. E., 1985. "Adaptive selection methods for genetic algorithms," *Proc. 1st Int. Conf. on Genetic Algorithms*, J. J. Grefenstette (Ed.), Lawrence Erlbaum, 101-111.

[17] Baker, J. E., 1987. "Reducing bias and inefficiency in the selection algorithm," *Proc. 2nd Int. Conf. on Genetic Algorithms*, J. J. Grefenstette (Ed.), Lawrence Erlbaum, 14-21.

[18] Ritzel, B. J., Eheart, J. W., and Ranjithan, S., 1994. "Using genetic algorithms to solve a multiple objective groundwater pollution containment problem," *Water Resources Research*, Vol. 30, 1589-1603.

[19] Goldberg, D. E. and Segrest, P. 1987. "Finite Markov chain analysis of genetic algorithms," *Proc. 2nd Int. Conf. on Genetic Algorithms*, J. J. Grefenstette (Ed.), Lawrence Erlbaum, 1-8.

[20] Goldberg, D. E. and Richardson, J., 1987. "Genetic algorithms with sharing for multimodal function optimization," *Proc. 2nd Int. Conf. on Genetic Algorithms*, J. J. Grefenstette (Ed.), Lawrence Erlbaum, 41-49.

[21] Silverman, B. W., 1986. *Density Estimation for Statistics and Data Analysis*, Chapman and Hall.

[22] Deb, K. and Goldberg, D. E., 1989. "An investigation of niche and species formation in genetic function optimization," *Proc 3rd Int. Conf. on Genetic Algorithms*, J. D. Schaffer (Ed.), Morgan Kaufmann, 42-50.

[23] Grefenstette, J. J., 1992. "Genetic algorithms for changing environments," *Parallel Problem Solving from Nature* 2, R. Männer and B. Manderick (Eds.), North-Holland, 137-144.

[24] Chipperfield, A. J. and Fleming, P. J., 1996. "Systems integration using evolutionary algorithms," *UKACC Control '96*, Exeter, U.K., Vol. 1, 705-710.

[25] Brown, H. and Elgin, J. A., 1985. "Aircraft engine control mode analysis," *ASME J. for Gas Turbines and Power*, Vol. 107, 838-844.

[26] Chipperfield, A. J., Pohlheim, H. P. and Fleming, P. J., 1994. "A genetic algorithm toolbox for MATLAB," Proc. *Int. Conf. Systems Engineering*, Coventry, U.K., 200-207.

Chapter 12:

Application of Genetic Algorithms in Telecommunication System Design

APPLICATION OF GENETIC ALGORITHMS IN TELECOMMUNICATION SYSTEM DESIGN

V. Sinkovic, I. Lovrek, and B. Mikac
Department of Telecommunications
Faculty of Electrical Engineering and Computing
University of Zagreb
Croatia

Many of design problems in telecommunications could be treated as optimization problems that include some kind of searching among a set of potential solutions. The choice of the method depends mostly on the problem complexity. If the number of possible solutions is not too big, one could enumerate them all, evaluate their goal functions, and select the best solution(s). If the function to be optimized is done by a derivative continuous function, analytical methods could be applied. In all other cases, where the problem space is too big and analytical methods are not applicable, some sort of heuristic search for optimal solution could be applied.

Genetic algorithms, to be presented in this chapter, could be classified as guided random search evolution algorithms that use probability to guide their search. Genetic algorithms are created by analogy with the processes in the reproduction of biological organisms. By natural selection or by forced selection in laboratories, new generations of organisms are produced. As a consequence of crossover and mutation processes on chromosomes and genes, the children could possess either better or worse features than their parents. The "better" organisms are those that have a greater chance than the "worse" ones to survive and to produce a new generation.

This chapter deals with the application of genetic algorithms in telecommunications. After a brief introduction into genetic algorithms in Section 1, three applications of genetic algorithms will be discussed. Section 2 describes a method for call and service process scheduling based on genetic algorithm. Section 3 deals with the analysis of call and service control in distributed environment, where a genetic algorithm is used to determine a response time. Section 4 presents an example of genetic algorithm application in optimization problem solving. An example is presented through the case study on availability–cost optimization of an all-optical network.

0-8493-9804-5/99/$0.00+$.50

1. Genetic Algorithm Fundamentals

The functioning of a genetic algorithm (GA) could be described by using specific data structure and by defining the set of genetic operators, as well as the reproduction procedure. In this section the explanations are mostly referred to the restricted class of simple genetic algorithms (SGA) including possible extensions to the more complex ones [1]. The terms used are a mixture taken from both biological and computational domains.

The biological organism to be specified is defined by one or by a set of chromosomes. The overall set of chromosomes is called genotype, and the resulting organism is called phenotype. Every chromosome consists of genes. The gene position within the chromosome refers to the type of organism characteristic, and the coded content of each gene refers to an attribute within the organism type.

In GA terminology, the set of strings (chromosomes) forms a structure (genotype). Each string consists of characters (genes). In SGA the character is two-valued, represented by a bit. In a search procedure, a string represents a full solution and a character (bit) represents a variable to be found, perhaps, as optimal.

The quality of a single solution in GA is determined by a fitness function f. The role of the fitness function could be twofold: to award the goodness of a solution on the one hand, and to punish the solution not satisfying constraints on the other. One solution is better than some other, if it has a higher fitness value. For an optimization procedure, the fitness function is equivalent to the goal function that could have a global optimum and a number of local optima.

In order to provide a reproduction process, producing a new generation of strings from the old one, in SGA the following genetic operators are used: reproduction, crossover, and mutation.

The reproduction operator, as the crucial one in SGA, is based on the string selection from roulette wheel. A string s_i among the set of all strings, with its cardinal number n_s, occupies an angle size α_i of the wheel, proportional to its fitness function value f_i.

$$\alpha_i = \frac{f_i}{\sum_{i=1}^{n_s} f_i} \cdot 2\pi$$

A new generation is produced by repeating gambling as shown in Figure 1, selecting a number at random in the interval $(0, 2\pi)$, until all strings are selected.

The option–elitism could be applied in which the best string offered to the selection process is transferred directly to the next generation. Note that in algorithms other than SGA, more sophisticated selection techniques are used, such as stochastic reminder selection technique and tournament selection.

Crossover operator provides the exchange of genetic material, represented by a set of characters, between two strings. The part(s) of a string to be exchanged are defined by one or more crossover points. For example, two strings are selected at random for crossing over in randomly selected crossover points (*), as shown in Figure 2. The occurrence of the crossover is defined by crossover probability.

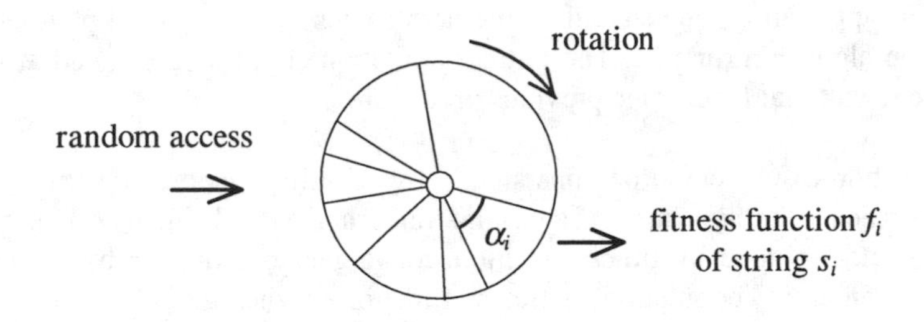

Figure 1. Roulette wheel selection.

two strings before crossover		two strings after crossover
100011*0100011*001	→	1000111001110001
crossover ↓ ↑		
001010*1001110*011	→	0010100100011011

Figure 2. Crossover example.

Mutation operator is related to a bit (gene) which could be changed according to the mutation probability. For example, some bits $\underline{x}$ in a string are selected at random to be mutated, as shown in Figure 3.

a string before mutation				the string after the mutation
1000111001110001	→	mutation	→	1000011001110101

Figure 3. Mutation example.

The set of strings in a generation is referred as a population. The population size could change, increase or decrease, from generation to generation, but, in most applications, it is taken to be constant. The population size is a function of string size and it should be big enough to avoid the "incest effect" in the reproduction process. If diversity of strings is small, there are few chances to change (improve) species using crossover operator. The diversity could be improved in this case by intensifying mutation, but, on the other hand, by increasing mutation probability, searching tends to be random. Generation gap is referred as the fraction of a population, in interval (0,1), which takes part in reproduction procedure creating a new generation.

In each GA application, one should define how to generate the initial population and how to stop algorithm running. The initial population could be generated at random, or as genetic material from some previous procedure.

The termination of GA could be done simply by counting if some prescribed number of steps is reached, or by testing, if a termination criterion is fulfilled. An autonomous stopping could be done by fitness function convergence testing or by homogeneity checking of an entire population. If fitness function reaches global or some of local optima, then all strings, because of preferring the best solution, tend to be equivalent. In this case, a high degree of homogeneity could stop the procedure, or adapt GA parameters in order to move searching of the solution to other areas of local optima as candidates for the global one. In this case, GA is considered to have adaptivity features.

A GA application to a specific problem includes a number of steps. Some of them are listed below:

- The problem definition – goal, assumptions, and constraints;
- Solution coding;
- Using SGA or an improved GA;
- Population homogeneity – testing or not testing;
- Adaptivity – including or not including;
- Selection technique definition;
- Elitism – including or not including;
- Defining the number of crossover points;
- Defining crossover and mutation probabilities;
- Defining population size;
- Defining fitness function;
- Generation of initial population;
- Defining termination criterion.

Three examples discussing GA applications in telecommunication system design follow.

2. Call and Service Processing in Telecommunications

2.1 Parallel Processing of Calls and Services

Call is a generic term related to the establishment, utilization, and the release of a connection between a calling user and the called user for the purpose of exchanging information. The call is also defined as an association between two or more users, or between a user and a network established by using network capabilities. Service is offered by a network to its users, in order to satisfy a specific telecommunication requirement.

Users initiate calls and services. Each call/service causes a number of processing requests. The call/service can be represented by a sequence of requests occurring in random intervals. The first request initiates the call, while further requests start different communication and information operations – named call phases. On this level of abstraction, unsuccessful calls shorten the call because some phases are skipped. The services shorten, modify (some phases are replaced by others), or enlarge the call (new phases are inserted), so the number of requests varies from case to case. The mean number of requests depends on implemented set of services as well as on traffic and other conditions in the network.

The call/service decomposition on phases depends on the interaction with the environment; information from the environment (request input data) is needed to start a specific call phase, and/or information (result) is sent to the environment after completing a call phase. Different types of calls and services represent different types of independent, cooperating, or mutually excluding tasks to be run on the network.

It should be pointed out that, when discussing call and service handling, control plane of the telecommunication network must be taken into consideration. Considering the aspect of control, the system is reactive. After receiving a request, and reflecting defined time conditions, a response has to be produced; otherwise, a call/service will be lost.

When discussing the decomposition of calls and services, the following features have to be taken into account:

- calls and services are real time processes with response time constraints and real time dependencies,
- calls and services are distributed processes, because telecommunication systems are built from a variable number of communicating nodes operating autonomously and weakly coupled,

- calls and services are processes with a potentially large number of parallel activities, called *elementary tasks* (ETs in this chapter).

There are different levels of parallelism to be considered in order to organize an efficient processing system:

- the simultaneous processing of different calls and services,
- the simultaneous processing of different/equal phases of different calls and services,
- simultaneous evaluation of ETs from different call/service and simultaneous evaluation of ETs from the same call/service.

The first two levels are related to the processing capacity, i.e., they influence the number of calls/services to be handled. The third level can improve the call/service processing time and, only in this case, a minimum processing time can be obtained. A parallel system should include all three levels [2, 3].

2.2 Scheduling Problem Definition

Each call/service processing task consists of a number of processes. The question is what could be considered as a basic call/service grain – what the elementary task is or how simple/complex it is. It is well known that telecommunication applications are programmed as if they were composed of a large number of small agents.

The solution with higher parallelism generally could be obtained with smaller ETs, but the proper boundary also depends on the way in which the concurrent programming language supports the creation and destruction of parallel activities, as well as interprocess communication. The programming reasons are against unconstrained parallelism, i.e., parallel activities at the instruction level. In fact, an individual ET consists of several sequential activities; it describes a sequential behaviour, but it is allowed to be concurrently executed [4].

This is the reason why the internal structures of call/service tasks have to be defined in order to arrange the parallel system, and to analyze it as well. It is assumed that all ETs have the same duration, i.e., each one can be handled in an elementary interval Δt. A task pattern associated to a processing request, r_i, is defined by the following parameters:

n_i the number of ETs,
p_i the maximum number of ETs which can be processed in parallel,
m_i the number of ET partitions which must be processed in sequence,

and an internal structure,

$$r_i \rightarrow \left(s_1, s_2, \ldots, s_j, \ldots, s_{m_i}\right),$$

where partition index j is referred as height, s_j represents the number of ETs in a partition j ($1 \le s_j \le p_i$), and $\Sigma s_j = n_i$. In that way, serial–parallel ordering of ETs defines a task pattern for a specific request. If $m_i = n_i$ and $p_i = 1$, ETs are evaluated in sequence; for $m_i = 1$ and $p_i = n_i$, all ETs could be simultaneously executed. All other cases define partial parallelism for a given set of ETs. The total request processing time, $m_i \times \Delta t$, cannot be decreased; it is determined by the inherent parallelism of ETs. An example is shown in Figure 4.

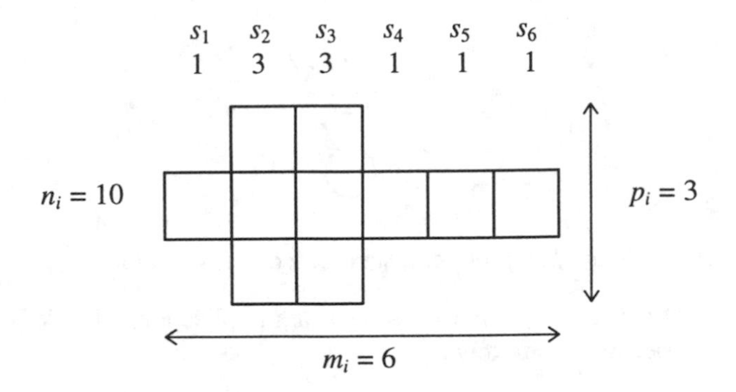

Figure 4. Task pattern example.

Note that the task pattern graphic representation corresponds to Gantt chart. The task pattern can also be defined as a set of partially ordered ETs and represented by a directed acyclic graph $G = (V, E)$. V is a finite set of vertices representing ETs. A set of edges E, where e_{ij} denote a directed edge from i to j, describes the partial ordering or precedence relation between ETs, denoted by >. If $i > j$ then i must be completed before j. We consider a graph with identical ET times (equals 1 Δt of time, each). The length of the critical path of the graph m_i is the minimum finishing time, i.e., the minimum time to complete the set of ETs.

The problem of determining optimum task pattern, i.e., the task pattern with minimum p_i satisfying the minimum processing time m_i is equivalent to the problem of determining the minimum number of processors required to evaluate the program in the shortest possible time. The problem of determining the task pattern with the shortest finishing time for some p_i corresponds to the classical scheduling problem based on a deterministic model, because the relationship between ETs and ET execution time are known [5, 6]. The goal is to assign ETs to the processors, so that precedence relations are preserved and the set of ETs is completed in the shortest possible time. The problem is known as computationally intractable and some heuristic algorithm should be used.

A graph representing the task pattern with ET1 - ET10, similar to the previous one, is shown in Figure 5 (both patterns are equal with respect to the number of ETs, maximum parallelism, and finishing time, but precedence relations are different).

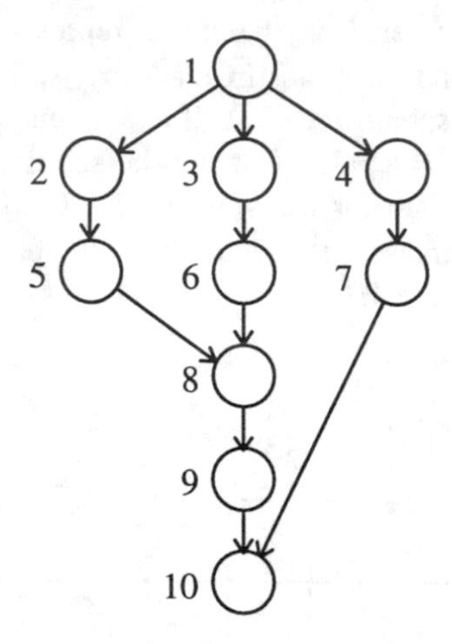

Figure 5. Graph representation of a task pattern.

An application of genetic algorithm for scheduling implies a careful definition of GA constructs [7], especially the following.

Schedule representation

The representation of a schedule must accommodate the precedence relation between ETs. The schedule is represented by means of several lists of ETs, one for each processor. The order of tasks in the list corresponds to the order of execution. Every task appears only once in the schedule. A schedule satisfying all these conditions is the legal schedule. Let us consider two processors, $P1$ and $P2$, and the graph representation of a task pattern shown in Figure 5. Legal schedules A and B, both with the finishing time FT, equal to 8 Δt, are, for instance,

	$P1$	$P2$	
A	[1 3 4 7 6]	[2 5 8 9 10]	$FT = 8$
height	0 1 1 2 2	1 2 3 4 5	
B	[1 2 4 6 7 10]	[3 5 8 9]	$FT = 8$
height	0 1 1 2 2 5	1 2 3 4	

Evidently, the schedule representation is more complex than string representation used in SGA, as described in Section 2.1. The genetic operators must maintain the precedence relation in each ET list (processor) and ensure a unique appearance of all ETs in the schedule.

Crossover

Crossover operator will produce a legal schedule, if crossover sites always lie between ETs with different heights, and if the heights of the tasks on the first right positions of the crossover sites are equal. The following example shows the result of a crossover operation between schedules A and B producing new schedules C and D. The crossover sites (*) are placed between following pairs of ETs: in processor $P1$, schedule A (ET4, ET7), in $P1$, B (ET4, ET6) (note ET7 and ET6 have equal heights), in $P2$, A (ET8, ET9), and in $P2$, B (ET8, ET9).

	$P1$	$P2$	
C	[1 3 4 6 7 10]	[2 5 8 9]	$FT = 7$
height	0 1 1 2 2 5	1 2 3 4	
D	[1 2 4 7 6]	[3 5 8 9 10]	$FT = 8$
height	0 1 1 2 2	1 2 3 4 5	

In this example, schedule C has a better fitness value than the parent schedules.

Mutation

Mutation operator will produce legal schedule, if ETs with identical heights are mutated, exchanging their positions in the schedule. For example, by using mutation operator on ET7 and ET6 of schedule A, a new schedule E is produced.

	$P1$	$P2$	
E	[1 3 4 6 7]	[2 5 8 9 10]	$FT = 7$
height	0 1 1 2 2	1 2 3 4 5	

In this example, the new schedule E has a better fitness value than old schedule A.

3. Analysis of Call and Service Control in Distributed Processing Environment

3.1 Model of Call and Service Control

According to the call and service processing concept described in Section 2.1, a model of call and service control is introduced. It consists of a processing system and its environment representing the telecommunication network and its users, as shown in Figure 6. Requests coming from the environment enter the request queue limited to N places. If the number of requests is greater than N, a loss occurs. Requests wait until the processing of the previous request has been completed. After that, all requests from the queue are moved to the task pattern formatting stage where a set of elementary tasks with their precedence relations is associated to every request. Further on, the elementary task scheduling is performed by using GA. Finally, elementary tasks enter processor queues.

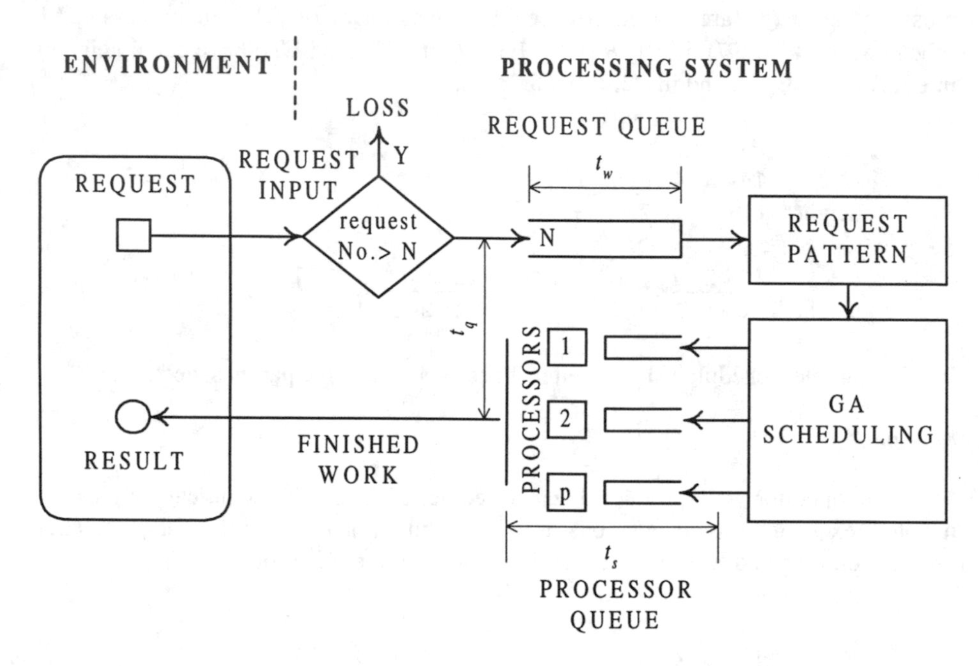

Figure 6. Model of call and service control.

The most important parameter describing control system performance is a response time, that is, the time needed to generate response to a processing request. To obtain response time t_q, two parameters have to be evaluated: finishing time T_s and waiting time t_w. The problem is similar to the problem of multiprocessor scheduling: t_s corresponds to the optimal finishing time for a given set of elementary tasks and t_w to the time needed to complete the previous set of elementary tasks.

3.2 Simulation of Parallel Processing

A genetic algorithm is applied for finishing time determination in a simulation method used for the analysis of parallel processing in telecommunications [8, 9, 10, 11]. The method includes three steps: request flow generation, finishing time determination, and response time calculation, that are repeated as many times as pointed out by defined number of simulation samples. The simulation is based on a generation of processing request flow with random structures with respect to the number of ETs and their precedence relations. A stochastic nature of processes is defined with probability functions for arrival characteristics of processing requests, and the number of ETs per request.

Each processing request consists of a certain number of ETs, each ET taking the same processing time Δt, as defined in the previous sections. For example, Figure 7 shows 10 requests represented by the sequence of numbers, where each number describes the number of ETs in each partition and determines how many tasks can be simultaneously processed in each processing phase. For the first request there is a sequence {3, 4, 4, 6, 6, 2}, which means that in the first Δt there are 3 ETs that could be processed in parallel, in the second one there are 4 ETs, etc. The whole request can be processed for at least 6 Δt if there is enough processor capacity in each Δt.

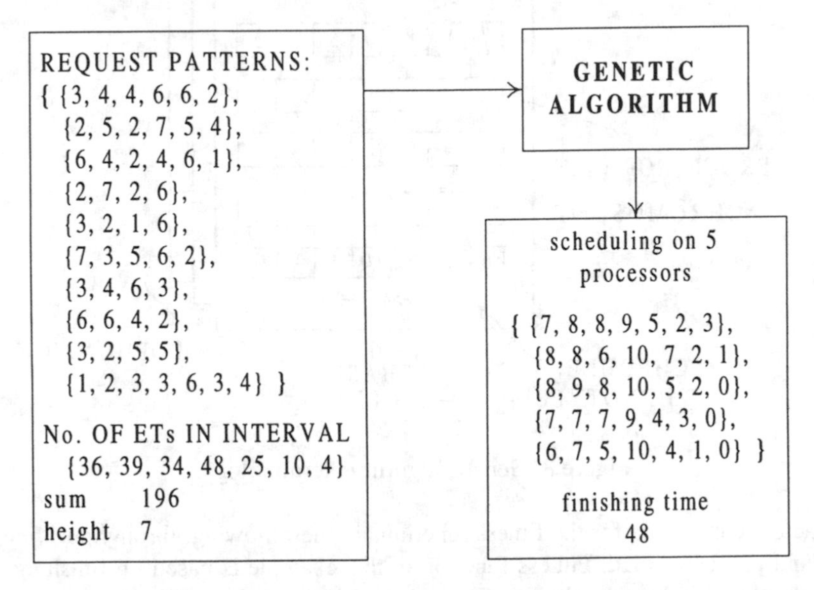

Figure 7. The survey of the genetic algorithm function.

The problem becomes more difficult if more demands with different patterns come at the same time as in this example, where the total number of ETs is given by the sequence {36, 39, 34, 48, 25, 10, 4}. Genetic algorithm will be used to distribute ETs on a certain number of processors (in this example, there are 5 processors) and to get the shortest time to finish all the tasks. One feasible schedule is shown in the same figure.

3.3 Genetic Algorithm Terminology

Before using genetic algorithm, we associate the meanings to the terms as shown in Figure 8.

Task schedule on the single processor forms a processor list that corresponds to the chromosome. The places in the processor lists representing processing phases correspond to the genes, and the number of ETs in the place represents the genetic code. One of possible schedules corresponds to the phenotype. Set of the schedules corresponds to the population on which genetic operations are applied in some steps of the genetic algorithm. In that way, a set of ETs defines a genetic code contained in a gene to be exchanged between schedules, i.e., genetic operations will operate on such ET sets and not on individual ETs. It should be noted that, in the example discussed in Section 2, each individual ET takes the notion of a gene.

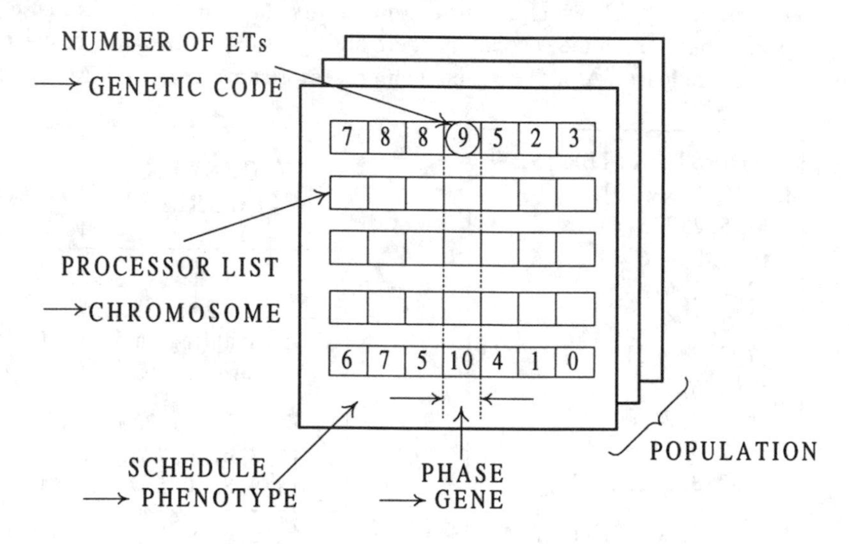

Figure 8. Genetic algorithm terminology.

The factors considered for the fitness function are the following: throughput, finishing time, and processor load. Fitness function in this example is based on finishing time. Finishing time *FT*, for the schedule *S* at a phase *h* is defined by

$$FT(S,h) = \max_{P_j}\{ftp(P_j)\},$$

where ftp (P_j) denotes finishing time of the last ET in the processor P_j. If n denotes schedule height, then each phase is denoted by $1 \leq h \leq n$, and the number of ETs associated to the processor P_j in the phase h is denoted by et (Pj, h), so the expression for the finishing time for the schedule S could be written in the form

$$FT(S) = \sum_h FT(S,h) = \sum_h \max_{P_j}\{et(P_j,h)\}.$$

In order to get the fitness function that is going to be maximized during the genetic algorithm running, the finishing time function has to be transformed into a maximized form by introducing the value $C_{\max}$ as the maximum value of the finishing time that occurs at any schedule until the determined moment during the algorithm run. The fitness function is defined as follows:

$$f = FF(S) = C_{\max} - FT(S).$$

Thus, the best schedule is the one with the shortest finishing time, i.e., the greatest value of the fitness function.

3.4 Genetic Operators

Reproduction

Reproduction is the basic genetic operator. Reproduction forms a new generation by choosing those individuals from the old population that have the highest value of the fitness function. The choice is based on the fact that the individuals with the highest value of the fitness function have greater chances to survive in the next generation. In our application it means that better task schedules have higher value of the fitness function and, because of that, they have to be preserved in the next generation. The choice is made by the roulette wheel selection technique. Since the schedules with a higher value of the fitness function take a greater part of the wheel, they have a higher probability to be chosen in the next generation. Also, an elitism procedure can be added, so that the best schedule from the old generation always moves into the new generation without a random choice.

Crossover

Two schedules A and B from the population are taken, as defined in Figure 9. Two new schedules C and D can be made by changing the parts of the schedules A and B according to the following crossover procedure:

A site (phase) h that divides processor lists into two parts is chosen randomly.

The left parts of the schedules remain unchanged, and the right parts are subjected to the crossover in such a way that the right part of the schedule A becomes the right part of the schedule B and vice versa.

Mutation

Mutation is considered as a stochastic alternation of the genetic code value in selected places in the schedule. Some modifications in the mutation operation defined in Section 1 are proposed because GA used for finishing time determination manipulate the ET sets. The modified procedure follows:

(1) The mutation is going to be performed by choosing a phase at random between the first and the last phase in the best schedule.

(2) The processors with the highest and the lowest load (number of ETs) are determined in the chosen phase.

(3) The ETs are redistributed between these processors in a way that half of them are allocated to each one.

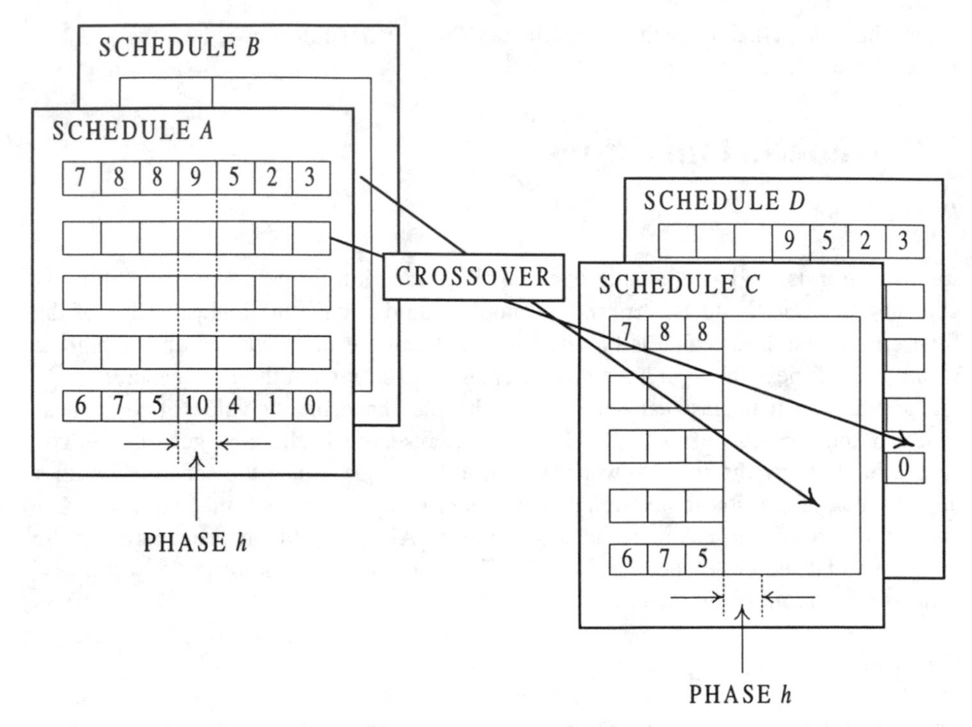

Figure 9. Crossover operation.

An example is shown in Figure 10. The second processor list has the lowest load (2 ETs) and the last one has the highest load (10 ETs). The total number of ETs (12) is redistributed on these processors by using mutation. The resulting schedule has 6 ETs in processors affected by this genetic operator.

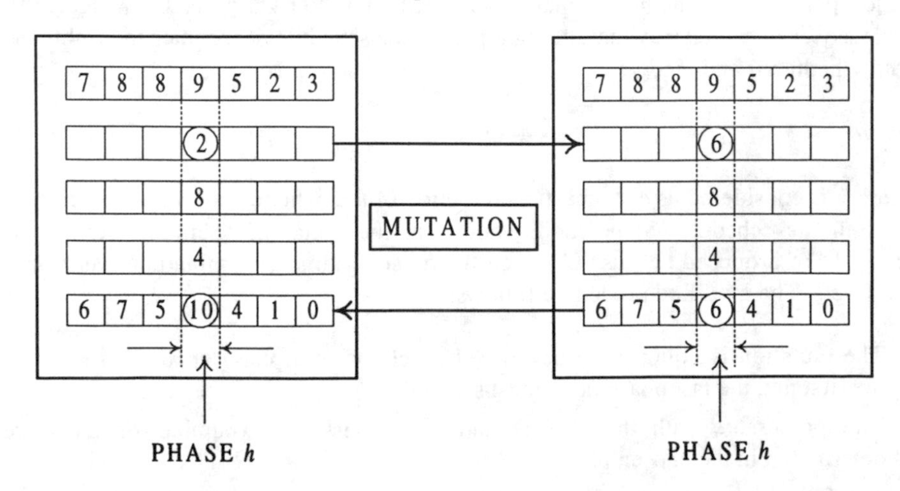

Figure 10. Mutation operation.

3.5 Complete Algorithm and Analysis Results

The flow chart of the complete genetic algorithm, whose parts have been described previously, is shown in Figure 11.

The start is defined by assigning the number of the generations explored by a genetic algorithm.

Afterward, the generation of the initial population is performed randomly, according to the assigned number of processors and the number of schedules in a population. Also, mutation and crossover probabilities, that are going to be used later, are defined.

After that, a fitness function is calculated for all schedules in the population.

For the schedule with the highest value of the fitness function, an operation of the modified mutation is done with the assigned probability that improves the best schedule.

The reproduction, the key operation of the genetic algorithm, follows. A new population, having the same number of schedules as the old one, is obtained. The best schedule is moved directly from the old generation into a new one.

Afterward, a new population is subjected to the crossover operation with assigned probability. Fitness function calculation shows the features of a new generation.

A resulting generation containing the best schedules is obtained when the defined number of generations is reached. After fulfilling convergence criterion, the algorithm terminates.

For the illustration of the genetic algorithm analysis in the given examples, an obtained finish time is considered as a complement of the fitness function in certain iterations.

A series of experiments was done to evaluate a performance of the genetic algorithm itself. Different problems, from regular to a heavy request flow, including request bursts were analyzed. Also, the initial population with respect to generated strings (schedules) and size has been evaluated. The comparison criterion was the best string in the population (the string with the lowest finishing time). An example of the analysis with different initial populations will be discussed here.

The analysis, shown in Figure 12, is done for three different initial populations (2, 10, and 20 strings) for a maximum number of generations, $GEN = 100$. The simulation method including GA has been programmed in Mathematica.

The best schedule and corresponding finishing time are simulation results. Note that the optimum ($FT_{opt} = 41\ \Delta t$) is not reached with the defined initial populations and the number of generations ($FT = 48\ \Delta t$).

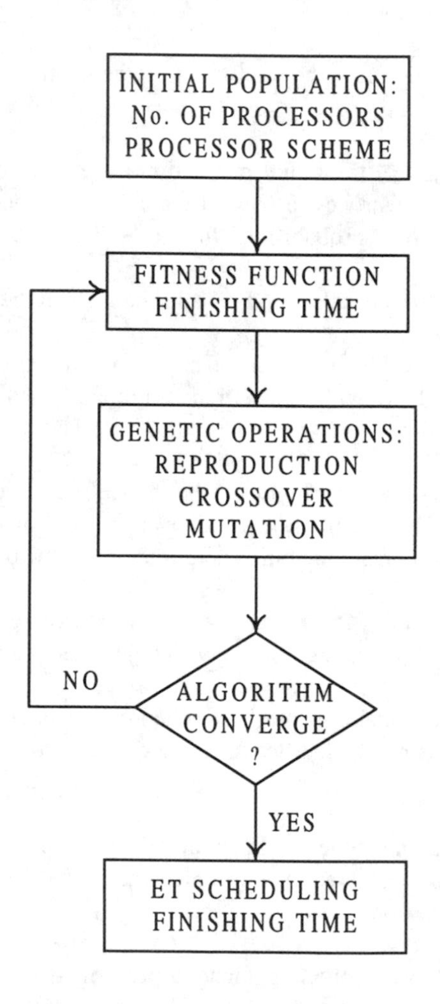

Figure 11. Flow chart of the genetic algorithm.

Parallel processing of calls and services in telecommunications is a stochastic process itself. This section presents the analysis of call and service control by using genetic algorithm – the method based on stochastic approach as well.

Systematic experiments have shown that GA approach offers a framework suitable for estimating performance of the parallel processing of calls and services in telecommunications and the associated control system.

request arrivals intensity 10
mean number of ETs 20
max. parallelism 8
number of processors 5
mutation probability 0.5
crossover probability 0.5

20 initial strings
best scheduling after 100 generations
{7, 8, 8, 9, 5, 2, 3},
{8, 8, 6,10, 7, 2, 1},
{8, 9, 8, 10, 5, 2, 0},
{7, 7, 7, 9, 4, 3, 0},
{6, 7, 5, 10, 4, 1, 0}
finishing time 48

generated request patterns
{3, 4, 4, 6, 6, 2},
{2, 5, 2, 7, 5, 4},
{6, 4, 2, 4, 6, 1},
{2, 7, 2, 6},
{3, 2, 1, 6},
{7, 3, 5, 6, 2},
{3, 4, 6, 3},
{6, 6, 4, 2},
{3, 2, 5, 5},
{1, 2, 3, 3, 6, 3, 4}
sum ETs 196
ETs per time slot
{36, 39, 34, 48, 25, 10, 4}
height 7
optimal finishing time 41

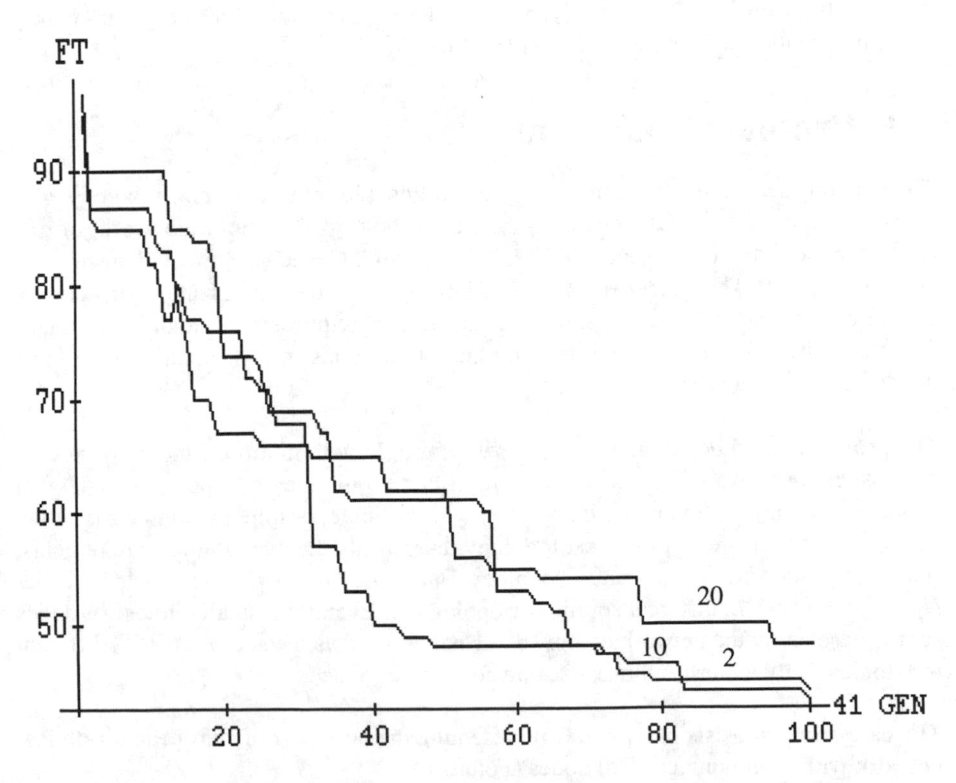

Figure 12. Finishing time vs. number of generations in GA application.

4. Optimization Problem - Case Study: Availability–Cost Optimization of All-Optical Network

The problem to be described is an example of genetic algorithm application in telecommunication network optimization. Note that this type of optimization problem could be solved by other methods as well, for example, simulated annealing and taboo search. Many of the design problems in telecommunications could be treated as optimization problems that include some kind of searching among the set of potential solutions. The choice of the method for solving the problem depends mostly on the problem complexity. In all cases, where the problem space is too big and analytical methods are not applicable, some sort of heuristic search for pseudo-optimal solution could be applied. For example, if one should find some kind of optimal network topology, among the set of n = 10 nodes, the number of possible links connecting predefined nodes is $n(n\text{-}1)/2$ = 45. Assuming that every candidate link could be present or not in the solution to be evaluated, the total number of topologies is 2^{45} = 3.5 10^{13}. If an enumeration method is used, assuming evaluation for one solution takes 1 ms, the solution will be reached in 1115 years.

4.1 Problem Statement

This section deals with the issues involved in generating an optimum topology of a European core all-optical network – a case study within the framework of the European Commission project COST 239 "Ultra-high Capacity Optical Transmission Networks" [12]. The objective of the optimization is the minimization of network unavailability and cost, while satisfying the traffic requirements among the major European cities, meeting current technological limitations in the optical domain, and the defined routing rules.

The problem could be defined in another way, too: how to minimize the network cost while keeping unavailability within the prescribed requirements, if possible. The goal is not only to have as minimum unavailability as possible, despite the high costs of the network, but to achieve a low-cost topology that fulfills the availability requirements, if any. In order to find an optimum topology for n nodes' network, one should consider $N_s = 2^k$, $k = n(n\text{-}1)/2$ different solutions (topologies). Even for a small number of nodes (in the case study the network comprises 11 nodes and has the set of 3.6 10^{16} different topologies) only a quasi-optimal solution could be obtained.

The case study consists of 11 nodes representing the core part of European all-optical network with total number of 20 nodes (Figure 13).

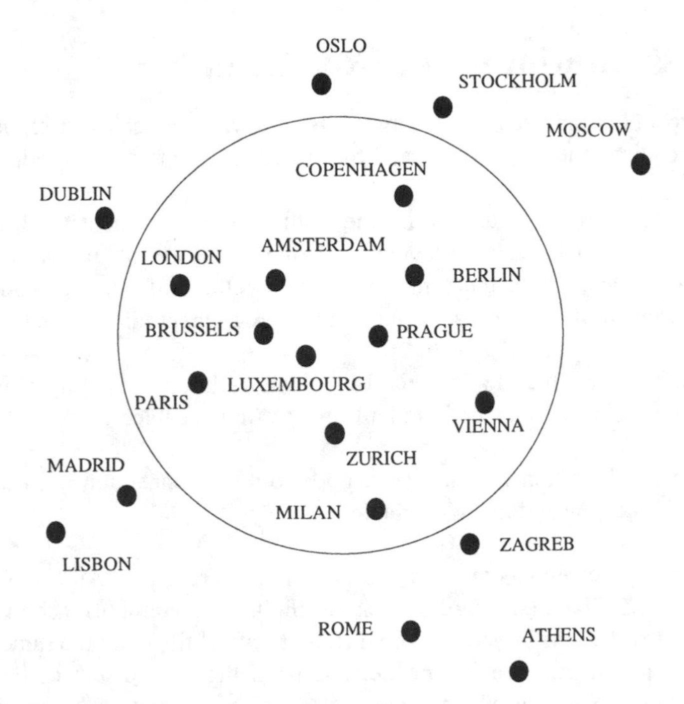

Fig 13. Case study: Core part of European all-optical network.

Every topology should fulfill symmetric traffic requirements (capacities) expressed by required bit rates, and take into account road distances between nodes (Table 1).

Table 1. Bit rate requirements and road distances

		Bit rates (Gbit/s)										
		0	1	2	3	4	5	6	7	8	9	10
		Par	Mil	Zur	Pra	Vie	Ber	Ams	Lux	Bru	Lon	Cop
Road distances x 10^3 (km)	0 Par	-	12.5	15	2.5	5	27.5	12.5	2.5	15	25	2.5
	1 Mil	0.82	-	15	2.5	7.5	22.5	5	2.5	5	7.5	2.5
	2 Zur	0.60	0.29	-	2.5	7.5	27.5	7.5	2.5	7.5	7.5	2.5
	3 Pra	1.00	0.87	0.62	-	2.5	5	2.5	2.5	2.5	2.5	2.5
	4 Vie	1.20	0.82	0.80	0.32	-	22.5	2.5	2.5	2.5	5	2.5
	5 Ber	1.09	1.01	0.90	0.34	0.66	-	20	5	15	20	7.5
	6 Ams	0.51	1.14	0.85	0.91	1.16	0.66	-	2.5	10	12.5	2.5
	7 Lux	0.34	0.71	0.38	0.73	0.93	0.75	0.39	-	2.5	2.5	2.5
	8 Bru	0.30	0.93	0.60	0.91	1.12	0.78	0.21	0.22	-	10	2.5
	9 Lon	0.45	1.22	1.00	1.31	1.51	1.17	0.55	0.60	0.39	-	2.5
	10 Cop	1.24	1.52	1.20	0.74	1.04	0.39	0.76	0.95	0.92	1.31	-

4.2 Assumptions and Constraints

In order to obtain an acceptable network topology, let us call it a regular topology, different requirements, limitations, and routing rules have to be fulfilled.

The network topology must fulfill the following requirements: all node-to-node connections should be established through the two shortest, mutually independent paths, primary and spare, the same for both directions of communication, ensuring network survivability in the case of single network element failure, the link or node.

A link failure is assumed to be caused by a failure in an optical amplifier, or in the fiber cable, causing an interruption of all services in the cable.

The following definition is assumed: a node-to-node connection is available if both directions of the connection are available.

The traffic requirements between all pairs of nodes are given. All link capacities are multiples of 2.5 Gbit/s (standard capacity in digital transmission), achieved through a number of wavelengths in one or more different optical fibers on the same optical link. The node pair direct distances are derived from the road distances between major European cities. Because of the accumulated noise and distortions in optical fibers, amplifiers, and node elements, the optical path length limitation is fixed at 2000 km. The distances between optical amplifiers are assumed to be maximum 100 km. Component failure and repair rate data for calculating the unavailability of the future all-optical network are taken from the existing data set for mature optical components, whereas, for new photonic components, the calculation is based on estimated data. Steady-state unavailability (the asymptotic value of unavailability if time tends to infinity) is considered, assuming constant failure and repair rates. In the total path unavailability calculation, the impact of node unavailabilities is negligible compared to the unavailabilities of optical links.

4.3 Cost Evaluation

The cost model applied in the network availability optimization was taken from [13]. The total network cost for the set of nodes N is a sum of all link and node costs is

$$C = \sum_{i,j\in N} C_{L_{ij}} + \sum_{i\in N} C_{N_i} = C_L + C_N,$$

where C_{Lij} is the cost of the link between nodes i and j, and C_{Ni} is the cost of the node i. Link cost is a function of link length L_{ij} (km) and link capacity V_{ij} (Gbit/s),

$$C_{L_{ij}} = L_{ij}\, V_{ij}.$$

The link capacity is determined for each link by summing up the contributions from all primary and spare paths that make use of it.

The node cost C_{Ni} is a function of node effective distance N_i (km) (N_i represents the cost of node in equivalent distance terms), and the total capacity of all links incident to the node – V_i (Gbit/s):

$$C_{N_i} = 0.5\, N_i\, V_i, \quad N_i = E + d_i\, F,$$

where d_i is node degree (the number of links incident to node i), E and F constants assumed to be 200 km and 100 km, respectively.

4.4 Shortest Path Evaluation

For each topology, the solution proposed by GA between all pairs of nodes – the first shortest path as the primary path, and the second shortest path as a spare path – have to be evaluated using Dijkstra algorithm. The weights W_{ij} of links to be used in shortest path evaluation reflect the influence of node parameters on the path "length".

$$W_{ij} = 0.5\, N_i + L_{ij} + 0.5\, N_j.$$

4.5 Capacity Evaluation

Superposing all traffic requirements between all pairs of nodes, using primary and spare paths, the capacities of links and nodes are obtained.

4.6 Network Unavailability Calculation

Network unavailability is defined as the worst case of all node-to-node connection unavailabilities (source-termination unavailability) [14]:

$$U = \max_{i,j}\left\{U_{ij}\right\}, \quad U_{ij} = \left(1 - \prod_{k \in pp}(1 - U_k)\right)\left(1 - \prod_{l \in sp}(1 - U_l)\right),$$

where U_k is the unavailability of a link from the primary path (*pp*), and U_l is the unavailability of a link from the independent spare path (*sp*). In other words, the unavailability model of a node-to-node connection could be described as a serial structure of two parallel.

Optical link is treated as a nonredundant structure comprising fiber in optical fiber cable and optical amplifiers. For small unavailability values of link elements, an approximate formula for the total link unavailability can be used.

$$U_{link} = \lambda_F \, L \, MTTR_F + N_{OA} \, \lambda_{OA} \, MTTR_{OA}$$

where λ_F is fiber cable failure rate per km, λ_{OA} is the failure rate of the optical amplifier (OA), N_{OA} is the number of optical amplifiers on the link, L is the link length, $MTTR_F$ and $MTTR_{OA}$ are mean times to repair of fiber (F) and OA, respectively (λ_F = 114 fit/km, $MTTR_F$ = 21 hours, λ_{OA} = 4500 fit, $MTTR_{OA}$ = 21 hours, fit = number of failures per 10^9 hours).

4.7 Solution Coding

Possible solutions are coded as binary strings with $n(n\text{-}1)/2$ bits. The position of every bit represents one direct link between two nodes. The value of the bit corresponding to 1 represents the existence of the link in the solution, while the value 0 stands for a missing corresponding link.

For instance, the case study network of 11 nodes should be coded by the string containing 55 bits. In the Figure 14 one random string is presented and corresponding network is shown in the Figure 15.

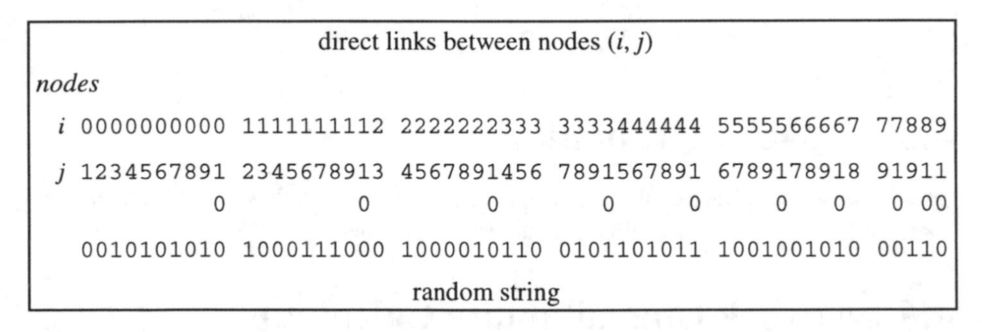

direct links between nodes (i, j)

nodes

```
i 0000000000 1111111112 2222222333 3333444444 5555566667 77889
j 1234567891 2345678913 4567891456 7891567891 6789178918 91911
           0          0          0     0    0     0    0  0 00
  0010101010 1000111000 1000010110 0101101011 1001001010 00110
```

random string

Figure 14. An example of coded topology in the case study.

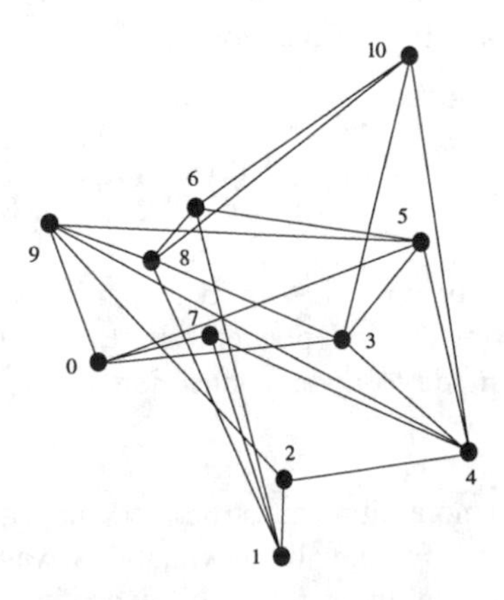

Figure 15. The topology representing a random string shown in Figure 14.

4.8 Selection Process

Different approaches could be applied in creating the selection process. Any selection principle reflects the definition of the fitness function. Here, two extreme approaches will be analyzed.

Approach 1: Preselection rejection

In the preselection process, all solutions not satisfying some easy-to-test fundamental requirements are rejected. For example, if a generated graph has a node degree less than 2, or if the number of branches in a topology is less than (N-1) (graph tree), it can surely be inferred that these solutions cannot satisfy the network requirement of two independent paths between all node pairs.

Fitness functions are very simple; in the case of cost minimization f_c and in unavailability minimization f_u, respectively,

$$f_c = \frac{1}{C}, \qquad f_u = \frac{1}{U}.$$

The advantage of this approach lies in reducing the number of topologies to be evaluated in detail (shortest path, capacity, cost, and unavailability calculation). For example, for 11 nodes, as used in the case study, the number of acceptable topologies is reduced to 10^{-4}% of all topologies, according to the "graph tree" preselection rule, as mentioned above.

On the other hand, the disadvantages of this approach are poor diversity of solutions in the population, and the very rough distinction between solutions – a solution is either regular, that is, acceptable, or irregular, that is, unacceptable. In the cases where solution limitations are very restrictive, the whole initial population could be rejected, disabling further search. Note that even a bad solution could produce a good offspring.

Approach 2: The use of fitness functions with penalizing

No topology is rejected but, is penalized, if assumptions or dynamic limitations are not satisfied.

The advantage of this approach lies in the great diversity of solutions to be evaluated, increasing the probability of finding different areas of local minima to be tested, in order to select the global one.

The disadvantage of this approach lies in an extensive evaluation time.

Despite higher time consumption than in Approach 1, Approach 2 is selected for optimization application as the more efficient one.

4.9 Optimization Procedure

In order to minimize unavailability–cost pairs, two types of optimization alternate. In odd optimization steps, the network cost is minimized. Fitness function for cost minimization is

$$f_c = \frac{1}{(C+PF)\,k\,(1+UP)},$$

$$UP = \frac{U - U_{lim}}{U_{lim}} \text{ for } U > U_{lim}, \text{ and } UP = 0 \text{ for } U \le U_{lim},$$

where k is the penalty slope and U_{lim} is the dynamic unavailability upper bound in an odd step, achieved as minimal in previous even step(s). PF is the penalty factor defined as follows:

$$PF = 2.5\ PathOver\ CapOver,$$

where *PathOver* is the sum of all excesses of path length limitation and distances between the node pairs without primary and/or spare paths. *CapOver* is the sum of capacity demands between the node pairs contributing to the *PathOver*. In even optimization steps, the unavailability is minimized. Fitness function for unavailability minimization is equal to

$$f_u = \frac{1}{U\,k\,(1+CP)},$$

$$CP = \frac{C + PF - C_{lim}}{C_{lim}} \text{ for } C + PF > C_{lim}, \text{ and } CP = 0 \text{ for } C + PF \le C_{lim}.$$

The penalty is effective for the costs higher than the cost limit C_{lim}, – the dynamic cost upper bound reached in previous odd optimization step(s).

Note that the genetic material is transferred from one step to the next one, forming initial population.

4.10 Optimization Results

The optimization results refer to the case study of European all-optical network.

The absolute minimum unavailability, as a reference value, was determined from the fully meshed network. The optimization target was to find the topology with the same or very close unavailability value and with cost as low as possible.

The genetic algorithm parameters are chosen as follows: population size = 100, string size = 55, crossover probability = 0.6, mutation probability = 0.05, two point crossover, roulette wheel selection scheme, generation gap = 1, the number of generations per step = 200, elitism.

As a result of optimization running, several quasi-optimal unavailability-cost pairs were obtained. Table 2 shows the results of two GA generated topologies, the minimum cost topology (MinC) and the minimum unavailability topology (MinU) (Figure 16), compared to the reference topology COST 239 (EON) and manually designed grid network (MG) (Figure 17) and fully meshed topology (FM) [15].

Table 2. The comparison of topology performances

	FM	EON	MG	MinC	MinU
U x10^{-5}	2.502	3.789	4.235	3.130	2.502
C x10^{6}	4.537	3.765	3.903	3.706	3.793
C_L x10^{6}	1.441	1.685	1.711	1.576	1.615
C_N x10^{6}	3.096	2.080	2.192	2.130	2.178
TFCL [km]*	44145	14775	11635	14610	19115
No. of links	55	25	22	25	29
d_{min}	10	4	2	3	4
d_{max}**	10	5	6	7	8
PathOver [km]	0	50	675	0	0

* *TFCL* - total fiber cable length
** d_{min}, d_{max} - the minimum and maximum node degrees.

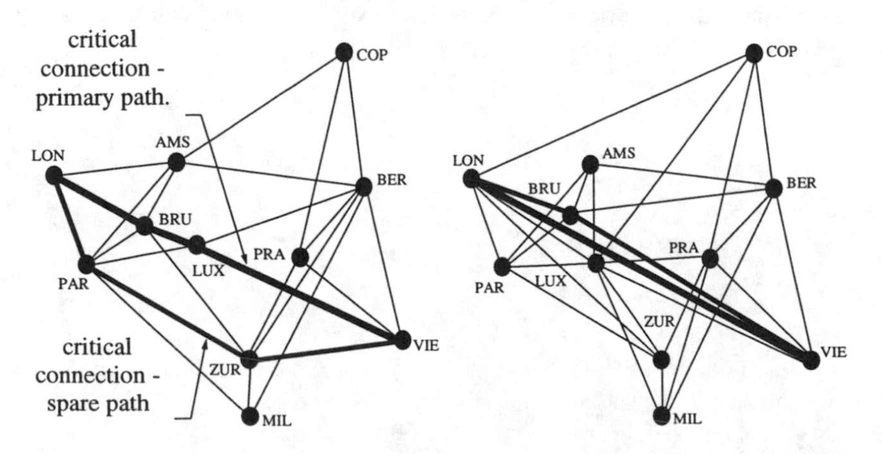

Figure 16. Minimum Cost (MinC) and Minimum Unavailability (MinU) topologies.

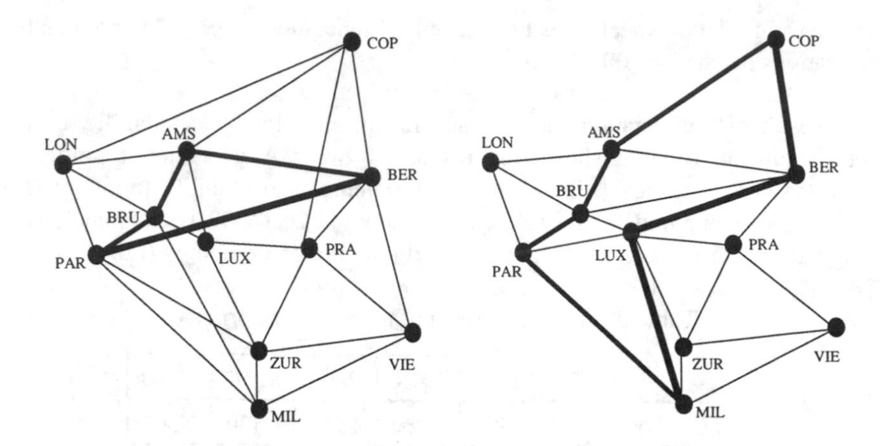

Figure 17. COST 239 case study (EON) and Manual–Grid (MG) topologies.

5. Conclusion

In this chapter an application of genetic algorithms in telecommunications is described. Genetic algorithms are based by analogy with the processes in the reproduction of biological organisms. These algorithms could be classified as guided random search evolution algorithms that use probability to guide their search. A genetic algorithm application to a specific problem includes a number of steps and some of them are discussed in three different telecommunication system design problems. Two of them are related to a method for call and service process scheduling and call and service control in distributed environment, where a genetic algorithm is used to determine a response time. Genetic algorithm application in optimization is presented through the case study on availability–cost optimization of an all-optical network.

References

[1] Goldberg, D.E. (1989), *Genetic Algorithms in Search, Optimization & Machine Learning*, Addison-Wesley, Reading.

[2] Sinkovic, V. and Lovrek, I. (1994), An Approach to Massively Parallel Call and Service Processing in Telecommunications, *Proceedings MPCS'94 Conference on Massively Parallel Computing Systems: the Challenges of General-Purpose and Special-Purpose Computing*, IEEE, Ischia, Italy, pp. 533–537.

[3] Sinkovic, V. and Lovrek, I. (1997), A Model of Massively Parallel Call and Service Processing in Telecommunications, *Journal of System Architecture - The EUROMICRO Journal*, Vol. 43, pp. 479–490.

[4] Dacker, B. (1993), Erlang - A New Programming Language, *Ericsson Review*, Vol. 70, No.2.

[5] Ramamoorthy, C.V., Chandy, K.M., and Gonzalez, M.J., Jr. (1972), Optimal Scheduling Strategies in a Multiprocessor System, *IEEE Transaction on Computers*, Vol. C-21, No. 2, pp. 137–146.

[6] Fernandez, E.B. and Bussell, B. (1973), Bounds on the Number of Processors and Time for Multiprocessor Optimal Schedules, *IEEE Transaction on Computers*, Vol. C-22, No. 8, pp. 745–751.

[7] Hou, E.S.H., Ansari, N., and Ren, H. (1994), A Genetic Algorithm for Multiprocessor Scheduling, *IEEE Transactions on Parallel and Distributed System*, Vol. 5, No. 2, pp. 113–120.

[8] Sinkovic, V. and Lovrek, I. (1995), Performance of Genetic Algorithm Used for Analysis of Call and Service Processing in Telecommunications, *Proceedings ICANNGA'95 International Conference on Artificial Neural Networks and Genetic Algorithms*, Ales, France, Springer Verlag Wien, New York, pp. 281–284.

[9] Lovrek, I. and Simunic, N. (1996), A Tool for Parallelism Analysis in Call and Service Processes, *Proceedings MIPRO'96 Computers in Telecommunications*, Rijeka, Croatia.

[10] Selvakumar, C. and Murthy, S.R. (1994), Scheduling Precedence Constrained Task Graphs with Non-negligible Intertask Communication onto Multi-processors, *IEEE Transactions on Parallel and Distributed Systems*, Vol. 5, No. 3, pp. 328–336.

[11] Lovrek, I. and Jezic, G. (1996), A Genetic Algorithm for Multiprocessor Scheduling with Non-negligible Intertask Communication, *Proceedings MIPRO'96 Computers in Telecommunications*, Rijeka, Croatia.

[12] O'Mahony, M.J., Sinclair, M.C., and Mikac, B. (1993), Ultra-high Capacity Optical Transmission Networks: European Research Project COST 239, *ITA - Information, Telecommunication, Automata*, Vol. 12, No. 1–3, pp 33-45.

[13] Sinclair, M.C. (1995), Minimum Cost Topology Optimisation of the COST 239 European Optical Network, *Proceedings ICANNGA'95 International Conference on Artificial Neural Networks and Genetic Algorithms*, Ales, France, Springer Verlag Wien, New York, pp. 26–29.

[14] Mikac, B. and Inkret, R. (1997), Application of a Genetic Algorithm to the Availability-Cost Optimisation of a Transmission Network Topology, *Proceedings ICANNGA'97 Third International Conference on Artificial Neural Networks and Genetic Algorithms*, Norwich, U.K., Springer Verlag Wien, New York, pp. 306–310.

[15] Inkret, R. (1995), All-optical Network Reliability Optimization by Means of Genetic Algorithm, *Project Report*, Department of Telecommunications, Faculty of Electrical Engineering and Computing, University of Zagreb (in Croatian).

INDEX

- E -

- F -

- G -

- H -

- I -

- K -

- L -

- M -

- N -

- O -